D0076045

INTRODUCTION TO

PROTEIN SCIENCE

Architecture, Function, and Genomics

ARTHUR M. LESK

INTRODUCTION TO

PROTEIN SCIENCE

INTRODUCTION TO PROTEIN SCIENCE

Architecture, Function, and Genomics

ARTHUR M. LESK

University of Cambridge

OXFORD
UNIVERSITY PRESS

Great Clarendon Street, Oxford OX2 6DP

Oxford University Press is a department of the University of Oxford.
It furthers the University's objective of excellence in research, scholarship,
and education by publishing worldwide in

Oxford New York

Auckland Bangkok Buenos Aires Cape Town Chennai
Dar es Salaam Delhi Hong Kong Istanbul Karachi Kolkata
Kuala Lumpur Madrid Melbourne Mexico City Mumbai Nairobi
São Paulo Shanghai Taipei Tokyo Toronto

Oxford is a registered trade mark of Oxford University Press
in the UK and in certain other countries

Published in the United States
by Oxford University Press Inc., New York

First published 2004

British Library Cataloguing in Publication Data
Data available

Library of Congress Cataloging in Publication Data
Data available

ISBN 0 19 926511 9

Typeset by Newgen Imaging Systems (P) Ltd., Chennai, India
Printed in Great Britain
on acid-free paper by
Ashford Colour Press Limited, Gosport, Hampshire

Dedicated to the memory of Max Ferdinand Perutz

— Mozart, K. 584

CONTENTS

PREFACE

This book completes a trilogy, joining *Introduction to Protein architecture: the structural biology of proteins*, and *Introduction to bioinformatics*, both of which are more specialized—*Bioinformatics* to sequences, and *Architecture* to structures. This one is intended as an introduction to, and synthesis of, the other two. It also complements them, with a focus on protein function, including the integration and regulation of function: this book is as much about the logic of life as about its chemistry. These three components—sequence, structure, and function—define the significance of proteins in biology.

Sydney Brenner once stated the questions to ask about any living system:

1. How does it work?
2. How does its structure form?
3. How did it evolve?

Protein science has appealed to many scientific disciplines to address these problems:

- Parts of protein science have developed out of the field of *Chemistry*, including the use of physical methods such as spectroscopy, kinetics, and techniques of structure determination; and organic/biochemical methods for working out the mechanism of enzymatic catalysis. Much of classical protein chemistry and enzymology studied purified molecules in isolation, revealing detailed information about the structures and activities of individual molecules.
- Parts of protein science have developed out of *Molecular and cellular biology*. The challenge is to coordinate our knowledge of the properties and functions of isolated molecules, into an understanding of the biological context and integration of protein activity.

  High-throughput methods of **genomics** and **proteomics** provide very large amounts of data about sequences and expression patterns. We now know the full DNA sequences of many organisms, and from these genome sequences we can infer the amino acid sequences of an organism's full complement of proteins. Proteomics tells us how the expression patterns of these proteins vary. We have pieces of a jigsaw puzzle that extends in both space and time. How can we fit the pieces together to understand the complex and delicate instrument that is the living cell?
- Parts of protein science have developed out of *Evolutionary biology*. Evolution takes place on the molecular scale, exploring variations in sequences, protein structures and functions, and patterns of expression. Much of what we see in living organisms depends on, and reveals, historical events no longer observable directly.
- Parts of protein science have emerged from *Computing*. Computer science is a flourishing field with the goal of making the most effective use of information technology hardware. Computers are essential for storing and analysing the very large amounts of data that have been coming onstream. Databanks archive and organize access to sequences

of genomes; to amino acid sequences, structures, expression patterns, and interaction networks of proteins; to metabolic pathways; to the molecular correlates of diseases; and to the scientific literature. Distribution of the information requires the facilities of computer networks and the World Wide Web. Computer programs provide access to the data, and tools for their analysis. *Bioinformatics* is a hybrid of biology, chemistry, physics, and computer science. Its goal is to bring the data to bear on biological problems, including applications to clinical medicine, agriculture, and technology.

Some readers may wonder why nucleic acids receive substantial attention in a book on protein science. The answer is that they belong here. It is impossible to make sense of the biology of proteins without discussion of the ribosome and control of gene expression, to pick just two examples. Nucleic acids provide the logical framework of the stage on which proteins act.

Whole-genome sequencing projects have transformed the field of molecular biology. Before, we were limited to studying selected available examples. The new results lay bare the choices that Nature has made. The three-dimensional counterpart of genome sequence determination, the provision of *structures* of all the proteins in an organism, is known as **Structural genomics**.

When we know the sequences and structures of all macromolecules in an *Escherichia coli* cell, we shall have a complete but *static* knowledge of the components of a living object. The next step will be to learn how the components are assembled and how their individual activities are integrated.

The most profound significance of proteins lies in their collective properties as a class of biomolecules. These properties include all their potential, as well as their actual, characteristics: proteins have an underlying chemical unity, they have the ability to organize themselves in three dimensions; and the system that produces them can create inheritable structural variations, conferring the ability to evolve. Recent developments in genomics, proteomics, and structural genomics permit study of the entire complement of proteins of an organism and their coordinated activities, and give glimpses of the *proteosphere*—the entire spectrum of the proteins in Nature.

Proteins are where the action is

Proteins are fascinating molecular devices. They play a variety of roles: there are structural proteins (molecules of the cytoskeleton, epidermal keratin, viral coat proteins); catalytic proteins (enzymes); transport and storage proteins (haemoglobin, myoglobin, ferritin); regulatory proteins, (including hormones, many kinases and phosphatases, and proteins that control gene expression); and proteins of the immune system and the immunoglobulin superfamily (including antibodies, and proteins involved in cell–cell recognition and signalling).

Amino acid sequences are now available for over 1 million proteins, and detailed atomic structures for over 24000. These data support a number of fascinating and useful scientific endeavours, including:

1. **Interpretations of the mechanisms of function of individual proteins**. The catalytic activities of enzymes can be explained in terms of physical–organic chemistry, on the basis of the interactions of substrates with residues in the active site of the protein.

2. **Patterns of molecular evolution.** For many families of proteins, we know tens or even hundreds of amino acid sequences, and many structures. We can observe the structural and functional roles of the sets of residues that are strongly conserved. We can describe the consequences of mutations, insertions, and deletions in the amino acid sequence that perturb protein conformation, and identify constraints that selection imposes on the structure to preserve function. When proteins evolve with changes in function, these constraints on the structure are relaxed—or rather, replaced by alternative constraints—and the sequences and structures change more radically.

On a broader scale, *comparative genomics* traces the changes and distribution of proteins and regulatory sequences in different species, and similarities and differences in genome organization.

3. **The amino acid sequences of proteins dictate their three-dimensional structures, and their folding pathways.**

(a) Under physiological conditions of solvent and temperature, proteins fold spontaneously to an active native state. Although we cannot yet confidently predict the conformation of a novel protein structure from its amino acid sequence in all cases, many principles of protein structure and folding are now understood.

(b) Relationships between the evolutionary divergence of amino acid sequence and structure in protein families do make it possible to predict a protein structure from an amino acid sequence, whenever we know the sequence and structure of one or more sufficiently closely related proteins. As we come closer to having a structure of every type that appears in Nature, then the nearest relative of a known structure will provide a reasonable quantitative model for almost all unknown proteins. In the future, this will give us access to structures of the proteins encoded in any genome.

(c) The amino acid sequence of a protein must not only preferentially stabilize the native state, it must contain a road map telling the protein how to get there, starting from the many diverse conformations that comprise the unfolded state. A combination of physicochemical measurements and studies of the effects of mutations on the kinetics of folding and unfolding are illuminating the mechanism of the folding process.

4. **Mapping the logic and the mechanisms of integration and control of life processes:**

(a) We can measure and compare patterns of biochemical activity in different tissues, or under different growth conditions. Comparisons of protein expression patterns, or of genotypes embedded in the DNA, using **microarrays**, have direct clinical applications, for many diseases involve failures in control mechanisms. Therefore differences in expression patterns in normal and disease states permit the precise diagnosis of disease, producing a more accurate prognosis and choice of treatment. Prediction, from details of genotypes, of enhanced susceptibility to disease may suggest preventive measures, including alteration of lifestyle.

(b) We can trace pathways of information flow in regulatory networks. Proteins receive and transmit signals, often using conformational changes to control activity. Proteins 'talk' to one another, and also to DNA. DNA–protein interactions control gene expression.

(c) We can study patterns in time as well in space. The cell cycle, and the unfolding of developmental programmes, are coordinated with changes in protein expression. Development of the nervous systems of higher organisms present great challenges to diversity plus accuracy in molecular signalling and recognition.

Advances in protein science have spawned the biotechnology industry:

5. **Protein engineering**. It is now possible to design and test modifications of known proteins, and to design novel ones. Applications include:

(a) Attempts to enhance thermostability, for example by introducing disulphide or salt bridges, or by optimizing the choice of amino acids that pack the protein interior.

(b) Clinical applications, for instance therapeutic antibodies. Rats can raise antibodies against human tumours. Transfer of the active site from a rat antibody to a human antibody framework can produce a molecule that retains therapeutic activity in humans, but reduces the side effects arising from patients' immunological response against the antigenic rat protein.

(c) Modifying antibodies to give them catalytic activity. Two features of enzymes are: (1) the ability to bind substrate specifically, and (2) the juxtaposition of bound substrate with residues that catalyse a chemical change. Antibodies can provide the binding and discrimination; the challenge to the chemist is to introduce the catalytic function.

(d) Modifications to probe mechanisms of function or folding, such as the identification of properties of folding intermediates by the effects of mutations on stability and folding kinetics.

6. **Drug discovery**. There are many proteins specific to pathogens that we want to inactivate. By knowing the structure of the AIDS proteinase, or the neuraminidase of influenza virus, it should be possible to design molecules that will bind tightly and specifically to an essential site on these molecules, to interfere with their function. Knowing the structures of receptors should make it possible to design agonists and antagonists of the signals to which they respond.

These and other topics comprise a new scientific speciality, *Protein science*. Using a combination of approaches and methods, gathered from many other fields of science, we are gaining insight into the principles governing this fascinating class of molecules. Let us apply this knowledge in medicine, agriculture and technology to improve the lot of mankind.

I thank F. Arnold, M. Bashton, S.E. Brenner, C. Chothia, J. Clarke, D. Crowther, T.J. Dafforn, A. Doherty, D.S. Eisenberg, I. Fearnley, J.R. Fresco, A. Friday, M. Gait, E. Gherardi, W.B. Gratzer, R. Henderson, O. Herzberg, T.J.P. Hubbard, J. Janin, T. Kiefhaber, G. Kleywegt, P.A. Lawrence, P.A. Lenton, E.L. Lesk, V.E. Lesk, V.I. Lesk, L. Liljas, L. Lo Conte, D.J. Lomas, J. Löwe, P.A. Lyons, A.J. McCoy, J.D. Mollon, S. Moran, J. Moult, A.G. Murzin, D. Neuhaus, M. Oliveberg, V. Ramakrishnan, R. Read, G.D. Rose, B. Rost, K. Scott, L. Serpell, R. Staden, R. Tregear, M. Vendruscolo, C. Vogel, G. Vriend, A.G. Weeds, J.C. Whisstock, A. Yonath, and L. You. I thank the staff of Oxford University Press for their skills and patience in producing this book.

Chapter 1

THE RIBOSOME—THE FULCRUM OF GENOMICS

LEARNING GOALS

1 **To understand the biological context of protein science,** including the key role of the ribosome as the machinery of protein synthesis.

2 **To know how DNA sequences code for amino acid sequences.** The genetic code is a triplet code. In all, 64 triplets of nucleotides, or **codons**, are translated to 20 amino acids, plus Stop signals. There is redundancy in the genetic code, several codons corresponding to one amino acid. The genetic code is universal among living things, with only a few exceptions.

3 **To learn the names and abbreviations of the nucleotides and amino acids,** and to be able to deduce the effect on the amino acid sequence of a protein of a mutation in the nucleotide sequence of a gene.

4 **To appreciate the 'central dogma' describing information transfer and expression,** in its original statement by Francis Crick and subsequent extensions.

Introduction

Where do proteins come from? Ribosomes are subcellular particles that synthesize proteins. The ribosome is the point of translation of genetic information into the amino acid sequences of proteins, and thereby into protein structures and functions. The ribosome makes possible the separation of *heritable genetic information*—in the form of nucleic acid—from the *agents of biochemical activity*—primarily proteins. It is this separation that makes evolution possible, and without evolution life would be severely impoverished, if not impossible. The ribosome is the point of contact between genes and proteins—it is the fulcrum of genomics.

Ribosomes are large nucleoprotein particles, approximately 250 Å in diameter. They are composed of two subunits, that dissociate reversibly upon lowering the Mg^{2+} concentration. Originally, the masses and sizes of particles were measured by their sedimentation constants in the ultracentrifuge. The nomenclature survives: the

Relative molecular mass, or r.m.m.—colloquially, 'molecular weight', expresses mass on a scale in which the most common isotope of carbon has a mass of exactly 12.

complete *Escherichia coli* ribosome is called the '70S particle'; it dissociates into 30S and 50S subunits. (Sedimentation constants are not additive!) Now we know the molecular sizes precisely: the full *E. coli* ribosome has a relative molecular mass of 2.5×10^6 and the two subunits are 1.65×10^6 and 0.85×10^6. The cytoplasmic ribosomes of eukaryotes are somewhat larger, and vary in size from species to species.

Ribosomes contain proteins and RNA. In prokaryotic and mitochondrial ribosomes the ratio by mass is about 2 : 1 RNA:protein. Eukaryotic cytoplasmic ribosomes have gained both additional protein and RNA, but the enrichment in protein is greater, giving a 1 : 1 RNA : protein ratio. In prokaryotic ribosomes, the large subunit contains two molecules of RNA, one with about 2900 bases and another with 120 bases. (Not base pairs—RNA is single-stranded, although it can form intramolecular stretches of double helix.) The large subunit also contains 31 different proteins. The small subunit contains one RNA molecule of 1500 bases, and 21 proteins. In eukaryotic cytoplasmic ribosomes, the large subunit contains three RNA molecules of approximately 4700, 156, and 120 bases; and 50 proteins, most of which are present in only one copy. The small subunit contains a 1900-base RNA molecule and 32 proteins.

Cells take ribosomes very seriously. An *E. coli* cell contains about 15 000 ribosomes, comprising approximately 25% of the cell mass! Conversely, protein synthesis is so vital a process that ribosomes are the targets of several powerful antibiotics, including streptomycin and tetracycline.

Ribosomes were implicated in protein synthesis very early on, in the 1950s. It was observed that radioactive amino acids first appeared in protein associated with ribosomes. Preparations of purified ribosomes capable of cell-free protein synthesis proved the function. Recognition that, in eukaryotes, genes were in the nucleus and ribosomes in the cytoplasm suggested the idea of messenger RNA to link them. The inability to detect specific and selective amino acid–nucleic acid interactions suggested the adaptor hypothesis: a special form of RNA would interact with both protein and nucleic acid and achieve the physical association of messenger RNA with specific amino acids. Transfer RNAs fill precisely this role.

These ideas and the cell-free protein synthesis systems came together to allow decipherment of the genetic code. The use of synthetic RNAs of known sequence (or known simple compositions) as artificial messengers, and observation of which amino acids were taken up into the polypeptides synthesized, permitted the assignment of triplets of nucleotides—**codons**—to amino acids.

A general picture of protein synthesis emerged with the metaphor of the ribosome as a 'tape reader'. A ribosome 'reads' codons sequentially from messenger RNA (mRNA). To process each codon, an appropriate transfer RNA (tRNA), charged with a specific amino acid, brings its amino acid into proximity to the growing polypeptide chain on the ribosome. The peptide bond is synthesized enzymatically; then the transfer RNA is released, and the 'tape advanced' to the next codon. One worrying portent: despite extensive investigation, no ribosomal protein that catalysed peptide bond synthesis was ever found.

Until the 1980s structural studies to provide physical details of the intact ribosomal machinery were limited to low-resolution electron microscopy. Some crystal structures of individual proteins were solved and the RNAs were sequenced, but it was not possible to coordinate this information into a complete detailed structure. Electron microscopy did show a rough general picture of the layout and topography of the ribosome, including the presence of a tunnel through the large subunit, through which, it later turned out, the nascent polypeptide chain passes.

Structural studies of ribosomes by X-ray crystallography and electron microscopy

A major triumph of structural biology in this new century has been the determination of the structures of ribosomal subunits and intact ribosomes to high resolution, achieved through a combination of X-ray crystallography and electron microscopy. These structures are the culmination of two decades of work since the first diffracting crystals were obtained in 1983.

The results came as a shock.

The cores of the subunits are nucleic acid, with the proteins mostly peripheral (Fig. 1.1). The subunit interface is also nucleic acid. Indeed, the enzymatic activity resides in the RNA: The ribosome is a **ribozyme**.

There are three major tRNA-binding sites on the ribosome:

- **The A (aminoacyl) site** binds the tRNA charged with the next amino acid to be added.
- **The P (peptidyl) site** binds the tRNA to which the growing polypeptide chain is attached.
- **The E (exit) site** binds the final, deacylated, tRNA before it dissociates from the ribosome.

Each tRNA first binds to the ribosome at the A-site, then occupies the P-site, then moves to the E-site, from which it dissociates (see Fig. 1.2).

Steps in protein synthesis include:

1. **Charging of the tRNAs:** A set of enzymes, tRNA synthetases, couples each tRNA to the appropriate amino acid.
2. **Initiation** of protein synthesis involves the formation of a complex that includes a formylmethionyl-tRNA, messenger RNA, and the ribosome, with the tRNA occupying the P-site.
3. **Elongation** of the nascent polypeptide chain requires: (1) selection of the proper aminoacyl-tRNA with the next amino acid, and its binding to the A-site; (2) formation of the new peptide bond, leaving the polypeptide chain attached to the last tRNA, bound to the A-site; and (3) a general shifting, of:
    - the deacylated tRNA from the P-site to the E-site and its dissociation from the ribosome;
    - translocation of the peptidyl-tRNA, still bound to the mRNA, from the A to the P-site; and
    - advance of the mRNA by three bases.

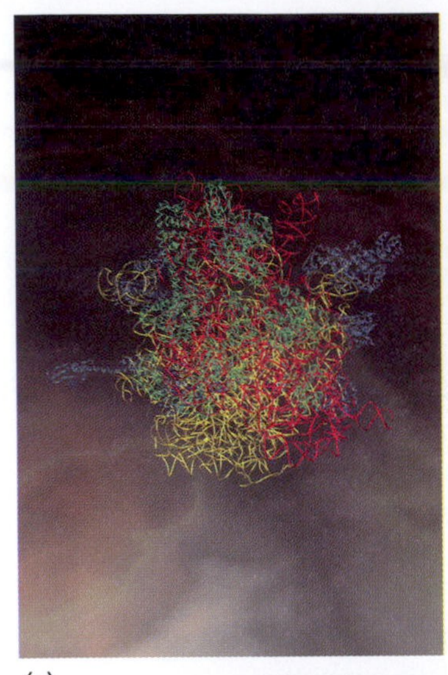

(a)

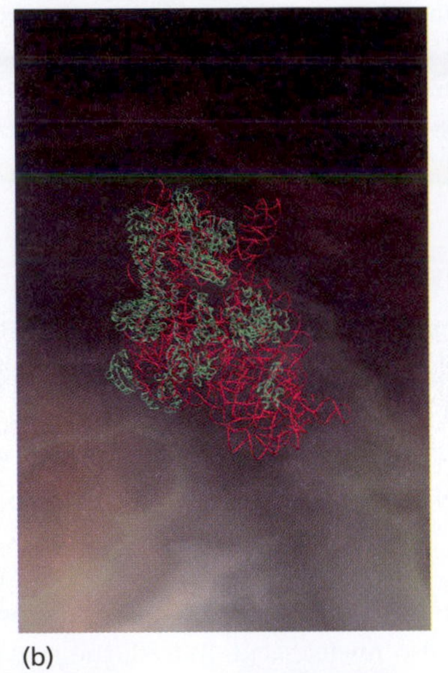

(b)

(c)

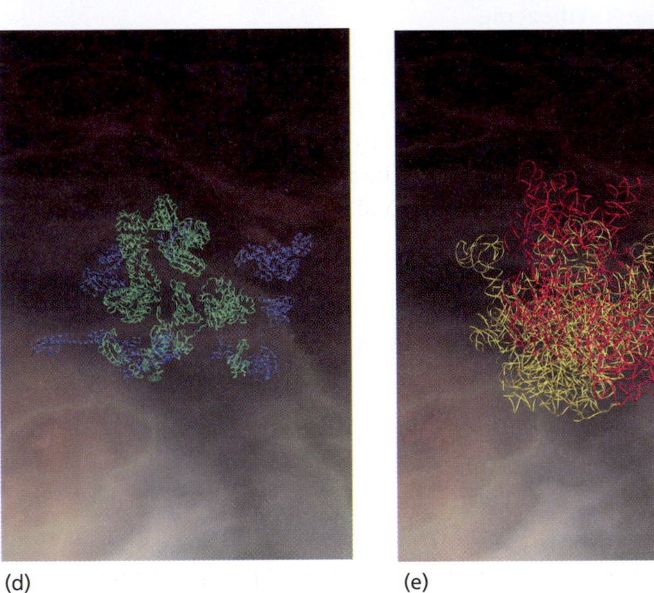

(d) (e)

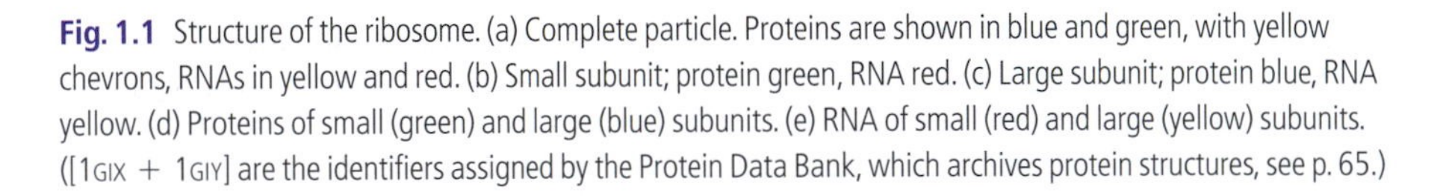

Fig. 1.1 Structure of the ribosome. (a) Complete particle. Proteins are shown in blue and green, with yellow chevrons, RNAs in yellow and red. (b) Small subunit; protein green, RNA red. (c) Large subunit; protein blue, RNA yellow. (d) Proteins of small (green) and large (blue) subunits. (e) RNA of small (red) and large (yellow) subunits. ([1GIX + 1GIY] are the identifiers assigned by the Protein Data Bank, which archives protein structures, see p. 65.)

The elongation step is carried out for each successive codon in the mRNA, until a Stop codon is encountered.

4. **Termination** of protein synthesis occurs when the appearance of a Stop codon leads to hydrolysis of the peptide–tRNA complex, followed by general dissociation, with release of the protein product.

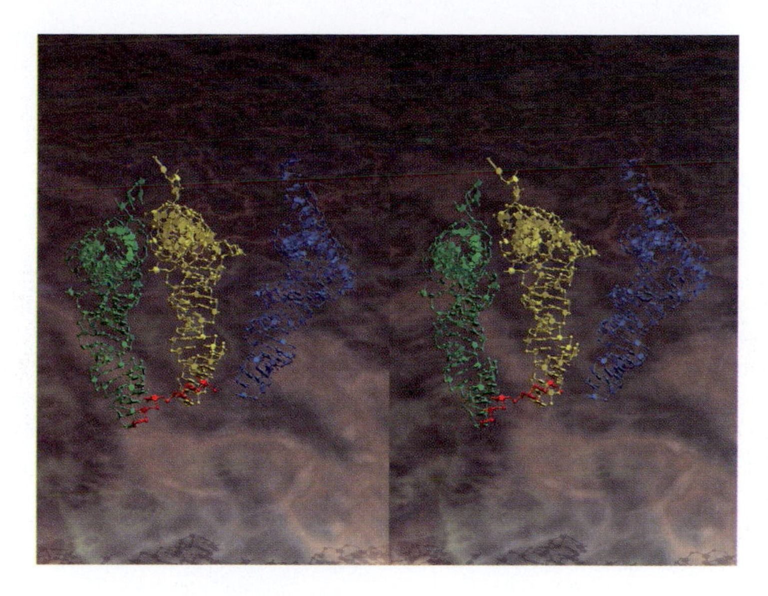

Fig. 1.2 Three tRNA molecules extracted from ribosome structure, in the relative positions of the A (green), P (yellow), and E (blue) sites, with a fragment of mRNA modelled in [1GIX + 1GIY].

Discovery of the ribosome structure was also an archaeological find of major significance. There is now consensus that before the current roster of biological molecules was established, there was a world of life based on RNA. Certainly the activity of the RNA in ribosomes is consistent with this. One puzzle about the relationship between the RNA world and our own is how the transition was achieved. The ribosome provides a crucial link.

The genetic code

The genetic code is the translation table between the nucleic acid sequences of genes and amino acid sequences of proteins (see Box). The unit of coding is a triplet of nucleotides. Four nucleotides—adenine, thymine, guanine, and cytosine—can combine to form a total of 64 possible triplets or codons.

More about amino acids in Chapter 3.

There is an excess of codons over amino acids: 64 codons are translated into only 20 amino acids, plus Stop signals. Most amino acids correspond to more than one codon. Five of the amino acids are encoded by blocks of four codons, which have the same nucleotides in the first two positions and all four possibilities in the third. Nine are encoded by blocks of two codons, with the same nucleotides in the first two positions and two nucleotides in the third position. Three amino acids are encoded by more than four codons, one amino acid by only three codons, and two amino acids by only one codon. The variability in the third position is related to the base pairing between the codon and the anticodon in the corresponding tRNA, which is more tolerant at the third position according to the **'wobble' hypothesis**.

The 4 naturally occurring nucleotides in DNA (RNA)

a	adenine	g	guanine	c	cytosine	t	thymine	(u	uracil)

The 20 naturally occurring amino acids in proteins

Non-polar amino acids

G	glycine	A	alanine	P	proline	V	valine
I	isoleucine	L	leucine	F	phenylalanine	M	methionine

Polar amino acids

S	serine	C	cysteine	T	threonine	N	asparagine
Q	glutamine	H	histidine	Y	tyrosine	W	tryptophan

Charged amino acids

D	aspartic acid	E	glutamic acid	K	lysine	R	arginine

Other classifications of amino acids can also be useful. For instance, histidine, phenylalanine, tyrosine, and tryptophan are aromatic, and are observed to play special structural roles in membrane proteins.

Amino acid names are frequently abbreviated to their first three letters—for instance Gly for glycine—except for isoleucine, asparagine, glutamine, and tryptophan, which are abbreviated to Ile, Asn, Gln, and Trp, respectively. The rare amino acid selenocysteine has the three-letter abbreviation Sec and the one-letter code U.

It is conventional to write nucleotides in lower case and amino acids in upper case. Thus atg = adenine–thymine–guanine and ATG = alanine–threonine–glycine.

The genetic code shown in the Box on p. 7 is the standard version used by the nuclear DNA of eukaryotes and by most prokaryotes. Other versions appear in mitochondria, chloroplasts, and sporadically in a few species. These alternative genetic codes show only a few differences from the standard one. For example, in the yeast mitochondrial genetic code cug is translated to Thr instead of Leu. In most other cases one or two of the usual Stop codons are translated into amino acids.

A mutation is a change in a genetic nucleotide sequence. Single-nucleotide polymorphisms (SNPs), or base substitutions, are a class of mutations important as genetic markers. SNPs can produce changes in amino acids, but do not necessarily do so. Many substitutions at the third positions of codons are silent, leaving the protein invariant. The assignments of codons to amino acids is such that many substitutions that change the amino acid make conservative changes in physicochemical class and volume. For instance, a single base substitution, in one of the four codons for Val, can change a Val to any of the following: Phe, Leu, Ile, Met, Ala, Asp, Glu, or Gly. Of these, Phe, Leu, Ile, Met, and Ala are physicochemically similar to Val and would be regarded as conservative substitutions, which might not do too much damage to a protein unless they were to occur at a particularly sensitive site. Of the other amino acids that can be produced from Val by single base substitutions, Asp and Glu are charged, and Gly is a special residue, lacking a side chain. Changing a Val to any of these would not be considered a conservative substitution.

The standard genetic code

uuu	Phe	ucu	Ser	uau	Tyr	ugu	Cys
uuc	Phe	ucc	Ser	uac	Tyr	ugc	Cys
uua	Leu	uca	Ser	uaa	STOP	uga	STOP
uug	Leu	ucg	Ser	uag	STOP	ugg	Trp
cuu	Leu	ccu	Pro	cau	His	cgu	Arg
cuc	Leu	ccc	Pro	cac	His	cgc	Arg
cua	Leu	cca	Pro	caa	Gln	cga	Arg
cug	Leu	ccg	Pro	cag	Gln	cgg	Arg
auu	Ile	acu	Thr	aau	Asn	agu	Ser
auc	Ile	acc	Thr	aac	Asn	agc	Ser
aua	Ile	aca	Thr	aaa	Lys	aga	Arg
aug	Met	acg	Thr	aag	Lys	agg	Arg
guu	Val	gcu	Ala	gau	Asp	ggu	Gly
guc	Val	gcc	Ala	gac	Asp	ggc	Gly
gua	Val	gca	Ala	gaa	Glu	gga	Gly
gug	Val	gcg	Ala	gag	Glu	ggg	Gly

Single-nucleotide changes are either **transitions** = purine to purine (A → G or G → A) or pyrimidine to pyrimidine (U → C or C → U), or **transversions** = purine to pyrimidine (A or G → U or C), or pyrimidine to purine (U or C → A or G). Transitions are more common mutations than transversions. A single transition could change a Val to an Ile, Met, or Ala. These substitutions are *all* conservative, with respect to their expected effect on a protein structure. Such relationships within the genetic code help to make proteins robust to mutation.

The biological context of protein synthesis—the basis of evolution

In 1958, as the basic paradigm of protein synthesis was emerging, Francis Crick brought the ideas together into a statement he called the central dogma of molecular biology: **DNA makes RNA makes protein**. He emphasized that the pathway of information flow is *unidirectional*: from DNA, the archival genetic material in most organisms → by transcription to messenger RNA → and then translation into the specific amino acid sequences of proteins. Reverse transcriptases were not known then.

See Box on p. 9 for an expanded version of the central dogma.

Most people would now add another step: amino acid sequence determines protein structure (Fig. 1.3). Even when purified in dilute salt solutions, proteins adopt specific three-dimensional conformations dictated by their amino acid sequences alone. These

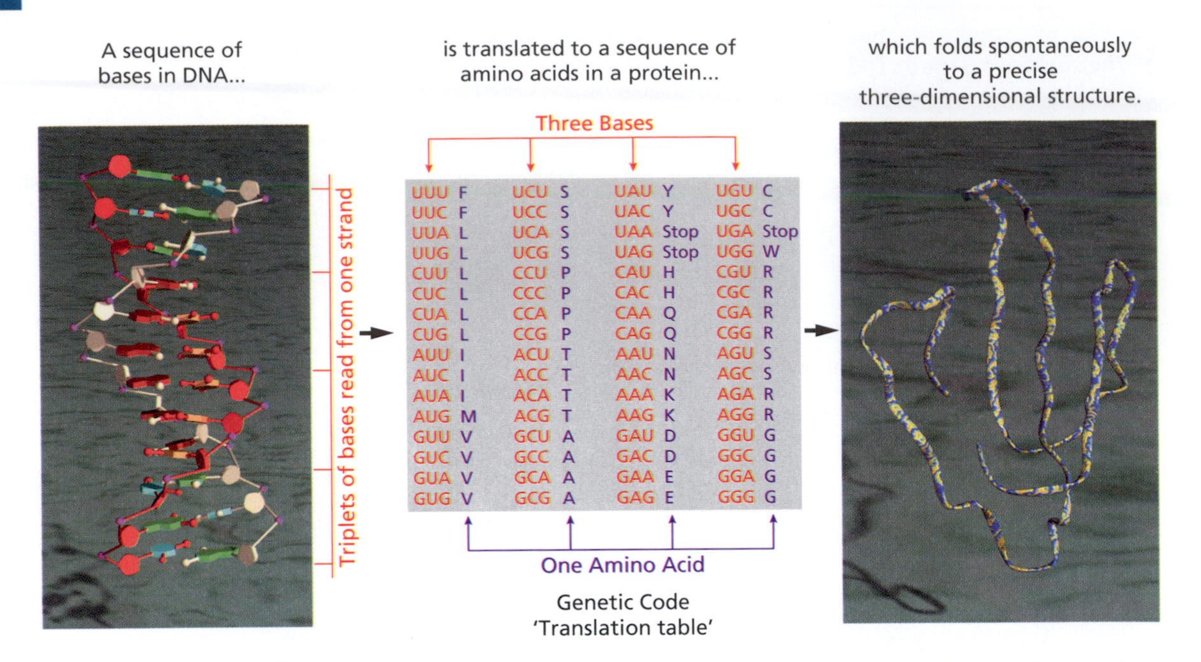

Fig. 1.3 Expression of gene sequences as three-dimensional structures of proteins. The three-dimensional structure is implicit in the amino acid sequence.

Logically, gene sequences are one-dimensional. Amino acid sequences are also one-dimensional. The spontaneous folding of proteins to form their native states is the point at which Nature makes the giant leap from the one-dimensional world of genetic and protein sequences to the three-dimensional world we inhabit.

There is a paradox: the translation of DNA sequences to amino acid sequences is very simple to describe logically; it is specified by the genetic code. The folding of the polypeptide chain into a precise three-dimensional structure is very difficult to describe logically. However, translation requires the immensely complicated machinery of the ribosome, tRNAs, and associated molecules; but protein folding occurs spontaneously.

naturally stable active conformations are the **native states** of proteins. Native protein structures are the primary agents of biochemical activity. Just as protein sequence determines protein structure, protein structure determines protein function. Perhaps a quick statement of the dogma should now read:

DNA makes RNA
makes amino acid sequence
makes protein structure
makes protein function.

The one-way flow of information imposes the need for Darwinian natural selection as a mechanism of evolution. Reverse transcriptases aside, the only way to change individual DNA sequences is by mutation, which generates variation but does not itself move a *population* to a better adapted state. Selection acts on the functions produced by the proteins encoded in the genome, and on regulatory mechanisms. Evolution governs inherited DNA sequences—including, but not limited to, alterations in **allele**

An expanded statement of the central dogma

- DNA is the archival genetic material in most organisms; some viruses have RNA genomes.
- Gene sequences from DNA are transcribed into messenger RNA. In eukaryotes, at this stage exons are spliced together, often in alternative ways. Alternative splicing patterns generate proteins with different amino acid sequences from the same gene.
- Messenger RNA is translated into the polypeptide chains of proteins, by synthesis on ribosomes.
- Amino acid sequences of proteins determine three-dimensional protein structures.
- Protein structures determine protein functions.
- Mutations change gene sequences, sometimes altering amino acid sequences, protein structures, and protein functions. Mutations in non-coding regions may affect regulation or splicing. Mutations create variation, providing the raw material on which evolution acts.
- Control mechanisms govern different rates of transcription and translation, altering the repertoire of proteins active at different stages of cell cycles and life cycles, and in different cells and tissues of higher organisms.
- Selection acts on function—including control mechanisms—effecting changes in the distributions of genome sequences within populations.

frequencies—to alter the nature and expression patterns of the proteins in the individuals of a population.

Regulation

Even the extended dogma is deceptively simple. We understand fairly well the mechanisms of many biochemical processes, in isolation. What is required for life is a set of mechanisms to integrate and control these activities.

Life as an integrated set of molecular activities involves regulation at the cellular level at a single time and at different stages of the cell cycle, and also mechanisms for unfolding programmes of development, differentiation, and ageing in higher organisms:

- In any cell at any time, there must be an appropriate distribution of the traffic through metabolic pathways to maintain overall harmony. For instance, unless successive steps in a reaction sequence have matched 'throughputs', some intermediate will accumulate.
- As a cell goes through the stages of its cycle, certain processes must be turned on and off and others adjusted. For instance, cyclin-dependent regulation of protein activity is essential for the control of cell replication and division.
- In a multicellular organism, different amounts of different proteins must be synthesized at different stages of life and in different tissues. For one example, most

humans produce a succession of five types of haemoglobin during their lifetimes: three embryonic forms, produced in the yolk sac for 5–6 weeks after conception; a fetal form, produced in the liver between 5–6 weeks after conception and birth; and an adult form afterwards. The preadult forms have properties that facilitate oxygen transfer from mother to child *in utero*. For another example, in fruit flies, genes involved in eye differentiation and vision are expressed only in the adult.

USEFUL WEB SITES

Primer on molecular genetics: **www.ornl.gov/TechResources/Human_Genome/publicat/primer/prim1.html**
Human genome project information: **www.ornl.gov/hgmis/**
Genome sequencing project information:
www.nslij-genetics.org/seq/
www–biol.univ-mrs.fr/english/genome.html **http://megasun.bch.umontreal.ca/ogmpproj.html** (organelles)

The following four sites are general ones, good starting points for 'surfing':

Home page of National Center for Biotechnology Information: **www.ncbi.nlm.nih.gov/**

Home page of the European Bioinformatics Institute, an outstation of the European Molecular Biology Laboratory: **www.ebi.ac.uk**

Home page of the Expasy molecular biology site of the Swiss Institute of Bioinformatics: **http://expasy.hcuge.ch/**

Home page of the GenomeNet WWW server, based at the Institute for Chemical Research, Kyoto University, and the Human Genome Center, Institute of Medical Science, University of Tokyo, Japan. In addition to access to standard databases, there are links to specialized databases and information-retrieval tools developed in Japan. **www.genome.ad.jp/**

RECOMMENDED READING

Alberts, B. (1998). The cell as a collection of protein machines: preparing the next generation of molecular biologists. *Cell,* **92**, 291–4.

Brenner, S. (1999). Theoretical biology in the third millenium. *Philos. Trans. R. Soc. London*, B354.

Brenner, S. (2003). Nature's gift to science (Nobel Lecture). *ChemBiolChem.*, **4**, 683–7.

Janssen, P., Audit, B., Cases, I., Darzentas, N., Goldovsky, L., Kunin, V., Lopez-Bigas, N., Peregrin-Alvarez, J.M., Pereira-Leal, J.B., Tsoka, S. and Ouzounis, C.A. (2003). Beyond 100 genomes. *Genome Biology*, 4, 402.

Knight, R.D., Freeland, S.J. and Landweber, L.F. (2001). Rewiring the keyboard: evolvability of the genetic code. *Nat. Rev. Genet.*, **2**, 49–58.

EXERCISES, PROBLEM, AND WEBLEM

Three types of questions appear at the ends of chapters. Exercises are short and straightforward applications of material in the text. Problems also depend only on information contained in the text, but require lengthier answers or, in some cases, calculations. The third category, 'Weblems', require access to the World Wide Web, preferably through a graphical interface. Weblems are designed to give readers practice with tools required for further study and research.

Exercises

1. (a) To what other amino acids (or Stop signal) can an Ala be changed by a single nucleotide substitution?

(b) To what other amino acids (or Stop signal) can a Ser be changed by a single nucleotide substitution?

(c) To what other amino acids (or Stop signal) can a Lys be changed by a single transversion? A transversion is a substitution of a purine by a pyrimidine (A or G $\rightarrow$ U or C), or a pyrimidine by a purine (U or C $\rightarrow$ A or G).

2. What is the minimum number of nucleotide substitutions required to change a codon for Phe to a codon for Lys?

3. What physicochemical class of amino acids tends to correspond to codons with a pyrimidine (u or c) in the second position?

4. During protein synthesis by the ribosome, is the tRNA charged (that is: is an amino acid bound to it) when it is in (a) the A site, (b) the P site, (c) the E site?

Problem

1. There are 4×64 possible single point mutations. (a) How many of them, if they occur within coding regions, leave the amino acid sequence unchanged? (b) How many possible transitions—purine to purine (A $\rightarrow$ G or G $\rightarrow$ A) or pyrimidine to pyrimidine (U $\rightarrow$ C or C $\rightarrow$ U) substitutions—are there? How many of them, if they occur within coding regions, leave the amino acid sequence unchanged? (b) How many possible transversions—purine to pyrimidine (A or G $\rightarrow$ U or C), or pyrimidine to purine (U or C $\rightarrow$ A or G) substitutions—are there? How many of them, if they occur within coding regions, leave the amino acid sequence unchanged?

Weblem

1. What would be the translation gene sequence of yeast hexokinase according to the yeast mitochondrial genetic code? How many amino acids would change physicochemical class from the normal protein? What are the active site residues of yeast hexokinase? Which are directly involved in catalysis? Is there any obvious reason to think that the molecule would not be functional?

Chapter 2

GENOMICS AND PROTEOMICS

LEARNING GOALS

1 **To know about the basic types of data supporting contemporary research in protein science**, including nucleic acid sequences of genomes, amino acid sequences of proteins, three-dimensional structures of proteins, metabolic pathways and their variations among different organisms, and expression patterns of proteins.

2 **To be able to chart the full-genome sequencing projects completed and in progress**, and their taxonomic distribution.

3 **To know the basic vocabulary (see Box on page 20).**

4 **To understand the concept and properties of the native state of proteins**: a compact, stable state of unique conformation, required for biological function. To know the meaning and significance of protein denaturation and renaturation.

5 **To appreciate the relationships among gene sequences, amino acid sequences, protein structures, and protein functions**. Gene sequences encipher amino acid sequences. Amino acid sequences determine the folding of proteins to their native structures. The sequence and structure of a protein determine its function(s).

It is easy to predict amino acid sequences from gene sequences, although in eukaryotes, multiple splicing patterns create a serious complication. It is difficult to predict protein structure and function from amino acid sequence.

6 **To understand the distinctions between**:

- primary structure—the chemical bonding of the atoms of a protein;
- secondary structure—the formation of helices and sheets from particular regions of a protein;
- tertiary structure—the way the helices and sheets are organized and interact in space; and
- quaternary structure—the formation of multisubunit proteins from individual polypeptide chains.

7 **To recognize the effectiveness of building up proteins from independent modules, as a means of generating variety of structure and function.**

8 **To be able to recognize, in a drawing of a protein structure**: helices, parallel and antiparallel sheets, β-hairpins, β-barrels.

9 **To appreciate some of the great variety of protein structures and functions;** understanding the purpose of comparative analysis of protein folding patterns.

10 **To distinguish various types of regulatory mechanisms, and the possible states of a network of activities that they can produce.**

11 **To understand the experimental techniques and results of measurements of protein expression patterns,** including mass spectrometry and microarrays.

12 **To develop skills in using sites on the World Wide Web to gain access to major databases in protein science,** including databases of nucleotide and amino acid sequences, protein structures, metabolic pathways, and the scientific literature.

Genome sequences

A genome is a specification of a potential life. The genome of an organism is the sequence of its main genetic material, usually one or more molecules of DNA. Organelles—mitochondria and chloroplasts—and plasmids contribute small amounts of additional genetic material. We now know the full genomes of approximately 100 prokaryotes, many viruses, organelles and plasmids, and over 12 eukaryotes, representing all major categories of living things (see Box and Figs 2.1 and 2.2).

Completed eukaryotic nuclear genomes

Type of organism	Species	Genome size (10^6 base pairs)
Primitive microsporidian	*Encephalitozoon cuniculi*	2.5
Fungi	*Saccharomyces cerevisiae* (Baker's yeast)	12.1
	Schizosaccharomyces pombe	13.8
	Neurospora crassa	40
Nematode worm	*Caenorhabditis elegans*	100
Insects: fruit fly	*Drosophila melanogaster*	180
mosquito	*Anopheles gambiae*	278
Malarial parasite (carried by *A. gambiae*)	*Plasmodium falciparum*	22.8
Plants: thale cress	*Arabidopsis thaliana*	116.8
rice	*Oryza sativa*	400
Human	*Homo sapiens*	3400
Mouse	*Mus musculus*	3454
Rat	*Rattus norvegicus*	2556
Chicken	*Gallus gallus*	1200

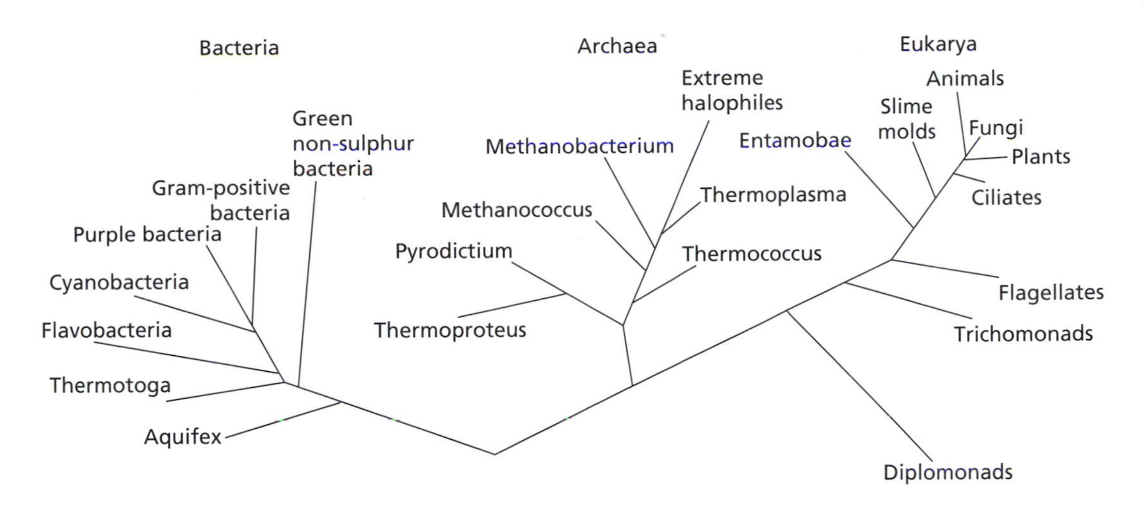

Fig. 2.1 Bacteria, Archaea, and Eukarya (or Eukaryotes) are the three major divisions of living things.

Many other full-genome projects are in progress. Figure 2.2 indicates completed and planned full-genome sequences of major eukaryotic lineages. The reader is urged to keep this chart up to date.

The ENCODE project (ENCyclopedia of DNA Elements) has the goal of determining the functions of all significant regions of a selected portion of the human genome, including coding and regulatory regions. For a selected 1% of the human genome (about 30 Mb), the corresponding regions in 30 vertebrate genomes will be sequenced. These data will illuminate each other. ENCODE will apply a variety of experimental and computational techniques, including comparative genomics. Lessons learned from the selected subset will guide the scaling up of successful methods to analysis of the entire genomes.

Gene sequences determine amino acid sequences

A gene is a segment, of DNA or viral genomic RNA, that encodes the sequence of a protein or structural RNA. In prokaryotes, genes for proteins are contiguous sequences of the DNA. In eukaryotes, genes may be split into **exons** (**ex**pressed regi**ons**) and **introns** (**int**ervening regi**ons**). Introns are transcribed but not translated: following synthesis of a full-length messenger RNA from the gene—introns plus exons—the introns are spliced out to produce a mature messenger RNA to be translated by the ribosome. The splicing is carried out in the nucleus by a large ribonucleoprotein particle called the **spliceosome**.

See Box: an example from rice.

In eukaryotes alternative splicing combinations can produce more than one mature message from the same region of the DNA. Each of these messages produces a different protein. Alternative splicing patterns thereby give genes an extra dimension of

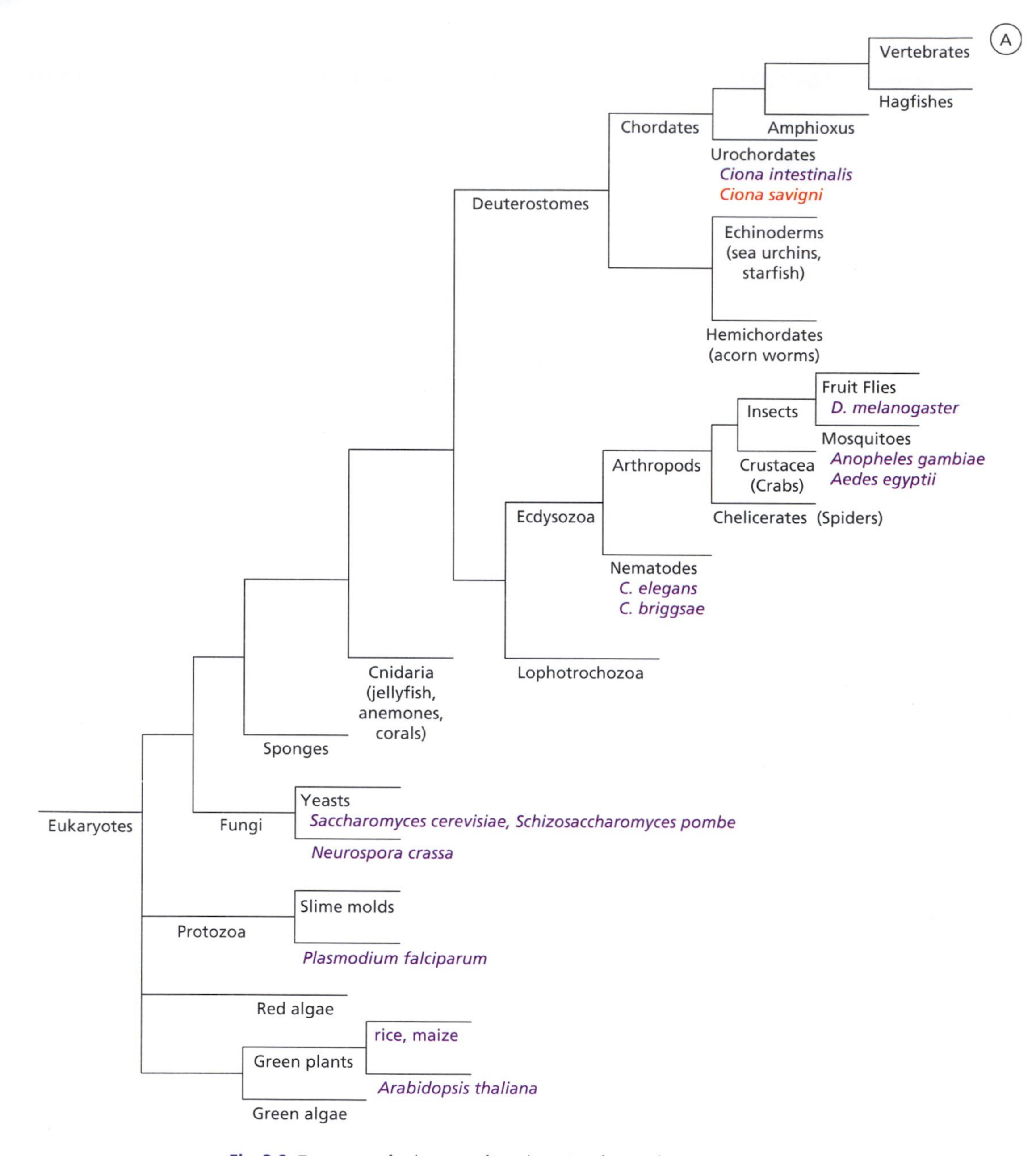

Fig. 2.2 Taxonomy of eukaryotes, from the point of view of genomic sequencing. Species for which complete genome sequences are known appear in blue. Species for which the determination of complete genome sequences is in progress appear in red.

versatility. They also make it more difficult to interpret and annotate a genome sequence! It is estimated that about half the genes in the human genome show alternative splicing.

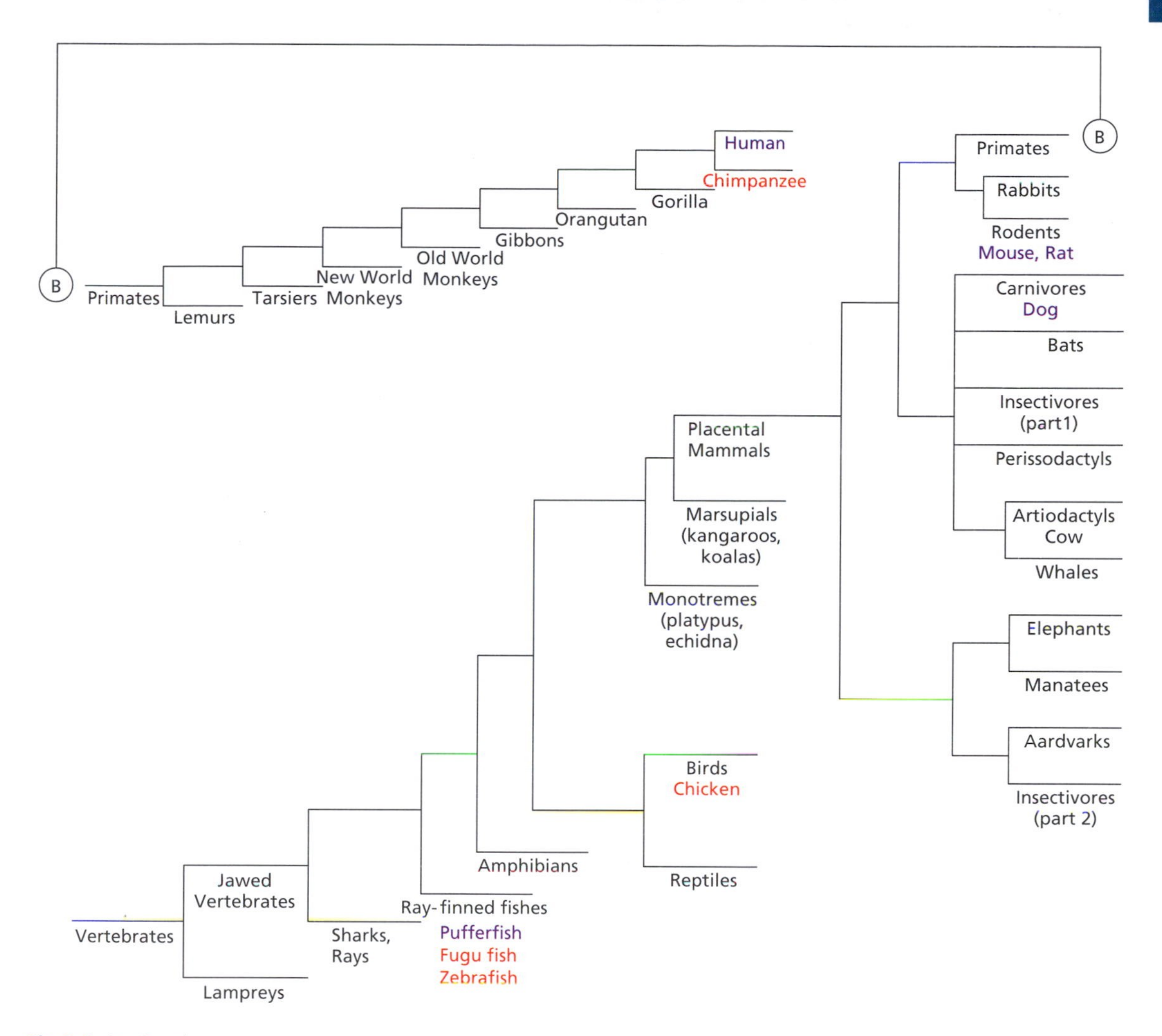

Fig. 2.2 Continued

Reminiscent of Saul Steinberg's 1976 New Yorker cover, 'View of the World from 9th Avenue', groups more closely related to the human appear in greater detail. This reflects the focus of attention of the genomic sequencing community.

The generation of the very great number and variety of antigen-binding segments of antibodies by the immune system also arises from the combinatorial splicing of genetic segments, but this is done at the DNA level, during the differentiation of B cells. Splicing of the antigen-binding segments with constant segments of immunoglobulins is done at the mRNA level.

Another kind of splicing is done at the protein level. **Inteins** are proteins that catalyse the excision of a stretch of residues within their own sequences.

The ribosome reads the messenger RNA and synthesizes a polypeptide chain at its direction. Usually, the protein is a faithful translation of the messenger RNA sequence. However, there are exceptions. In some cases, one genomic DNA sequence encodes parts of more than one polypeptide chain. For example, in *E. coli* one gene codes for

An example of alternative splicing in rice *(Oryza sativa)*

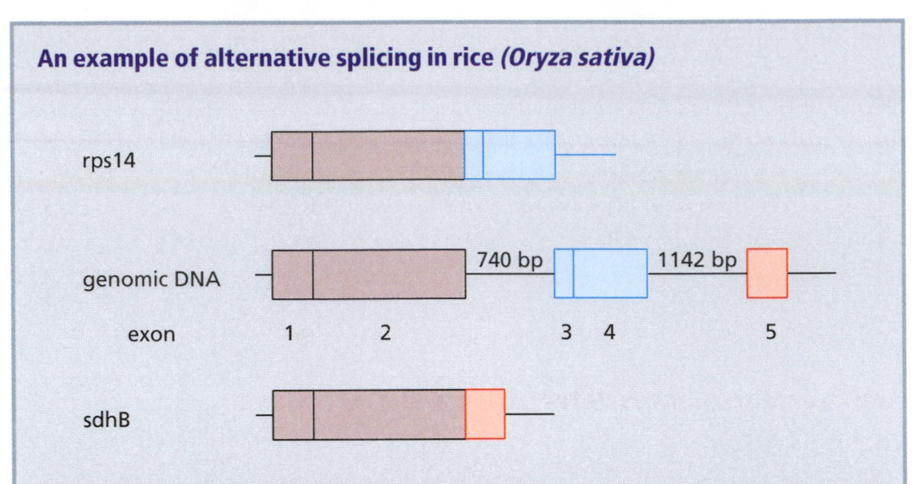

A region of chromosome 8 in the nuclear genome of rice contains five exons (centre strip), encoding two *mitochondrial* proteins: rps14 (ribosomal protein 14 of the small subunit) and sdhB (the B subunit of succinate dehydrogenase). rps14 is the translation of exons 1, 2, 3, and 4. sdhB is the translation of exons 1, 2, and 5.

These proteins are synthesized in the cytoplasm and transported into the mitochondria. It is likely that the genes encoding these proteins were originally in the mitochondrial genome. They illustrate the traffic between organelle and nuclear genomes. A region in the rice mitochondrial genome similar to *rps14* is translated, but has become non-functional as a result of four single-nucleotide deletions that destroy the **reading frame**. Certain other higher plants have functional mitochondrial genes for rps14 (broadbean, rapeseed), whereas others resemble rice in containing non-functional mitochondrial *rps14* genes but functional nuclear genes encoding the mitochondrial protein (potato, *Arabidopsis*).

Compared to other *rps14* genes, the protein encoded by the rice nuclear gene for mitochondrial rps14 has an N-terminal extension derived from the *sdhB* exons. Incorporation of the additional exons provides the protein with a targeting signal directing the protein to mitochondria. This region of the protein is cleaved off in the mitochondria.

Genes similar to *sdhB*, in contrast, have not been observed in plant mitochondrial genomes. It is likely, therefore, that the move of the *sdhB* gene from the mitochondrial → nuclear genome is an old event, and the move of the *rsp14* gene a relatively recent one.
(Names of genes are in italics, names of proteins are in roman type.)

both the τ-and γ-subunits of DNA polymerase III. Translation of the entire gene forms the τ-subunit. The γ-subunit corresponds approximately to the N-terminal two-thirds of the τ-subunit. A frameshift on the ribosome leads to premature chain termination 50% of the time, yielding a 1:1 expression ratio of τ-and γ-subunits.

Another mechanism breaking the fidelity of mRNA → protein translation is the enzymatic modification of certain bases between transcription and translation, a process known as 'mRNA editing'. In the human intestine, two isoforms of apolipoprotein B appear. To produce one of these forms, mRNA editing alters one nucleotide, changing a C to a U. This turns a Gln codon (caa) into a Stop codon (uaa), truncating the altered isoform to half the size of the longer form which is synthesized from the unedited mRNA.

Post-translational modifications alter many proteins, often with important effects on function. Some of these are additions of a variety of chemical groups, for example glycosylation. Others involve removal of peptides. (1) Some proteolytic enzymes are synthesized as inactive **zymogens**. Cleavage of a 16-residue, N-terminal peptide from trypsinogen results in a conformational change to active trypsin. (2) Active insulin contains two polypeptide chains, of 21 and 30 amino acids. The precursor proinsulin is a single polypeptide chain from which excision of an internal peptide produces the mature protein.

The proinsulin stage appears to be essential for proper folding—see Chapter 6.

Amino acid sequences determine protein structures

Proteins are polymers of amino acids containing a constant **main chain** or **backbone** of repeating units, with a variable **side chain** attached to each:

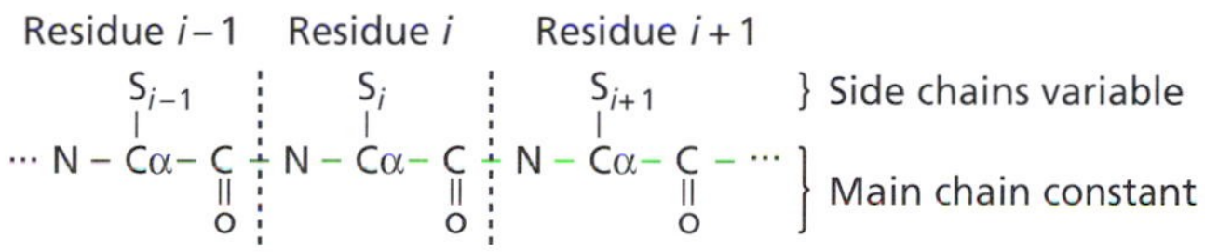

```
Residue i−1   Residue i      Residue i+1
     S_{i−1}  :    S_i    :      S_{i+1}       } Side chains variable
     |        :    |      :      |
··· N − Cα− C : N − Cα− C : N − Cα− C − ···    }
            ||:         ||:          ||         } Main chain constant
            O :         O :          O         }
```

Think of strings of Christmas-tree lights: the wire is like the repetitive backbone; the order of colours of the bulbs is variable. Each protein has a unique sequence of the side chains that determines its individual characteristics.

Every protein has a free amino terminus at one end and a free carboxy terminus at the other (except for a few cyclic polypeptides). Amino acid sequences are always given in order from the N-terminal to the C-terminal. This is also the order in which ribosomes synthesize proteins; ribosomes add amino acids to the free carboxy terminus of the growing chain. The *directionality* of the chain allows us to distinguish parallel and antiparallel strands in sheets (see Fig. 2.3).

The amino acid sequence of a protein, together with any post-translational modifications, specify the **primary structure** of a protein, the fixed chemical bonds. Because the chain is flexible, the primary structure is compatible with a very large number of spatial conformations of the main chain and side chains. However, each protein, under suitable conditions of solvent and temperature, will *spontaneously* adopt a unique structure, the **native state** (see Box, p. 21).

See Box for a general Glossary.

The conformation of the native state is dictated by the amino acid sequences.

For a protein to take up a unique conformation means that evolution has produced a set of interresidue interactions that stabilize the desired state, and that no *alternative* conformation has comparable stability. How is this achieved? Any possible conformation of the chain places different sets of residues in proximity. The interactions of the side chains and main chain, with one another and with the solvent and with ligands, determine the stability of the conformation. Proteins have evolved so that one folding pattern of the chain produces a set of interactions that is significantly more favourable than all others. This corresponds to the native state.

Protein structure—basic vocabulary

Polypeptide chain	Linear polymer of amino acids.
Main chain	Atoms of the repetitive concatenation of peptide groups ... N–Cα–(C=O)N–Cα–(C=O) ...
Side chains	Sets of atoms attached to each Cα of the main chain. Most side chains in proteins are chosen from a canonical set of 20.
Primary structure	The chemical bonds linking atoms in a protein.
Hydrogen bond	A weak interaction between two neighbouring polar atoms, mediated by a hydrogen atom.
Secondary structure	Substructures common to many proteins, compatible with main-chain conformations free of interatomic collisions, and stabilized by hydrogen bonds between main-chain atoms. Secondary structures are compatible with all amino acids, except that a proline necessarily disrupts the hydrogen-bonding pattern (see Chapter 3).
α-Helix	Type of secondary structure in which the chain winds into a helix, with hydrogen bonds between residues separated by four positions in the sequence (Fig. 2.3(a)).
β-Sheet	Another type of secondary structure, in which sections of main-chain interact by lateral hydrogen bonding (Figs 2.3(b),(c)).
Folding pattern	Layout of the chain as a curve through space.
Tertiary structure	The spatial assembly of the helices and sheets, and the pattern of interactions between them. (Folding pattern and tertiary structure are nearly synonymous terms.)
Quaternary structure	The assembly of multisubunit proteins from two or more monomers.
Native state	The biologically active form of a protein, compact and low-energy. Under suitable conditions proteins form native states spontaneously.
Denaturant	A chemical that tends to disrupt the native state of a protein.
Denatured state	Non-compact, structurally heterogeneous state formed by proteins under conditions of temperature extremes, or high concentrations of denaturant.
Post-translational modification	Chemical change in a protein after its creation by the normal protein-synthesizing machinery.
Disulphide bridge	Sulphur–sulphur bond between two cysteine side chains. A simple example of a post-translational modification.

Characteristics of the native states of proteins

- They have a definite three-dimensional structure, dictated by the amino acid sequence.
- Many similar amino acid sequences produce similar structures. That is, protein structures are robust to mutation—not to *all* possible mutations, but to many. This provides pathways for evolution to explore.
- The stabilization of native states is only marginal, about 20–60 kJ mol^{-1}—the equivalent of about two hydrogen bonds.
- Native states are compact, well-packed structures. Only rarely are there holes or channels.
- The conformations of most of the individual residues are unstrained.
- Burial of hydrophobic, or oil-like, side chains removes them from unfavourable contacts with water. (See Chapter 3: The **hydrophobic effect**.)
- Burial of main-chain polar groups requires satisfying their hydrogen-bonding capacity. This is achieved primarily by forming **secondary structures**: α-helices and β-sheets (see text).

Secondary structure

Helices and sheets are favourable conformations of the chain that recur in many proteins

α-Helices and **β-sheets** are two conformations of the polypeptide chain that appear in many proteins. They: (1) keep the main chain in an unstrained conformation, and (2) satisfy the hydrogen-bonding potential of the main-chain N–H and C=O groups.

Helices and sheets are like 'Lego' pieces, standard structural units which can be assembled in different ways, from which many proteins are built. They were predicted by Linus Pauling and Robert Corey in 1950 from model building. Upon reading their report, Max Perutz went into his laboratory and photographed the X-ray diffraction pattern of a horse hair. The keratin in the hair contained the predicted helices! Later he and his colleagues found them to be the basis of the structures of myoglobin and haemoglobin, the first globular protein structures determined.

α-Helices are formed from a single consecutive set of residues in the amino acid sequence (Fig. 2.3(a)). They are a *local* structure of the polypeptide chain; that is, they form from a set of residues consecutive in the sequence. The hydrogen-bonding pattern links the C=O group of residue i to the H–N group of residue $i + 4$.

Proteins also contain 3_{10} helices, in which hydrogen bonds form between residues i and $i + 3$; and π-helices, in which hydrogen bonds form between residues i and $i + 5$. The 3_{10} and π-helices are much rarer than α-helices.

β-Sheets form by the lateral interactions of several independent sets of residues (Figs 2.3(b),(c)). They can bring together sections of the chain widely separated in the amino acid sequence. *Strands* are *local* structures, *sheets* can be *non-local* structures. Figure 2.3(b) shows an ideal β-sheet, with all strands *parallel*. The strands are not fully extended, but accordion-pleated. The sheet is not flat, but each strand is rotated from its neighbours so that the sheet appears twisted in propeller fashion. Each of the three central strands in Fig. 2.3(b) has two neighbouring strands in the sheet. The two edge strands have only one neighbour in the sheet.

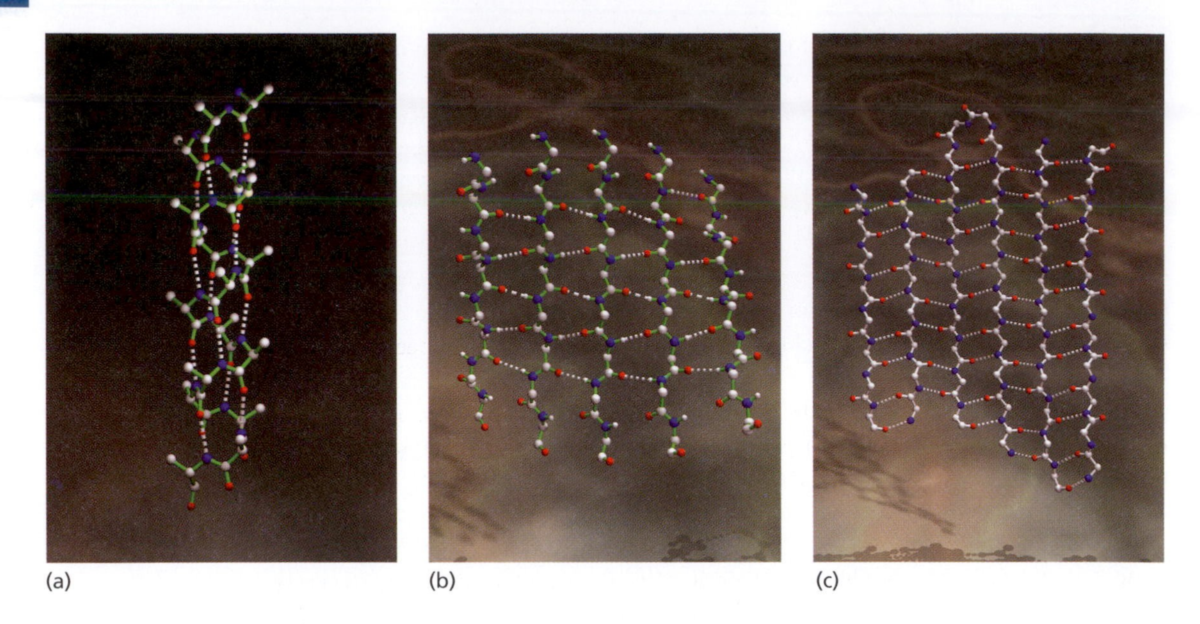

Fig. 2.3 α-Helices and β-sheets are two 'prefabricated' structures that appear in many proteins.

Figure 2.3(c) shows an *antiparallel* β-sheet, from bacteriochlorophyll a protein, in which every strand points in a direction opposite to its neighbours. β-Sheets can be parallel, antiparallel, or mixed, with respect to the relative directions of the strands. In Fig. 2.3(c), the central pair of adjacent strands is connected through a short loop called a **β-hairpin**.

The strands of some sheets form a closed structure, with no edge strands, called a **β-barrel**. The strands lie roughly on the surface of a cylinder.

Tertiary and quaternary structure

Many proteins contain helices and sheets. Different proteins combine helices and sheets in different ways, to create different spatial arrangements of the chain, and different patterns of interaction between the helices and sheets. This is called the **tertiary stucture**, or **folding pattern**.

Many proteins contain more than one subunit, or monomer. There may be multiple copies of the same polypeptide chain, or combinations of different polypeptide chains. The assembly of the subunits is called the **quaternary structure**.

Protein stability and denaturation

The native structure can be broken up, by heating or by high concentrations of certain chemicals such as urea—a process called **denaturation**. Denaturation destroys the secondary, tertiary, and quaternary structure, but leaves the polypeptide chain intact.

If the original conditions are restored the protein will spontaneously *renature*, re-forming its original structure and recovering its biological activity.

Proteins are only marginally stable, and achieve stability only within narrow ranges of solvent and temperature conditions. Overstep these boundaries, and proteins denature: they lose their definite compact structure, and take up states with disorder in the backbone conformation, and few, if any, specific interactions among residues: the pieces of the jigsaw puzzle have been pulled apart and deformed. Homeostatic mechanisms that keep the internal environment of our bodies at a relatively constant temperature, salt concentration, and acidity are essential to health, if only because proteins lose the effectiveness of their function if any of these conditions varies beyond limits that are in many cases quite narrow.

Formation of the native state is a *global* property of the protein. In most cases, the entire protein (or at least a very large part) is necessary. This is because many of the stabilizing interactions involve parts of the protein that are very distant in the polypeptide chain, but brought into spatial proximity by the folding. The β-sheets shown in Fig. 2.3(c) are good examples.

The structure of the native state is essential for biological activity. The residues forming the active site of an enzyme may be distant in the sequence, but juxtaposed in the native structure. Typically, an active site may involve only 10% of the residues of a protein, or even fewer. The rest of the structure is required as scaffolding to create the exquisite spatial relationships among the active residues. (Later we shall see that still more is involved: (1) The protein may provide more than a *fixed* scaffolding. In some cases, conformational changes are an essential part of the mechanism of activity. (2) Some residues in active sites are in strained conformations, which play a role in catalysis. The scaffolding 'pays for' the energy required to allow these residues to adopt unfavourable conformations.)

In the very crowded regime inside cells, the smooth process of protein folding as observed in dilute solution is threatened by aggregation. Protein aggregates are associated with several diseases, including Alzheimer, Huntington, and the **prion** diseases, the **spongiform encephalopathies**. Misfolded or partially folded proteins are particularly prone to aggregation. Many mutant proteins show an increased tendency to aggregate.

See Chapter 7.

Cells defend themselves against aggregation of misfolded molecules with **chaperone proteins**. (The metaphor was suggested by the idea that chaperones inhibit the formation of improper alliances.) Chaperones unfold misfolded proteins and give them another chance to fold properly. But chaperones do not themselves contain any information about the correct structure of any protein. The fact that they can catalyse the folding of many different proteins proves this.

The relationship between amino acid sequence and protein structure is robust

The rules relating protein sequence to structure can tolerate many, but not all, mutations. This is of the utmost significance for evolution, because it permits the exploration of sequence variations. A protein structure that required a unique amino acid sequence—so that any mutation would destroy it—would not be observed in Nature because evolution could never find it. (If any mutation would destroy the structure, it could not have any precursor.) If such a protein were designed and its gene artificially synthesized and released into Nature, it would be unstable to mutation.

If we compare corresponding proteins from related species, we find that the sequences have diverged. However, in many instances the structures and functions have stayed very much the same. As an example, histidine-containing phosphocarrier

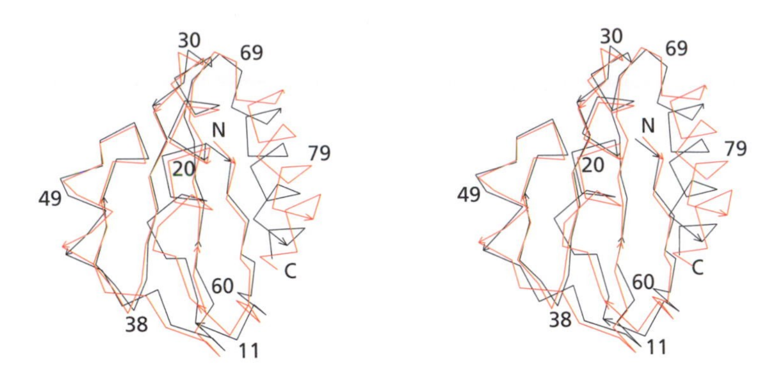

Fig. 2.4 Superposition of structures of histidine-containing phosphocarrier proteins from *Escherichia coli* [1POH] (black) and *Streptococcus faecalis* [1PTF] (red). Although over 60% of the residues have mutated, the folding pattern is completely intact.

proteins from *Escherichia coli* and *Streptococcus faecalis* have only 38% identical residues in an alignment of their sequences:

```
                     10        20        30        40
E. coli      MFQQEVTITAPNGLHTRPAAQFVKEAKGFTSEITVTSNGKSASA
             |   |  | |  | | |||   |  |  | | |     |||
S. faecalis  MEKKEFHIVAETGIHARPATLLVQTASKFNSDINLEYKGKSVNL

                     50        60        70        80
E. coli      KSLFKLQTLGLTQGTVVTISAEGEDEQKAVEHLVK-LMAE-LE
             ||      ||  ||   |||   | ||       |  |  | |
S. faecalis  KSIMGVMSLGVGQGSDVTITVDGADEAEGMAAIVETLQKEGLA
```

However, their structures retain the same general spatial course of the main chain (Fig. 2.4).

A survey of protein structures and functions

General classes of protein function, and the types of structures that provide them, include:

General classes of protein structures and functions

Fibrous proteins provide mechanical functions, including (but not limited to) building connective tissue. Most form long, thin, insoluble rope-like structures.

Enzymes catalyse chemical reactions. Most enzymes are globular molecules with a cleft in their surface that contains an active site.

Antibodies are responsible for recognizing and defending the vertebrate body against foreign molecules. Proteins related to antibodies carry out more general functions involving molecular recognition, for instance, target identification of cell interactions in neural development.

continues ...

Inhibitors interact with enzymes to reduce their activity. (Many non-protein molecules are also inhibitors.)
Carrier proteins bind specific metabolites, and transport them around the body.
Membrane proteins form structures and carry out their activities at least partly within the lipid bilayer environment of a membrane. Transmembrane proteins contain some segments that pass through the membrane, and other segments that protrude from either side of the membrane.
Receptors are responsible for detecting signal molecules, and initiating an appropriate response. Receptors that detect molecules at the outer surfaces of cells, nuclei, or organelles, and transmit the signal to the interior, are transmembrane proteins.
Regulatory proteins control the activity of other molecules. Some bind DNA sequences to regulate transcription; others phosphorylate or dephosphorylate target proteins to change their state of activation. Of course, protein inhibitors are regulatory proteins.
Motor proteins transduce chemical energy to mechanical energy. Some move themselves along fibrous protein tracks; others exert force on other molecules, as in muscle contraction.

Fibrous proteins

Traditionally, textbook presentations of fibrous proteins begin with hair and silk. These are familiar from daily life, and have the longest history of scientific study, stimulated by the textile industry. Hair and silk give the impression that fibrous proteins have purely structural roles. However, *intracellular* fibrous proteins show that the biological roles of fibrous proteins should be thought of as mechanical, not structural—in many cases dynamic rather than passive.

Eukaryotic cells contain three types of internal fibrous structures:

1. **Actin-containing microfilaments** are thread-like fibres, 3–6 nm in diameter, composed predominantly of polymers of actin. Microfilaments associated with myosin are responsible for muscle contraction.
2. **Microtubules** are cylindrical tubes ~25 nm in diameter, composed of polymers of α- and β-tubulin. They are involved in mitosis, motility, and transport. Cilia and flagella are assemblies of microtubules. Microtubules provide 'tracks' for the movements of motor proteins such as dynein and kinesin (see p. 39).
3. **Intermediate filaments** create an intracellular, three-dimensional network between the nucleus and outer cell membrane, determining the shape, form, and mechanical properties of the cell. They are a dynamic assembly, breaking up and re-forming during the cell cycle.

FtsZ is a filamentous protein involved in cell division in prokaryotes. It is the major component of the septum, that appears between the nascent progeny cells and pinches off to separate them. FtsZ occurs in all Bacteria and Archaea, and in chloroplasts but not mitochondria. It is related to the eukaryotic protein tubulin.

The FtsZ protofilament contains a linear array of monomers (Fig. 2.5). Protofilaments combine to form filaments.

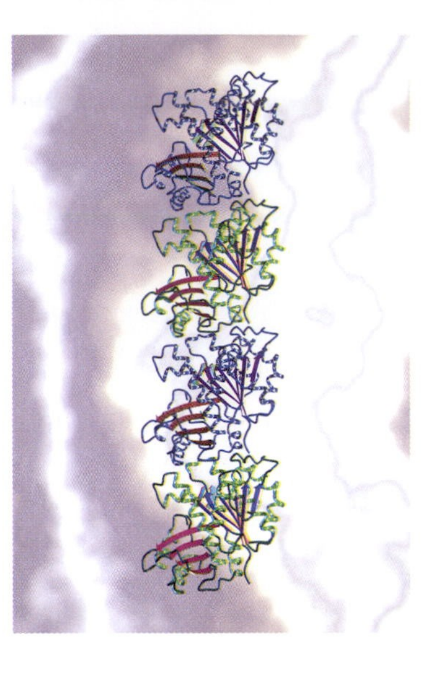

Fig. 2.5 Model of the FtsZ protofilament, based on the crystal structure of the *Methanococcus jannaschii* monomer [1FSZ], showing four subunits assembled by homology with tubulin. (I thank J. Löwe and L. Amos for the coordinates.)

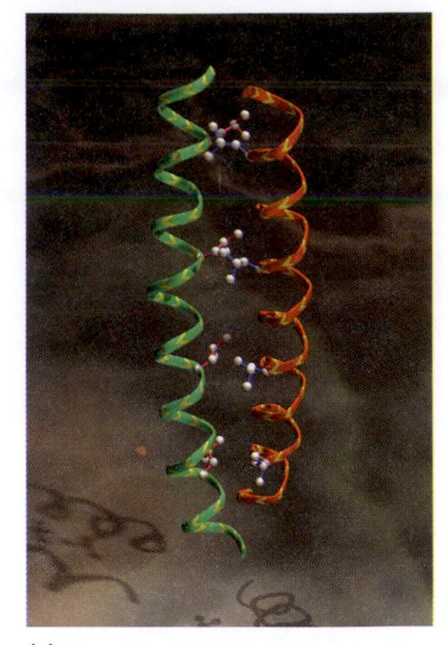

(a)

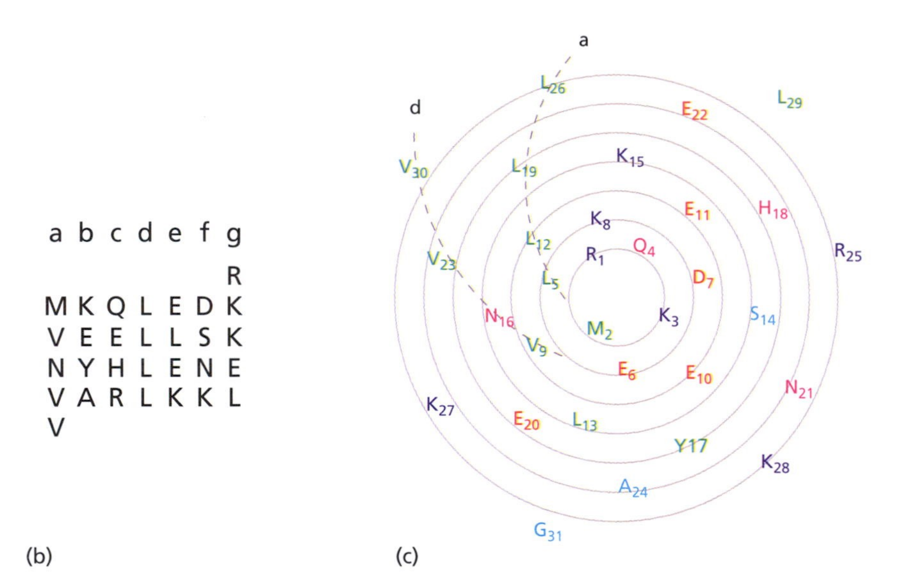

Fig. 2.6 (a) The coiled-coil structure of α-keratin also appears in the eukaryotic transcriptional activator GCN4 [2ZTA]. This structure contains two helices coiled around each other. It is known as the 'leucine zipper' because of the leucine repeats every seven residues (shown in ball-and-stick representation). The pitch is 14.7 nm. (b) The sequences of most proteins containing coiled-coils show *heptad repeats*—seven-residue patterns, the positions labelled **abcdefg**, in which the first and fourth positions (**a** and **d**) are usually hydrophobic. (c) If the sequence of GCN4 is plotted on a *helical wheel* corresponding to the geometry of a straight α-helix, the hydrophobic positions form a stripe up one face of the helix. (Colour coding: green, medium-sized and large hydrophobic; cyan, small; magenta, polar; blue, positively charged; red, negatively charged.) The **a** and **d** positions are indicated in this figure. The supercoiling of the helices around each other brings the **a** and **d** positions into even better register than suggested by this figure (see Problem 1).

Traditional fibrous proteins

When you look at other people, what you see is mostly fibrous protein. Their hair contains α-keratin, a 'rope' assembled from a **coiled-coil** of α-helices (Fig. 2.6). The helical section of a single subunit typically contains ~300 residues in each of the two strands, and is 48 nm long. There are also small capping domains at the head and tail. These units assemble into large aggregates, embedded in a matrix of cysteine-rich proteins to form a network, extensively crosslinked by disulphide bridges. The 'permanent wave' treatment of hair first breaks the crosslinks; then after setting the hair into the desired conformation, the links are allowed to re-form.

Helices are a common structural theme in biology: see Box.

The horny outer layer of the skin, and fingernails, are other forms of keratin, differing in amino-acid composition and sequence. Fingernails are less flexible than skin or hair because of a greater abundance of cysteine residues forming crosslinks. The layer of the skin just beneath the surface contains **collagen**, a glycoprotein (Fig. 2.7 and Box). The cornea of the eye is another form of collagen.

Describing the geometry of a helix

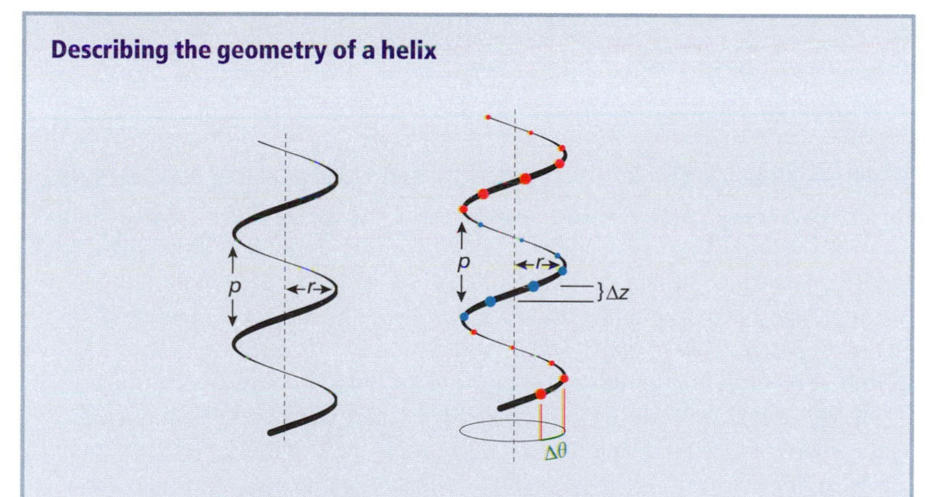

(Left) A helix formed by the coiling of a continuous line. The broken line indicates the helix axis. The structure is described completely by two numbers: the pitch p = the distance along the axis between successive turns, the radius r; and the enantiomorph or 'hand': right- or left-handed. (Right) A helix formed by the assembly of discrete units, such as the bases in one of the strands of the DNA double helix, or the amino acids in a polypeptide helix. The structure is defined by: the pitch, p; the radius, r; and the rise per residue, Δz. The number of residues per turn is equal to $p/\Delta z$. The angular difference between successive residues, $\Delta\theta = 360° \times \Delta z/p$. In this diagram there appear to be seven residues per turn (count the light blue dots). However, the number of residues per turn need not be a whole number.

More complicated aspects of helix geometry include two or more helices wound around each other, as in DNA; or supercoiling, as in leucine zippers (see Fig. 2.6) or collagen.

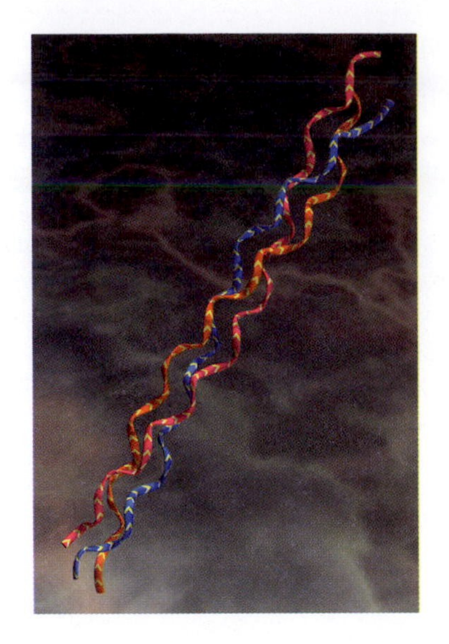

Fig. 2.7 The structure of collagen, a three-stranded supercoil formed by braiding together three polypeptide chains. In each strand, the rise per residue is approximately 0.3 nm. Each polypeptide chain itself forms a helix, with approximately 3.3 residues per turn. The repeat distance of the supercoil is approximately 1 nm [1BKV].

Many traditional clothing materials—before the development of synthetic fibres—are also protein. Wool consists of α-keratin—from the hair of sheep or other animals. (Cotton and linen, in contrast, are plant fibres, formed of cellulose.) Animal hair is, of course, chemically similar to human hair, but varies in its physical properties. (Paintbrushes have been made from the fur of many animals, including sable, ox, squirrel, pony, goat, hog, camel, and mongoose. Artists are sensitive to the variations in stiffness, and also to the differences in the retention and delivery of paint.)

Silk is β-fibroin, with the repetitive sequence ... Gly–(Ala or Ser)–Gly–(Ala or Ser) ... The cocoons of moths contain fibres of β-fibroin glued together by a second protein, sericin; adult moths secrete the proteolytic enzyme cocoonase to dissolve the sericin and let themselves out. It is a paradox that moths can digest the keratin of woollen sweaters, and the fabric of their own cocoons, but not silk scarves. The explanation is that most silk cloth contains the β-fibroin but not the sericin.

Fibrous proteins are also ubiquitous beneath the skin. Connective tissue, such as tendons, also contains collagen—indeed, collagens make up about one-third of the protein content of the human body (see Box).

Enzymes—proteins that catalyse chemical reactions

For many proteins, the focus of interest is on their interactions with other proteins or small metabolites.

Over a century ago Emil Fischer enunciated a fundamental principle of structural complementarity: 'Only with a similar geometrical structure can molecules approach

Collagens

The many different types of collagen have a common basic structure, a triple-stranded helix, with three polypeptide chains coiled around one another in a braid or plait (see Fig. 2.7). Individual chains are each ~1000 amino acids long. They have a glycine every *third* residue—the sequence scheme is $(Gly–X–Y)_n$. X and Y are mostly alanine, but are also special modified amino acids. X is often a 3- or 4-hydroxyproline, and Y a hydroxyproline or hydroxylysine. The prolines effectively prevent the chains from forming α-helices (see Chapter 3).

Individual molecules assemble into fibrils in different ways that are suitable for the differing mechanical requirements of the tissues in which they appear. So far, 12 types of human collagen have been distinguished, encoded by 28 genes on 12 different chromosomes. The differences in amino acid composition and sequence govern their modes of processing and assembly. Hydroxylysines provide sites of linkage to sugars or short polysaccharides. These carbohydrate adducts crosslink collagen molecules, contributing to the mechanical strength of the fibre. Differing amounts of hydroxylysine account for the different carbohydrate content of different collagen types. Indeed, not all types of collagen are fibrillar—for instance, type IV collagen, found in basement membrane, is not. It contains smaller stretches of repeated Gly–X–Y triplets.

Several genetic abnormalities affect the structure of connective tissue. In one form of *Ehlers–Danlos* syndrome, mutations in the gene for the enzyme lysine hydroxylase causes defective post-translational modification of lysine to hydroxylysine. This reduces carbohydrate binding, and the consequent loss of crosslinking lowers the mechanical strength. The symptoms of Ehlers–Danlos syndrome include spidery fingers and unusually high flexibility of the joints. The nineteenth-century violinist Paganini exhibited these anatomical characters—and very good use he put them to! It has been speculated that he had either Ehlers–Danlos syndrome or Marfan syndrome, a condition affecting another connective tissue protein, fibrillin.

The symptoms of scurvy, the deficiency disease caused by a lack of vitamin C, result from defective hydroxylation of prolines and lysines, weakening collagen. In severe cases, loss of integrity of gum tissue and of dentin inside teeth (dentin forms by mineral deposition on a collagen matrix) leads to loosening and ultimately falling out of teeth.

each other closely, and thus initiate a chemical reaction. To use a picture, I could say that the enzyme and substrate must fit each other like a lock and key…' Figure 2.8 gives an example of enzyme–substrate complementarity.

The rate of a reaction is typically limited by the highest energy point on a trajectory from substrate to product, called the **transition state** (Fig. 2.9). One way in which enzymes speed up reactions is by altering the energy barrier between the substrate and the product, by binding transition states more tightly than they bind substrates.

The functions of many enzymes are well understood, in terms of the relationship between the three-dimensional conformation and the physical–organic chemistry of their mechanism. (Biochemists have had an advantage over cell biologists in that many enzymes can be purified and studied in isolation from the cellular context.)

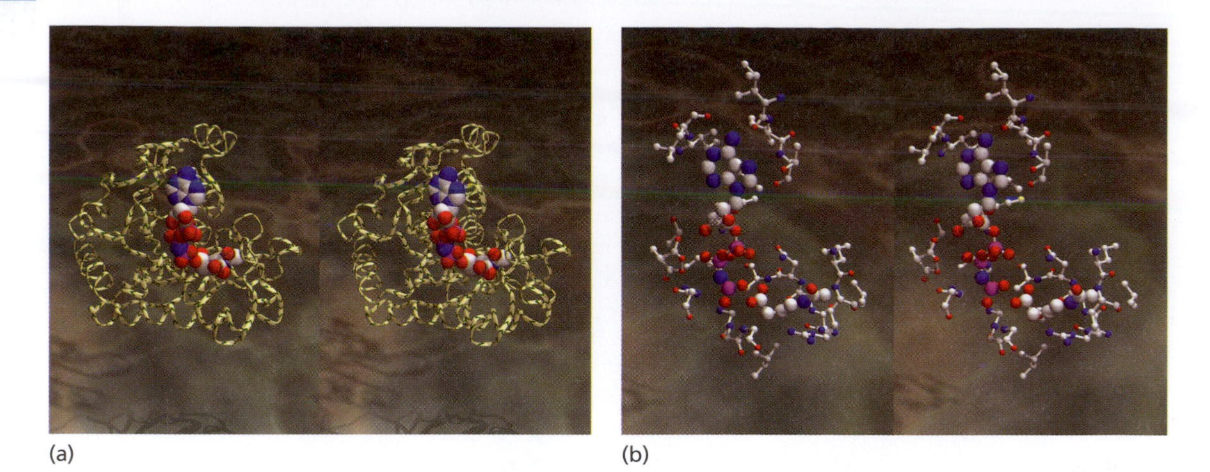

Fig. 2.8 An enzyme–substrate complex: *E. coli* *N*-acetyl-L-glutamate kinase binding the substrate *N*-acetylglutamate and the inhibitory cofactor analogue AMPPNP (instead of the natural cofactor ATP) [1GS5]. The substrate and inhibitor nestle snugly into the enzyme, which holds them in proper proximity for phosphate transfer. (a) The substrate and cofactor analogue occupy a crevice in the molecule. (b) The main chain and side chains that surround the ligands. *N*-acetyl-L-glutamate kinase catalyses a step in the biosynthesis of arginine:

$$\begin{array}{lcll} COO^- & & COO^- & \\ | & & | & \\ CHNHCOCH_3 & & CHNHCOCH_3 & \\ | & & | & \\ CH_2 & + ATP \rightarrow & CH_2 & + ADP \\ | & & | & \\ CH_2 & & CH_2 & \\ | & & | & \\ COO^- & & COOPO_3H^- & \end{array}$$

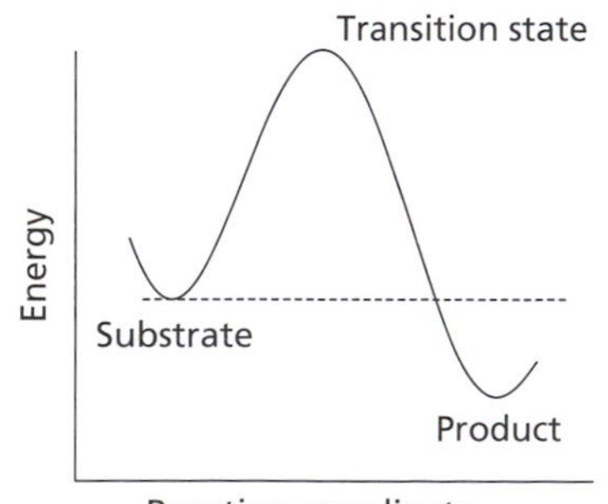

Fig. 2.9 Simplified diagram of the energetics of a reaction. Typically, substrates and products are stable species, as shown by their positions at local minima of the energy. To change substrate to product, the system must pass through intervening states of higher energy. The highest-energy peak in the trajectory is called the **transition state**. The height of this energy barrier determines the rate of reaction. To speed up reactions, enzymes can lower energy barriers by stabilizing transition states. In some cases, coupling of reactions to enzymes can provide alternative pathways of reaction, with lower transition-state energies.

Antibodies

These are a family of proteins from vertebrates that function by binding molecules foreign to the organism, notably the surface proteins of pathogens, and triggering cellular mechanisms for killing them.

Enzymes bind substrates, and juxtapose them with catalytic residues. Most antibodies show binding alone. For this reason, some attempts to design artificial enzymes have started with antibodies raised against transition-state analogues. The antibody provides the binding; the chemist can supply catalytic residues if necessary. In fact, some natural antibodies have proteolytic catalytic activity based on a mechanism similar to that of chymotrypsin.

The immune system is discussed in Chapter 7.

Catalytic antibodies are called abzymes.

Inhibitors

Inhibitors are common weapons in the biochemical wars between species. Many drugs are based on adaptations of natural inhibitors; others are the products of chemists' ingenuity.

- Leeches use the thrombin inhibitor hirudin to prevent their victims' blood from clotting (Fig. 2.10).
- Many drugs for the treatment of AIDS are inhibitors of the HIV-1 protease (Fig. 2.11). One general approach to inhibitor design is that of using **peptidomimetics**,

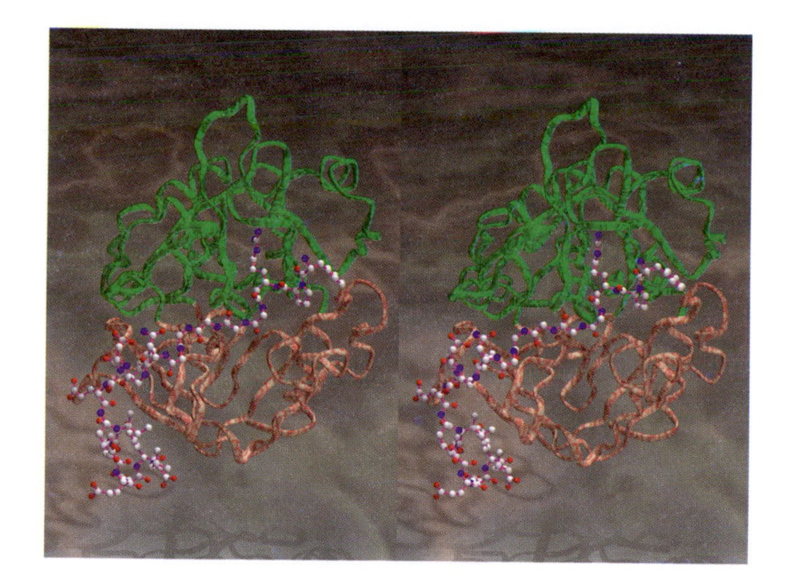

Fig. 2.10 Thrombin, a key player in the control of blood coagulation, is a member of the chymotrypsin family of serine proteases. The active site lies in a cleft between two domains. The two domains are homologues, that arose by gene duplication and divergence.

Human thrombin binds the synthetic inhibitor hirulog-3, a 20-residue peptide related to the natural inhibitor hirudin from the leech [1ABI]. Hirulog interacts with both the catalytic site and an anion-binding exosite specific to thrombin.

It is common for binding sites on proteins to appear at interfaces between domains or subunits.

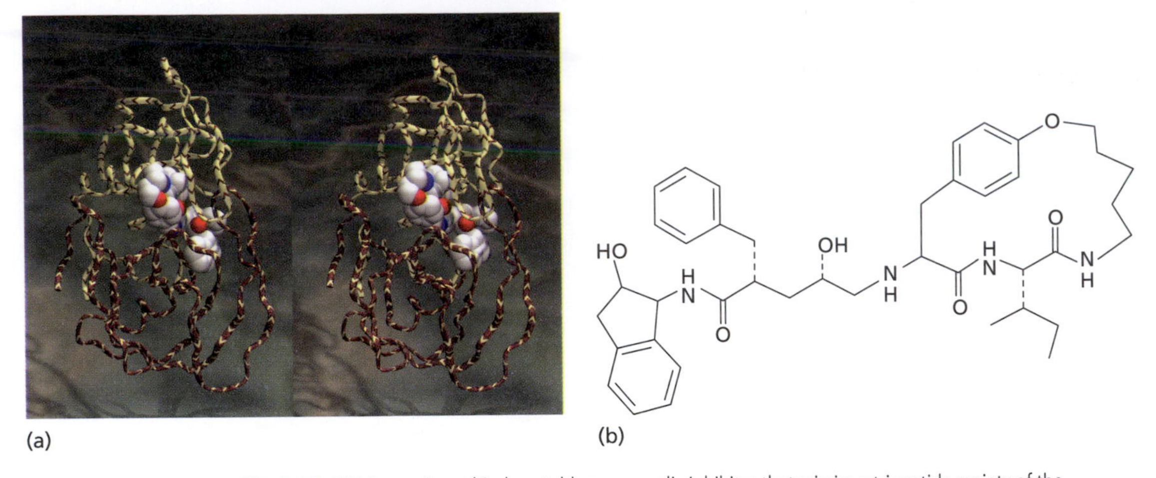

Fig. 2.11 HIV-1 proteinase binds a stable macrocyclic inhibitor that mimics a tripeptide moiety of the natural substrate [1D4K]. (a) The proteinase is a homodimer, with a binding site shared between the monomers. (In order to show the ligand, the orientation is oblique to the axis of symmetry.) (b) Chemical structure of inhibitor.

compounds that look enough like peptides to compete for a binding site, but cannot undergo reaction.

Carrier proteins

Retinol-binding protein circulates in the blood plasma in complex with transthyretin, to transport retinol (Fig. 2.12). Transthyretin is also a carrier protein, transporting thyroxine. Another example is the best-known transport protein, haemoglobin. Haemoglobin transports O_2 from the lungs where it is absorbed, to the tissues; and CO_2 from the tissues back to the lungs, where it is excreted.

Membrane proteins

Membranes are the wrapping around cells and subcellular compartments and organelles. They contain phospholipid bilayers (Fig. 2.13). The acyl tails of the phospholipids mimic a non-polar organic medium, about 3 nm thick. Between this layer and the surrounding aqueous medium, on either side, is a transitional layer approximately 1.5 nm thick.

It is estimated that approximately 30% of genes in the human genome encode membrane proteins.

As an environment for proteins, the membrane differs from a conventional aqueous solution not only in containing an organic medium, but in being **anisotropic**—that is, having a favoured direction. Proteins interacting with membranes have definite orientations, not only parallel to and perpendicular to the membrane, but with respect to the inside and outside of a cell or compartment.

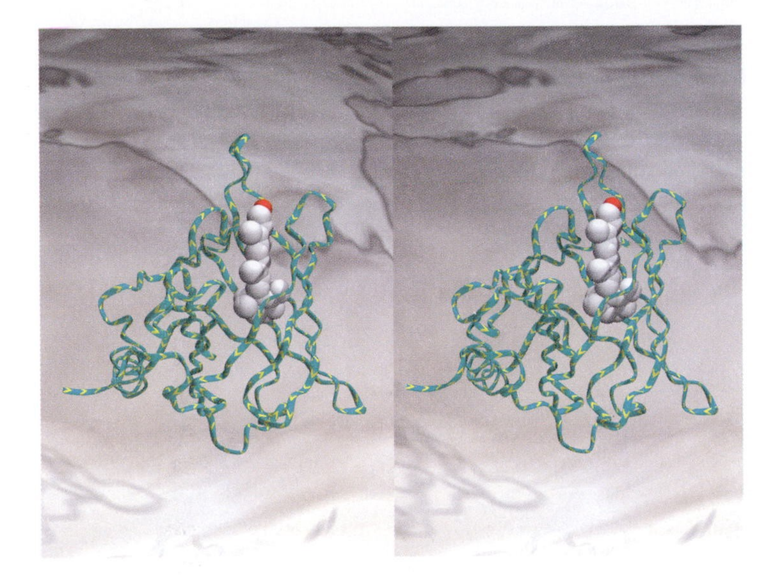

Fig. 2.12 Retinol-binding protein [1RBP] is an example of a β-barrel, in which the strands of a β-sheet are wrapped around into a cylinder, with continuous lateral hydrogen bonding around the circumference of the barrel. By forming barrels of different sizes, from different numbers of strands, proteins can create interior cavities of different sizes to bind different ligands.

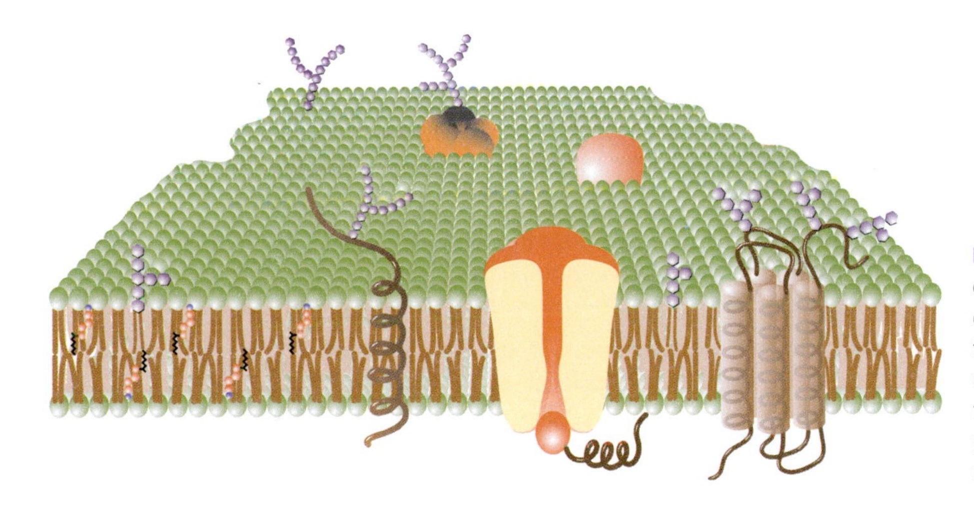

Fig. 2.13 Schematic diagram of the structure of a membrane. This figure gives an unwarranted impression that the membrane is a perfectly ordered and relatively rigid structure.

Many proteins are designed to sit within membranes. Their adaptations include regions containing only non-polar residues that interact with the organic layer of the membrane. These regions, which thread their way through the membrane, are connected through regions in the transitional layers that interact with the phospholipid head groups, and other regions entirely outside the membrane that have physicochemical properties similar to those of soluble proteins.

Structurally, membrane proteins fall into two major classes: (1) proteins containing transmembrane helices, such as bacteriorhodopsin (Fig. 2.14), and proteins containing

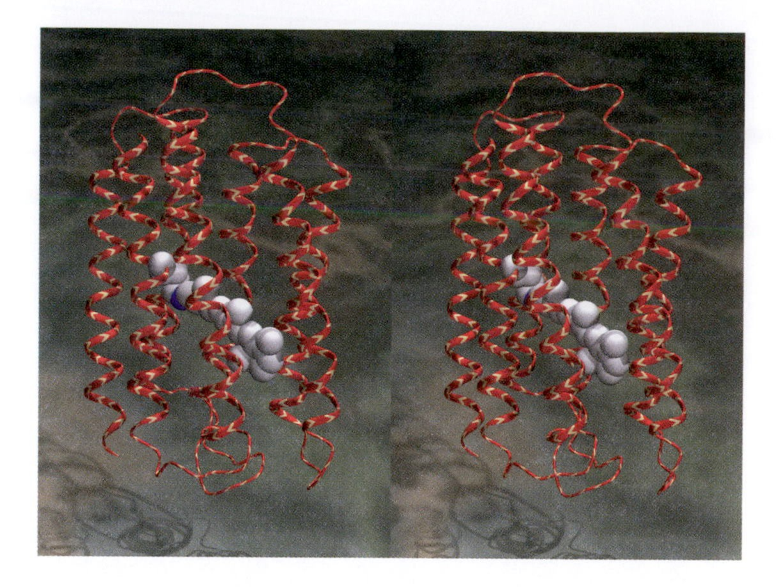

Fig. 2.14 The structure of bacteriorhodopsin [2BRD] from *Halobacterium salinarum*, illustrating the common theme of a 7-transmembrane helix structure. Bacteriorhodopsin is a light-driven pump, converting light energy absorbed by the chromophore, **retinal**, to a proton gradient across the membrane.

β-barrels, such as the transmembrane domain of *E. coli* outer membrane protein-A (ompa) (Fig. 2.15; compare this β-barrel with retinol-binding protein).

Structural characteristics of membrane proteins include:

- A typical helical membrane protein contains α-helices approximately 20 residues long, connected by loops typically 10–15 residues long (although there are many exceptions).
- The 'positive-inside rule': the loops between helices live either entirely inside or entirely outside the cell. Those inside contain a preponderance of positively charged residues.
- Segments that interact with head groups, in the transitional region, tend to be enriched in the large, polar aromatic residues tryptophan and tyrosine.

Membrane proteins mediate the exchange of matter, energy, and information between cell interiors and surroundings.

- **Channels, pores, pumps, and carriers** selectively control the import and export of molecules through membranes. (See Chapter 6.)
- **Energy transduction** is the function of a number of membrane-resident proteins and protein complexes, including opsins, photosynthetic reaction centres, and ATPases. A common feature of the mechanism is the generation or release of concentration gradients across cell or organelle membranes. The electrochemical potential energy of a concentration disequilibrium across a membrane is an intermediate in a number of energy transformations. (See Chapters 5 and 6).
- **Receptors detect and report the arrival of signals at the cell surface.** (See Figs. 2.14 and 2.15.)

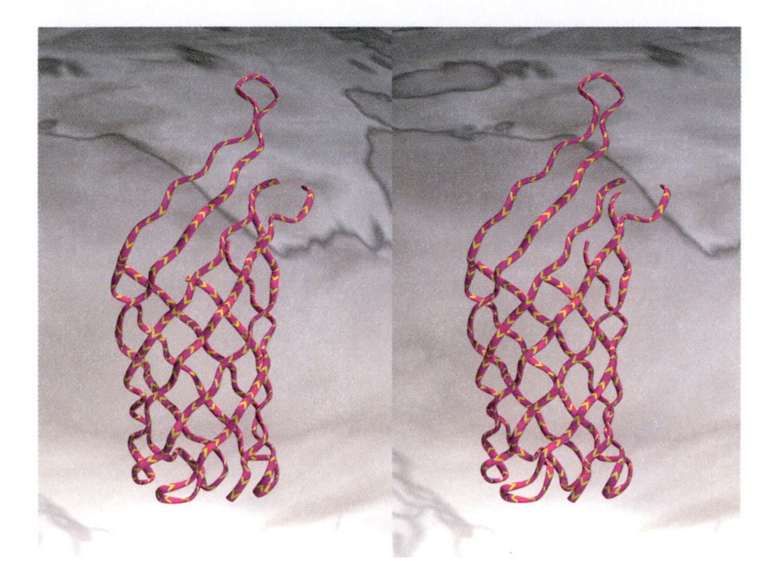

Fig. 2.15 The structure of *E. coli* outer membrane protein A (ompa) [1QJP], a β-barrel protein traversing the cell membrane. In Gram-negative bacteria, ompa appears as a structural membrane protein interacting with lipoproteins. It also serves as a docking site for the bacteriocidal protein colicin, and some phages, and is also involved in conjugation.

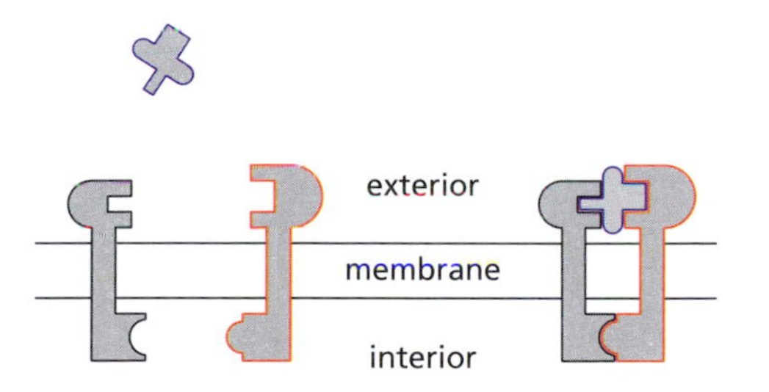

Fig. 2.16 Schematic diagram of the dimerization mechanism for transmission of a signal across a cell membrane. Receptor molecules contain exterior, transmembrane, and interior segments. In the absence of ligand they are monomeric (left). Binding of ligand brings together the exterior *and* the interior domains. Dimerization of the interior domains causes conformational changes that activate processes inside the cell, as a consequence of binding of the signal molecule outside the cell.

Receptors

How can a cell detect a signal molecule in the external medium and report its arrival to the cell interior, without the signal molecule itself ever needing to enter the cell? Many receptors use an ingenious dimerization mechanism (see Fig. 2.16). The receptor has external, transmembrane, and internal segments. An external ligand binds to *two* molecules of the receptor. The juxtaposition of the external portions also brings the internal portions together, because they are tethered to the external regions by the transmembrane segments. Interaction between the interior segments triggers a conformational change that activates a process such as the phosphorylation of a protein. This may initiate a signal transduction cascade that can amplify the original stimulus.

Fig. 2.17 Human growth hormone (blue) in complex with two molecules illustrating the dimerized exterior domain of its receptor (orange, green) [3HHR].

We do not have the structure of a complete receptor. Figure 2.17 shows the binding of one molecule of human growth hormone (blue) to two molecules of the external segment of the human growth hormone receptor (orange and green).

Regulatory proteins

Most proteins are regulatory proteins, in that they participate in the web of control mechanisms that pervade living processes. Control may be exerted:

- 'In the field': by several mechanisms, such as inhibitors; dimerization, ligand-induced conformational changes including, but not limited to, allosteric effects; GDP–GTP exchange or kinase–phosphorylase switches; and differential turnover rates.
- 'At headquarters': through control over gene expression.

Any regulatory action requires: (1) a stimulus; (2) transmission of a signal to a target; (3) a response; and (4) a 'reset' mechanism to restore the resting state. Some stimuli arise from genetic programmes. Some arise from changes in internal metabolite concentrations. Others originate outside the cell; the signal is detected by surface receptors, and transmitted across the membrane to an intracellular target. Individual control interactions are organized into linear signal transduction cascades, and reticulated into control networks.

One signal can trigger many responses. Each response may be stimulatory (increasing an activity) or inhibitory (decreasing an activity). Transmission of signals may damp out stimuli or amplify them. There are ample opportunities for complexity, opportunities of which cells have taken extensive advantage.

Examples of regulatory proteins appear throughout this book. The human growth hormone receptor appeared in Fig. 2.17. Bacteriorhodopsin (Fig. 2.14) is a member of a family of proteins called 'G-protein-coupled receptors'. Allosteric changes are discussed in Chapter 3. Proteins controlling transcription are discussed in Chapter 6.

GTP-binding proteins (or G-proteins) are an important class of signal transducers. One of them, p21 Ras (Fig. 2.18), is a molecular switch in pathways controlling cell growth and differentiation. Ras has two conformational states. The resting, inactive, state binds GDP. Membrane-bound, G-protein-coupled receptors—proteins related to bacteriorhodopsin—trigger a GDP–GTP exchange transition, associated with a conformational change (see Fig. 2.19). Ras thus activated binds Raf-1, a serine/threonine kinase. The Ras–Raf-1 complex then initiates the MAP kinase phosphorylation cascade. Ultimately the signal enters the nucleus, where it activates transcription factors regulating gene expression.

See Box: G-protein-coupled receptors and G-proteins.

Ras has a GTPase activity to reset it to the inactive state. Mutations that inactivate the GTPase activity are oncogenic. Such mutants are trapped in the active state, continuously triggering proliferation. Mutations in *ras* appear in 30% of human tumours.

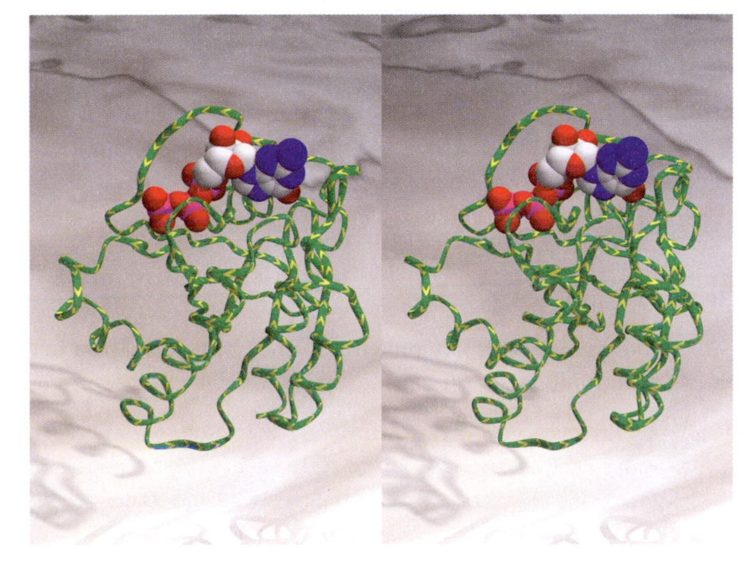

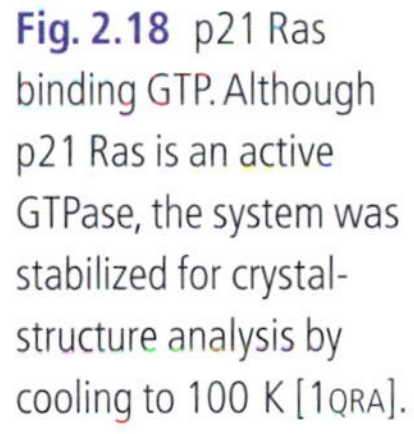

Fig. 2.18 p21 Ras binding GTP. Although p21 Ras is an active GTPase, the system was stabilized for crystal-structure analysis by cooling to 100 K [1QRA].

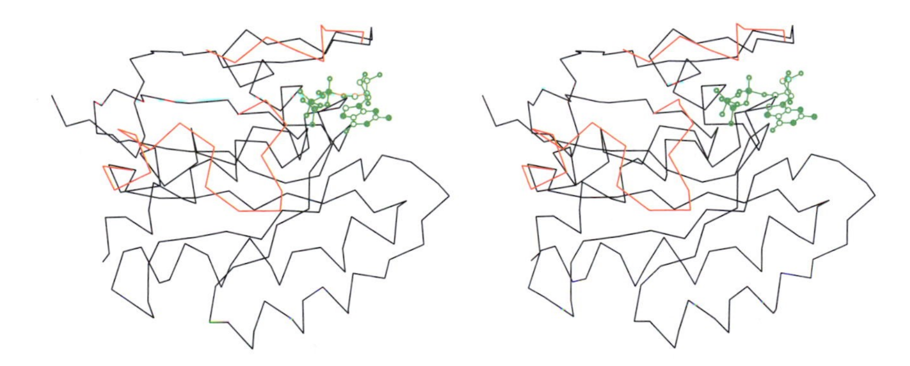

Fig. 2.19 The conformational change in p21 Ras from the inactive GDP-binding conformation to the active GTP-binding conformation primarily involves two regions (shown here in red), that form a patch on the molecular surface [1QRA, 1Q21].

G-protein-coupled receptors and G-proteins

G-protein-coupled receptors and G-proteins mediate the initial steps of many signal transduction cascades. In the specific details of their structure and interaction they illustrate general principles common to many regulatory and signal transduction processes.

G-protein-coupled receptors (GPCRs) are a large family of transmembrane proteins involved in signal transduction into cells. They share a substructure containing seven transmembrane helices, arranged in a common topology (see Fig. 2.14). The transmembrane part is generally flanked by N- and C-terminal domains. The N-terminal domain is always outside the cell, and the C-terminal domain always inside. Some GPCRs are involved in sensory reception, including vision, smell, and taste. Others respond to hormones and neurotransmitters. Some, like opsin and bacteriorhodopsin, bind chromophores. (Bacteriorhodopsin is not a signalling molecule but a light-driven proton pump.)

The downstream partners of GPCRs in signal transduction pathways are heterotrimeric G-proteins. These consist of three subunits: $G_\alpha G_\beta G_\gamma$. G_α is homologous to monomeric G-proteins such as p21 Ras. $G\alpha$ and $G\gamma$ are anchored to the membrane. G_β forms a tightly-bound complex with G_γ. In the resting, inactive state, $G\alpha$ binds GDP. Reception of a signal by a GPCR induces a conformational change, activating the GPCR. The activated GPCR binds to a specific G-protein, and catalyses GTP–GDP exchange in the $G\alpha$ subunit. This destabilizes the trimer, dissociating $G\alpha$:

$$G_\alpha(\mathrm{GDP})G_\beta G_\gamma \rightleftharpoons G_\alpha(\mathrm{GTP}) + G_\beta G_\gamma.$$

The two components G_α and $G_\beta G_\gamma$ activate downstream targets, such as adenyl cyclase.

An activated GPCR can interact successively with over 100 G-protein molecules, amplifying the signal. It is essential to turn them off—mutations that render a GPCR constitutively active cause a number of diseases, the symptoms emerging from a war between the rogue receptor and the feedback mechanisms that are unequal to the task of restraining its effects.

Different GPCRs have different mechanisms for restoring the resting state. Rhodopsin, for example, is inactivated by cleavage of the isomerized chromophore (see Chapter 5).

The heterotrimeric G-proteins are reset via the GTPase activity of G_α, converting $G_\alpha(\mathrm{GTP}) \rightarrow G_\alpha(\mathrm{GDP})$. $G_\alpha(\mathrm{GDP})$ does not bind to its receptors—therefore shutting down that pathway of signal transmission. $G_\alpha(\mathrm{GDP})$ rebinds the $G_\beta G_\gamma$ subunits. This resets the system. The GTPase activity of G_α is stimulated by a class of proteins called 'regulators of G-protein signalling' (RGSs).

GPCRs constitute the largest known family of receptors. The family is as old as the eukaryotes, and is large and diverse. Mammalian genomes contain ~1500–2000 GPCRs, accounting for about 3–5% of the genome. A similar fraction of the *Caenorhabditis elegans* genome codes for GPCRs. GPCRs are the targets for many drugs, used in the treatment of high blood pressure, asthma, and allergies.

Each GPCR interacts with a specific G-protein target. Mammalian genomes contain several homologues of G_α, G_β, and G_γ subunits, giving the potential for a large number of combinations. Although it is not known how many actually exist, it is likely to be less than the number of GPCRs. Therefore many GPCRs must target a single G-protein. For instance, all odorant receptors—several hundred in the human—target the same G_α.

Motor proteins

Motor proteins convert chemical energy to mechanical energy. There are two requirements: (1) coupling ATP hydrolysis to conformational change, to generate force; and (2) organizing a cycle of attachment and detachment to a mechanical substrate, to allow the force to generate movement.

Some motor proteins propel themselves—and their cargo—by exerting force against a stationary object, such as a cytoskeletal filament. Others remain stationary, and propel movable objects.

See Chapter 3, Conformational change.

- **Myosins** interact with actin during muscle contraction.
- **Kinesins and dyneins** interact with microtubules, mediating organelle transport, chromosome separation in mitosis, and movements of cilia and flagella.

Myosins, kinesins and dyneins are primarily *linear* motors. In contrast:

See Chapter 6 ATPase

- **ATPase** is a *rotatory* motor. The mechanical step is part of the mechanism for converting the free energy of the potential gradient across a membrane to the high-energy phosphate bond of ATP.

This section has sketched out broad categories of protein structure and function. One respect in which it is a simplification is that the categories overlap.

Protein folding patterns

Appreciation of protein structures, like connoisseurship in art, requires both a sensitive and trained eye, and technical scientific analysis.

The polypeptide chains of proteins in their native states describe graceful curves in space. These are best appreciated by temporarily ignoring the detailed interatomic interactions and focussing on the calligraphy of the patterns. The major differences among protein structures come not at the level of local interactions, which in most proteins are quite similar; but at this calligraphic level, in which similar substructures are differently deployed in space to give different *protein folding patterns*.

Study of protein folding patterns in terms of the curves the main chain traces out in space can: (1) expose recurring structural patterns, such as helices and sheets, and show the variety of their patterns of combination; and (2) provide an approach to a comparison of different protein folding patterns, to clarify evolutionary relationships. This approach can sacrifice the detailed representation of interatomic interactions.

Cow **acylphosphatase** is a small protein—only 98 residues long—but it contains many of the basic structural themes that also appear in larger proteins (Fig. 2.20(a)). Two regions of the chain (blue) curl up into α-helices. Five other segments, in which the chain is drawn out almost straight, assemble side-by-side to form a β-sheet.

Connecting the helices and strands of a sheet are regions called '*loops*', which exhibit much greater structural variety. Most loops appear on the surface of the structure, and effect a change in direction of the chain. The general idea is that helices and strands of

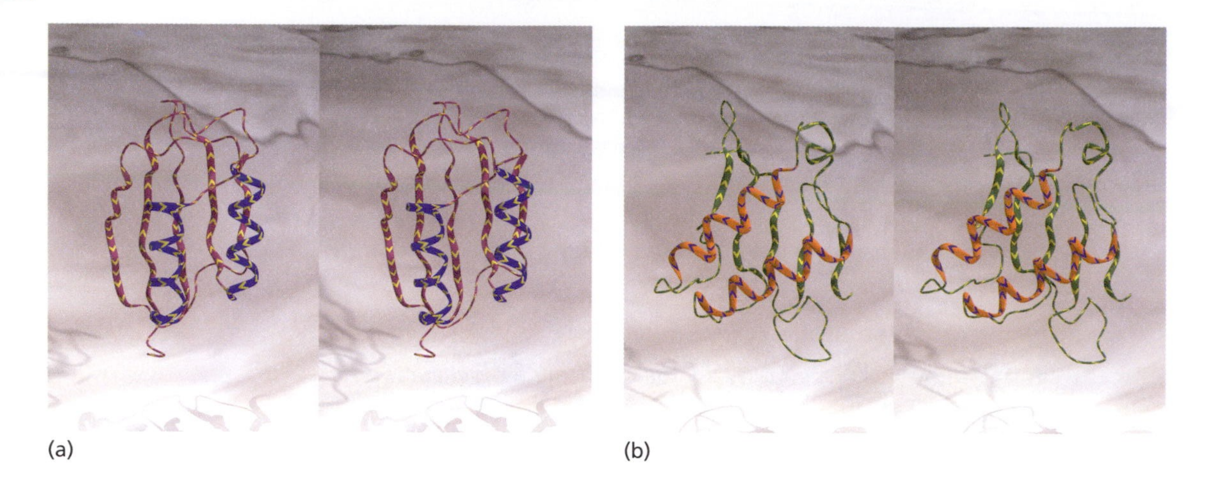

Fig. 2.20 Comparison of the folding patterns of two small proteins: (a) cow acylphosphatase [2ACY], (b) viral toxin from corn smut fungus (*Ustilago maydis*) [1KPT]. Although there are many superficial similarities between the folding patterns of these two proteins, they have different topologies (unlike the two related proteins shown in Fig. 2.4).

a sheet pass through the protein from one side to the other, then the chain turns around through a loop and begins another helix or strand of a sheet to pass through the structure again.

Figure 2.20(b) shows another small protein, KP4, a toxin from corn smut fungus, *Ustilago maydis*, encoded by a symbiotic virus. This 105-residue protein also contains two helices in front of a sheet. As in acylphosphatase, the helices are close enough to the sheet that side chains from the helices and strands of sheet are in contact; we say that the helices are packed against the sheet.

In both proteins the helices are about four turns long (~15 residues) and the strands about 10 residues long. These lengths are typical (although helices and sheets show great variation in length even in globular proteins—a helix in myosin is over 50 residues long, and helices in fibrous proteins can be much longer). Indeed, much of the *local* main-chain structure is very similar in these and other proteins. In fact, in these two structures even the *order* along the sequence of the elements of the secondary structure is virtually the same:

$$\beta_1\text{–}(\alpha)\text{–}\alpha_1\text{–}\beta_2\text{–}\beta_3\text{–}\alpha_2\text{–}\beta_4\text{–}\beta_5,$$

in which the helices and strands of a sheet are numbered according to their positions in the *sequence*, not the structure. The (α) in parentheses indicates a short turn of helix within the first loop of the toxin structure. This is followed, within that loop, by the full-blown helix α_1.

Can the reader suggest any reasons for the difference in orientation of the helices relative to the sheet? (See Case Study).

In acylphosphatase the axes of the helices are roughly parallel to the strands of the sheet. In the toxin, the helix axes are both tipped by about 45° to the strands of the sheet. The reason why the helices appear approximately parallel to each other in both cases is that this arrangement provides a large interface for the packing together of the helices.

In these pictures, because only the tracing of the main chain is shown the structures look empty. But a picture containing all the atoms makes clear that the structures are fairly compact globules of atoms (Fig. 2.21).

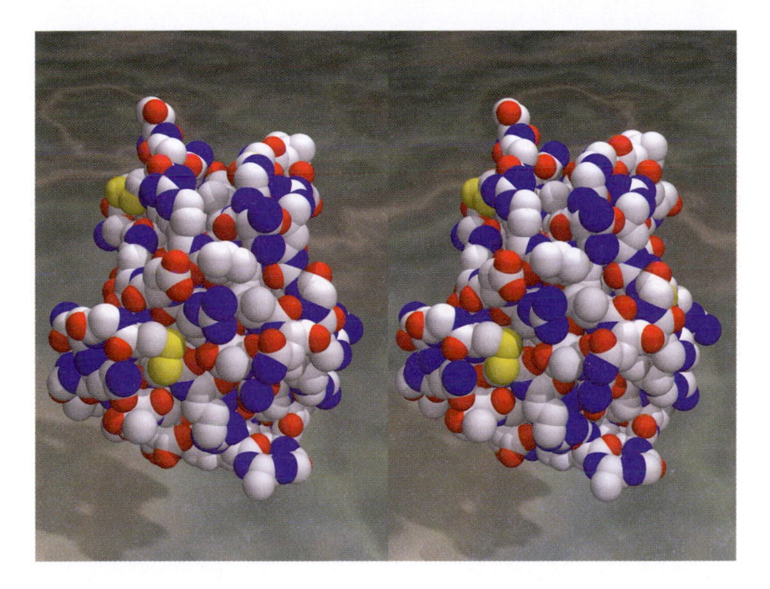

Fig. 2.21 All-atom representation of corn smut fungal toxin [1KPT]. In contrast to the representation of this protein in Fig. 2.20, this picture shows the compactness of the packing of the structure, and the topography of the surface; but it would be difficult to trace the chain in this picture. Can you see an α-helix? (Not easy, but it's there.)

CASE STUDY Comparison of the folding patterns of acylphosphatase and the fungal toxin

Let us trace the chains through the two structures in Fig. 2.20. As we examine the correlation between the order of secondary structural elements in the sequence and their relative disposition in space, we shall see that the helices and strands of sheet are 'wired up' differently.

In acylphosphatase, strand 1—the first strand in order of appearance in the *sequence*—is second from the left (in the orientation depicted. The reader will have to be careful to distinguish statements that refer only to the orientation illustrated, from those characterizing the intrinsic structure of the protein. The statement that the helices are in front of the sheet is true of this picture. The statement that the protein contains two helices packed against the same surface of a sheet would be true no matter how the structure were reoriented.) In the toxin the first strand is the rightmost strand. Following the first strand, in both proteins the chain passes through the first helix to the strand second in the sequence, which in acylphosphatase is the fourth strand from the left and in the toxin the first on the left. Note that the intrastrand region in the toxin has the longest distance to cover—from the right edge of the molecule to the left. Perhaps this is a partial explanation of why the helices are roughly parallel to the strands in acylphosphatase but oblique in the toxin.

In acylphosphatase the third strand is adjacent to the second strand, and occupies the position third from the left in the sheet. In the toxin the third strand is also adjacent to second but on its *right*, second from the left in the sheet. This combination of two adjacent antiparallel strands, connected by a loop, occurs frequently in β-sheets. It is called a '**hairpin**'.

Fig. 2.3 showed a hairpin.

continues ...

After the third strand, in both structures the chain passes through the second helix to the fourth strand. This is the leftmost strand in acylphosphatase, and fourth from the left in the toxin. Then in acylphosphatase the chain takes a long excursion to the fifth and last strand, the rightmost. In the toxin, the fourth strand is connected by a second hairpin to the fifth strand, the third from the left.

We conclude that acylphosphatase and the viral toxin have very nearly the same secondary structure but very different tertiary structure.

In the toxin every pair of adjacent strands points in opposite directions—for instance, strand 2 points up and strand 3 points down. This is an *antiparallel* β-*sheet* (see Fig. 2.3(c)). One way to construct an antiparallel β-sheet would be to assemble a succession of hairpins, in which the order of the strands in the sequence and the structure would match. (The sheet in the toxin, although antiparallel, is not constructed in this manner.) In other proteins, sheets occur with all strands pointing in the same direction, called *parallel* β-sheets. The sheet in acylphosphatase is neither parallel nor antiparallel. It is a *mixed* β-*sheet*. β-Sheets in proteins are free to mix parallel and antiparallel strands.

These relationships can be summarized in diagrammatic representations of the topology of (a) acylphosphatase and (b) the toxin.

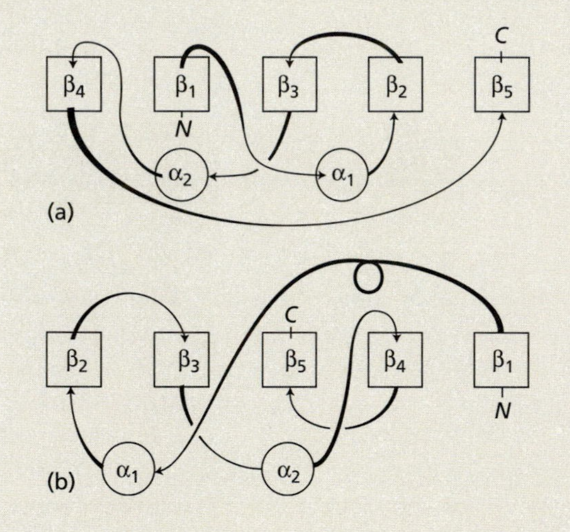

There is an interesting substructural symmetry in acylphosphatase. The first strand (in the sequence) is followed by a helix, followed by a second strand that is positioned two strands to the right of the first strand. The third strand (in the sequence) begins the same pattern, but upside down in this picture. A variant of this pattern appears in the toxin: a strand followed by a helix, followed by a strand positioned two strands to the *left*.

It would not be possible to overlay the structures of acylphosphatase and the toxin by rigid motions as in the superposition of the structures of *E. coli* and *S. faecalis* histidine-containing phosphocarrier proteins in Fig. 2.4. This implies that it would not be easy for evolution to interconvert these two structures, and therefore that they are not related. The reader is urged to study the figures, and verify, visually, the verbal description and comparison of these structures.

Folding patterns in native proteins—themes and variations

It is in their folding patterns that proteins show how versatile the polypeptide chain can be in forming structures adapted for different roles in living systems. It is very interesting to classify the different types of patterns observed in native proteins, and to try to correlate different structure types with different functions. At present, we know about 800 different folding patterns in the known protein structures. It has been estimated that only about 1000 different ones will be observed in the majority of proteins in Nature. When we know this repertoire almost completely—which will happen soon—we shall then be in a better position to appreciate the set of observed proteins in terms of variations on a fixed set of themes.

See Chapter 5.

A thousand folding patterns is much less than the number of proteins of known structure. Therefore many proteins must share the same folding pattern. This may come about in two ways:

1. **Evolutionary divergence**. When proteins evolve, they tend to maintain their folding pattern. We can therefore recognize families of proteins that share a folding pattern.
2. **Building up complex structures from different combinations of simpler ones**. Acylphosphatase and the fungal toxin form single compact assemblies of about 100 amino acids. These are relatively simple structures, as proteins go. There are

See Box: Similarity and homology.

(a) (b)

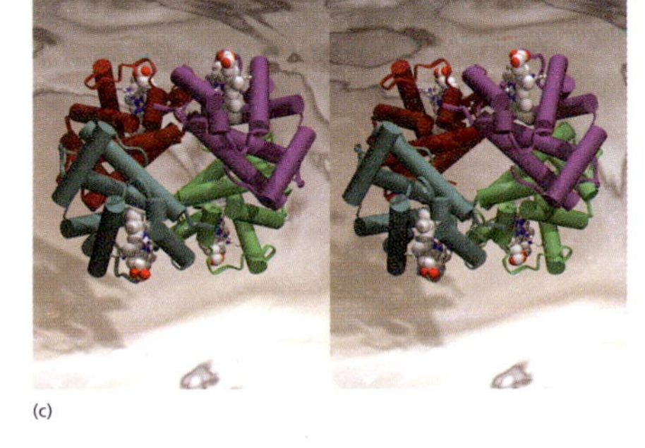

(c)

Fig. 2.22 Proteins from the globin family assemble different combinations of oligomers. (a) monomer: sperm-whale myoglobin [1MBO]; (b) dimer of two identical subunits: Ark clam globin [4SDH]; (c) mixed tetramer: human haemoglobin, containing two α-chains and two β-chains.

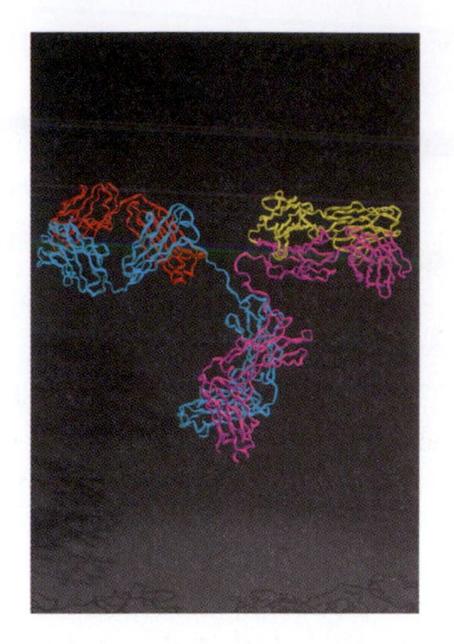

Fig. 2.23 Antibody molecule, illustrating both the concatenation of domains within each of the four chains, and the formation of a dimer. Like haemoglobin, this molecule is a dimer of dimers, containing two identical light chains (yellow and orange) and two identical heavy chains (blue and magenta). The antigen-binding sites are at the 'wingtips' [1IGT].

several ways to form more complex structures:

- Create a larger, but still single, compact unit by synthesizing a longer polypeptide chain.
- Form oligomers (*oligo* = few) by combining several monomers. Sperm-whale myoglobin is a monomer; the globin from the ark clam *Scapharca inaequivalvis* is a dimer formed by two identical monomers, and human haemoglobin is a tetramer (Fig. 2.22). Haemoglobin is an oligomer formed from non-identical monomers. Multisubunit proteins can contain very large numbers of subunits; for example, ribulose-*bis*-1,5-diphosphate carboxylase/oxygenase contains 16 subunits, pyruvate dehydrogenase contains over 100. Viral capsids may contain thousands of monomers.
- A kind of compromise between these two methods is to build up a protein from small compact units, formed from successive regions of a single, long, polypeptide chain. It is common for structures seen as monomeric proteins to appear as domains within longer ones. Such proteins are called **modular proteins**. One effective mechanism of evolution is to recombine modules in different orders (see next section.) Immunoglobulins illustrate *both* concatenation of domains and oligomer formation (Fig. 2.23).

Modular structure of proteins

Proteins such as acylphosphatase and the fungal toxin are single domains: individual compact units typically about 100 residues in length. The combination of domains into a modular protein, as in immunoglobulins, is an important way of creating complex proteins from simpler components (see Fig. 2.24).

Similarity and homology

The distinction between *similarity* and *homology* is subtle and important.

- **Similarity** is the observation or measurement of resemblance and difference, independent of the source of the resemblance.
- **Homology** means, specifically, that the sequences and the organisms in which they occur are descended from a common ancestor. The implication is that the similarities are shared ancestral characteristics.

Similarity of sequences or structures is observable *now*, and involves no historical hypotheses. In contrast, assertions of homology are statements about historical events that are almost always unobservable. Homology must be an *inference* from observations of similarity. Only in a few special cases is homology directly observable; for instance in pedigrees of human families carrying Huntington disease; or in laboratory populations of animals, plants, or microorganisms; or in clinical studies that follow sequence changes in viral infections in individual patients.

As proteins evolve, their amino acid sequences diverge, and the corresponding structures also diverge. Structures tend to change more conservatively than sequences. Figure 2.4 showed a typical example. Proteins that are very similar in sequence, structure, and function may be presumed to be homologous. For proteins with similar folding patterns, but for which no substantial similarities appear in the sequences, it is difficult to decide whether they are homologues or not.

In prokaryotes, approximately 2/3 of the proteins contain more than one domain. The several thousand proteins in a typical bacterium such as *E. coli* are made up of combinations of about 400 domains with different folding patterns. In eukaryotes, approximately 3/4 of the proteins are multidomain proteins, combinations of about 600–700 domains with different folding patterns. Many of these individual domains are common to Bacteria, Archaea, and Eukaryotes, but some are—as far as we now know—specific to each group.

Sometimes separate genes for proteins in prokaryotes become fused into a single gene for a modular protein in eukaryotes. For example, five separate enzymes in *E. coli*, that catalyse successive steps in the pathway of biosynthesis of aromatic amino acids, correspond to five regions of a single protein in the fungus *Aspergillus nidulans*.

Combination of domains in modular proteins does not occur at random. A few domains combine with many different partners, but most combine with only one or a few. Conversely, many domain combinations are common, and others rare. If domains exert their function independently, they can appear in different orders in different proteins. In other cases, the order of the domains is important for function. Tandem repeats of the same domain occur frequently, created by gene duplication and subsequent divergence. A giant muscle protein, titin, contains about 300 modules, including repeats of immunoglobulin and fibronectin type-3 domains.

Protein evolution thus appears to be going on at two levels. By variation of individual residues, mutations are exploring the immediate neighbourhoods of particular sequences, structures, and functions. Larger-scale evolutionary variation involves trying out different combinations of domains.

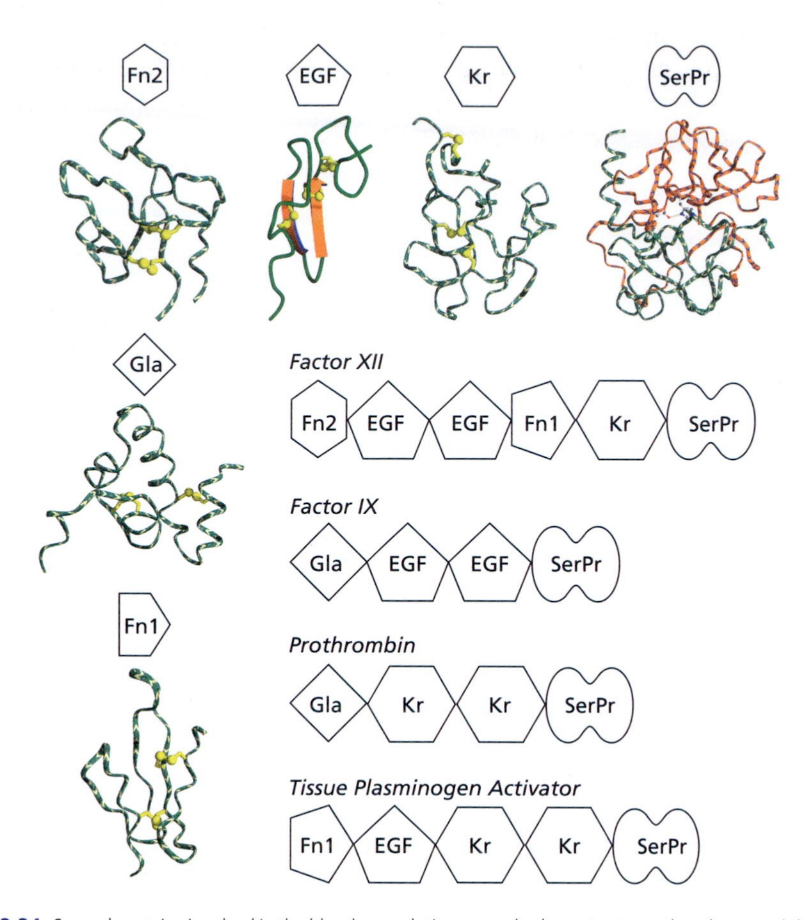

Fig. 2.24 Several proteins involved in the blood coagulation cascade show structures that share modules. The composition and the order of the modules is not preserved. Each module is a relatively small compact unit in its own right. The serine proteinases contain two halves with structural similarities, that arose by gene duplication and divergence, but are never seen separately. Abbreviations: Fn1, fibronectin type 1; Gla, γ-carboxyglutamic acid-rich; Fn2, fibronectin type 2; EGF, epidermal growth factor homologue; Kr, Kringle; SerPr, Serine Proteinase.

Protein evolution

The atomic event of protein evolution is:

- a change in the nucleic acid sequence of a genome,
- reflected in a change in the amino acid sequence or in the expression pattern of a protein,
- causing a change in the activity of the protein,
- which can produce a selective advantage or disadvantage.

Not all changes in DNA sequences alter the amino acid sequences of proteins. Some base substitutions are silent, including changes in the third position of many codons in exons, and many changes in untranslated regions. Other mutations alter regulatory sequences, affecting patterns of expression but not amino acid sequences.

The simplest change to a protein is the substitution of a single amino acid. What is the effect on the protein structure and function?

- In some cases there is no noticeable effect. We have seen that proteins are robust to mutation. Many corresponding enzymes in closely related species differ by only one or a few single-site mutations, and appear to be diverging at random rather than under a specific selective pressure. Genetic engineering experiments confirm that proteins can tolerate many substitutions with little change in structure and function.
- Conversely, some amino acid substitutions do entirely prevent the folding of a protein. Several such mutations in haemoglobin present clinically in a set of diseases called the thalassemias. For instance, the haemoglobin β-chain mutation β60 Val→Glu produces an unstable protein that is rapidly degraded, and the product of β106 Leu→Arg precipitates on the red-cell membrane (Fig. 2.25).
- A single point mutation, at an exquisitely sensitive position in the active site, can alter the function of a protein. Lactate and malate dehydrogenases are two related enzymes that carry out analogous reactions on similar substrates. A single Gln→Arg substitution is sufficient to convert a malate dehydrogenase to a lactate dehydrogenase.

See Chapter 5.

- Some point mutations leave the protein structure intact, but render the protein more prone to aggregation. This can lead to very serious clinical consequences. (1) Sickle-cell anaemia is caused by a single-site mutation in the β-chain of haemoglobin, which creates a sticky patch on the surface that leads to aggregation within red blood cells under conditions of low oxygen tension. These aggregates reduce the

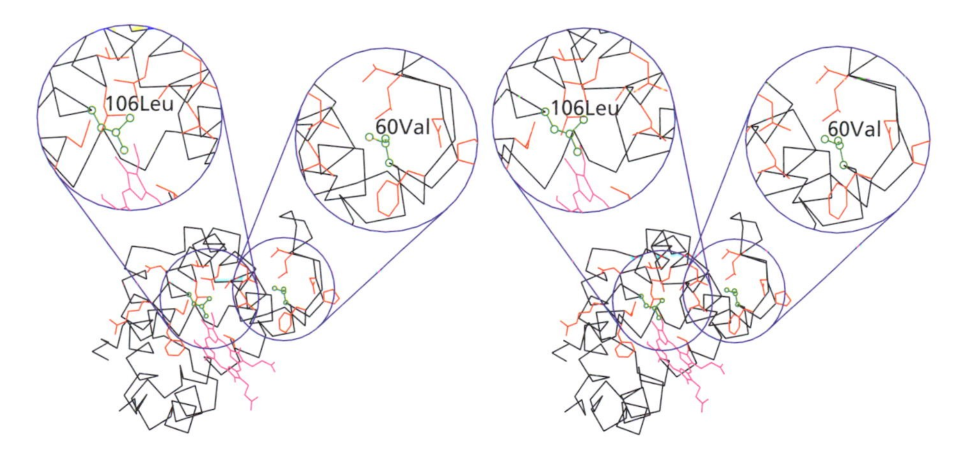

Fig. 2.25 Mutations in the β-chain of human haemoglobin that destabilize the structure include β60 Val→Glu and β106 Leu→Arg. These side chains (shown in green) are buried in the interior of the molecule, surrounded by other side chains (shown in red). Both mutations: (1) increase the volume of the side chain, which can no longer fit in the space available in the structure; and (2) change a buried hydrophobic residue to a charged residue—burial of a charged residue in a state inaccessible to solvent is energetically unfavourable.

deformability of the red cells, preventing them from passing smoothly through capillaries. (2) The Z-mutant of α_1-antitrypsin increases the risk of aggregation by destabilizing the structure.

Other types of changes in protein structures include insertions, deletions, and transpositions. (1) A common cause of cystic fibrosis is a three-nucleotide deletion in the *cystic fibrosis transmembrane conductance regulator* (CFTR) protein, deleting one amino acid—Phe508. The effect is defective translocation of the protein, which is degraded in the endoplasmic reticulum rather than being transported to the cell membrane. (2) The extension of polyglutamine tracts in Huntington disease leads to aggregates in nerve cells.

How do proteins develop new functions?

Evolution of function is discussed in Chapter 4.

Evolution has pushed the limits in its exploration of sequence–structure–function relationships. Observed mechanisms of protein evolution that produce altered or novel functions include: (1) divergence, (2) recruitment, and (3) 'mixing and matching' of domains, or modular evolution. Divergence involves progressive localized changes in sequence and structure, leading initially to changes in specificity and ultimately to changes in the nature of the reaction catalysed. In recruitment, one protein is adapted, sometimes with no change at all, for a second function. Modular evolution involves large-scale structural changes. Individual domains may retain their function in a new context, or their function may be modified by their new surroundings, or they may participate in entirely different processes.

One consequence of the robustness of protein structure to mutation is a maintenance of structure in spite of the divergence of sequences during evolution. But, although similar sequences determine similar structures, the converse is not true—proteins with very different sequences can have similar structures. Indeed, there are many cases of proteins with very similar structures with *no* easily recognizable relationship between the sequences.

The box summarizes the situation.

Even less reliable is the reasoning from sequence to function, or even from sequence + structure to function. Many proteins with similar sequences and structures have similar functions. But many do not, and conversely many functions can be carried out by proteins with unrelated sequences and structures.

Integration and control of protein function

Proteins are social animals, and life depends on their interactions. Because individual proteins have specialized functions, control mechanisms are required to integrate their activities. The right amount of the right protein must function in the right place at the right time. Failure of control mechanisms can lead to disease and even death.

Under unchanging environmental conditions, an organism's biochemical systems must be stable. Under changing conditions, the system must be robust, accommodating both neutral and stressful perturbations. Over long periods, the rates of processes

Relationships among sequence, structure, and function

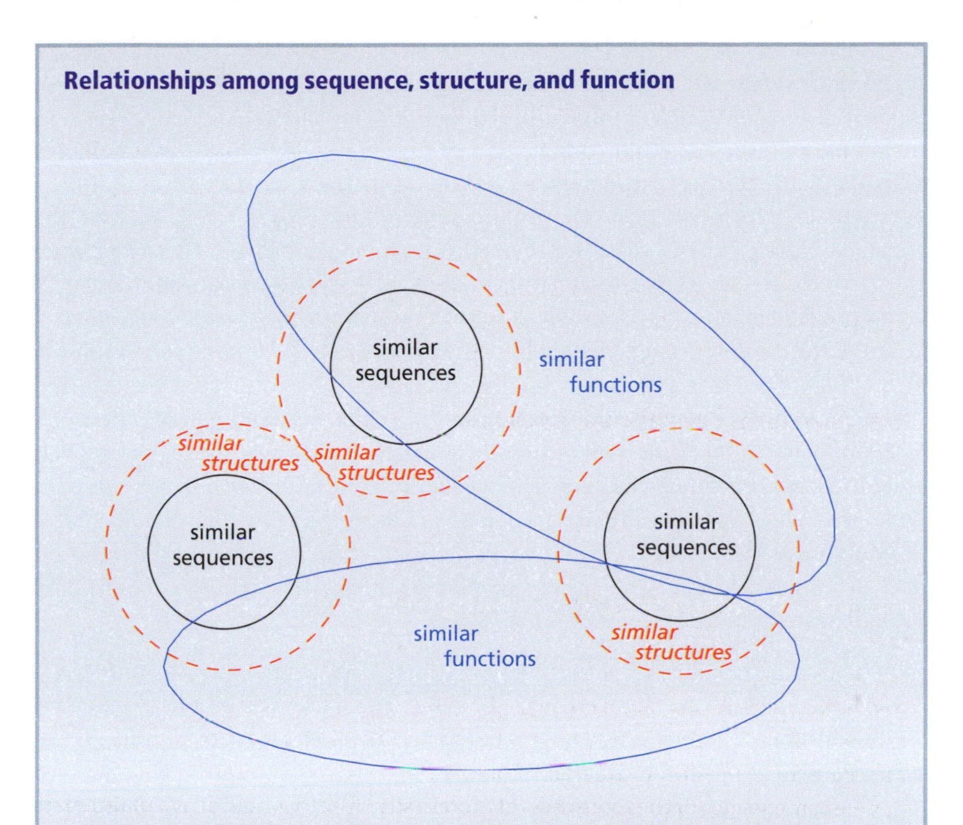

- **Similar sequences can be relied on to produce similar protein structures**, with divergence in structure increasing progressively with the divergence in sequence.
- Conversely, **similar structures are often found with very different sequences**. In many cases the relationships in a family of proteins can be detected *only* in the structures, the sequences having diverged beyond the point of our being able to detect the underlying common features.
- **Similar sequences and structures often produce proteins with similar functions**, but exceptions abound.
- Conversely, **similar functions are often carried out by non-homologous proteins with dissimilar structures**; examples include the many different families of proteinases, sugar kinases, and lysyl-tRNA synthetases.

must be altered, or even switched on and off. This regulation includes short-term adjustments, for instance in the stages of the cell cycle; or responses to external stimuli such as changes in the composition or levels of nutrients or oxygen. Longer-term regulatory activities include control over developmental stages during the entire lifetime of an organism.

In trying to understand regulation, let us start with the idea that the individual protein provides the unit of activity of the cell. (A reasonable approximation for purposes of this discussion.)

Metabolism can be thought of as the flow of molecules and energy through a *network of chemical reactions*. Of course the reactions involve proteins and nucleic acids as well as small compounds such as sugars. Individual proteins catalyse individual steps in metabolic pathways. The full complement of metabolic reactions forms a complex network. Some subsets of the network are linear pathways, such as the multistep synthesis of tryptophan from chorismate. Others form closed loops, such as the tricarboxylic acid (Krebs) cycle. But the pathways interlock densely. The structure of the network—its connectivity or topology—and its activity patterns can be analysed in terms of a mathematical apparatus dealing with graphs and flows and throughputs.

To control the flow through metabolic pathways, a distinct second, *regulatory network* also connects proteins and metabolite concentrations. The connectivity of the signalling or regulatory network is different from that of the metabolic pathway network. Corresponding to the succession of steps in a metabolic pathway, the regulatory network assembles signalling cascades. The regulatory network has two major components. One component deals with the regulation of existing proteins; for instance control of protein activity by feedback inhibition or allosteric effectors, or by phosphorylation and dephosphorylation. The other component affects transcription, controlling protein concentrations and expression patterns.

The regulatory machinery could achieve several types of states in different parts of both the metabolic transformation and regulatory networks (see Box):

- **equilibrium**
- **steady-state**
- **states that vary periodically**
- **unfolding of developmental programs**
- **chaotic states**
- **runaway or divergence**
- **shutdown.**

Much is known about the mechanisms of individual elements of control and signalling pathways. Understanding their integration is a subject of current research.

Underlying the intrinsic complexity of complex overlapping networks are fundamental mathematical problems. The idea that healthy cells and organisms are in stable states is certainly no more than an approximation (and in most cases an idealization). The description of the actual dynamic state of the metabolic and regulatory networks is a very delicate problem. Understanding *how* cells achieve even an apparent approximation to stability is also quite tricky. It is likely that a great redundancy of control processes lies at the basis of stability.

In most familiar organizations or pieces of machinery, control is imposed top-down—think of university departments, industrial companies, military hierarchies, airplanes, or dishwashers. In contrast, cellular regulation is based on the result of many individual control mechanisms—here a short feedback loop, there a multistep cascade. Somehow the independent actions of all the individual signals combine to achieve an overall, integrated result. It is like the operation of the 'invisible hand' that, according to Adam Smith, coordinates individual behaviour into the regulation of national economies.

States of a network of processes

- At **equilibrium** one or more forward and reverse processes occur at compensating rates to leave the amounts of different substances unchanging:

$$A \rightleftharpoons B.$$

Chemical equilibria are generally self-adjusting upon changes in conditions, or changes in concentrations of reactants or products.

- A **steady-state** will exist if the total rate of processes that produce a substance is the same as the total rate of processes that consume it. For instance, the two-step conversion:

$$A \rightarrow B \rightarrow C$$

could keep the amount of B constant, provided that the rate of production of B (the process $A \rightarrow B$) is the same as the rate of its consumption (the process $B \rightarrow C$). The net effect would be to convert A to C.

A cyclic process could maintain a steady state in all its components:

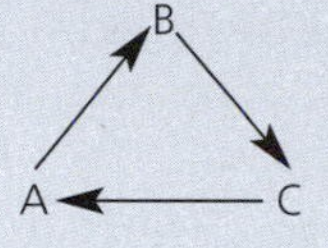

A steady state in such a cyclic process, with all reactions proceeding in one direction, is very different from an equilibrium state. Nevertheless, in some cases, it is still true that altering external conditions produces a shift to another, neighbouring, steady-state.

- **States that vary periodically** occur in the regulation of the cell cycle, in circadian rhythms, and in seasonal changes such as annual patterns of breeding in animals and flowering in plants. Circadian and seasonal cycles have their origins in the regular progressions of the day and year, but have evolved a certain degree of internalization.

- Many equilibrium and some steady-state conditions are **stable**, in the sense that concentrations of most metabolites are changing slowly if at all, and the system is robust to small changes in external conditions. The alternative is a **chaotic state**, in which small changes in conditions can cause very large responses. Weather is a chaotic system: the meteorologist Lorenz asked, 'Does the flap of a butterfly's wings in Brazil set off a tornado in Texas?' In a carefully regulated system, chaos is usually well worth avoiding, and it is likely that life has evolved to damp down the responses to the kinds of fluctuations that might give rise to such chaos. Chaotic dynamics does sometimes produce the approximations to stable states—these are called '**strange attractors**'. Understanding stability in dynamical systems subject to changing environmental stimuli is an important topic, but beyond the scope of this book. It is the subject of a new field, called '**Systems biology**'.

- **Unfolding of developmental programmes** occurs over the course of the lifetime of the cell or organism. Many developmental events are relatively independent of external conditions, and are primarily controlled by regulation of gene expression patterns.

continues ...

- **Runaway or divergence**. Breakdown in control over cellular proliferation leads to unconstrained growth, as in cancer.
- **Shutdown** is part of the picture. **Apoptosis** is the programmed death of a cell, as part of normal developmental processes, or in response to damage that could threaten the organism, such as DNA strand breaks. Breakdown of mechanisms of apoptosis—for instance, mutations in protein p53—is an important cause of cancer.

Protein expression patterns in space and time: proteomics

An integrated understanding of the roles of proteins in living systems can only emerge from analysis of the high-throughput, whole-organism data of **proteomics**. Here we discuss some of the background and methods that provide these data.

Subcellular localization

Proteins have adapted to a variety of working environments (see Box). Some are secreted and function outside the cell; others are deployed in specific intracellular compartments or organelles, or are integrated into membranes. Differences in the physicochemical properties of these environments are reflected in the properties of proteins. For instance, membrane proteins have surfaces designed to interact with the lipid bilayer of membranes. Disulphide bridges cannot form in the reducing environment of the cytosol of bacterial cells.

It is useful to know *where* a protein is active, for location is an important clue to determining the interaction networks in which a protein participates. It is thereby an aid to assigning function. Looking towards clinical applications, extracellular proteins provide more accessible drug targets. However, some proteins are active in more than one place; indeed, they may even have different functions at different sites.

Proteins are directed to their destinations by *sorting signals*. Short sequences, usually at or near the N-termini, serve as a kind of 'postcode' or 'zip code' specifying the destination. For instance, the N-terminal sequence KDEL specifies proteins retained in the endoplasmic reticulum. Proteins destined for the nucleus have a nuclear localization signal containing a stretch of 4–8 positively charged residues. We have seen that, in higher plants, the transfer of a mitochondrial gene to the nucleus required the gene to pick up a signal sequence to direct its protein product to the mitochondrion.

Signal sequences can interact with receptors on a target organelle directly, or with a carrier-protein complex. In eukaryotes, proteins destined for the *secretory pathway*—some, but not all, of which are exported from the cell—bind to a *signal recognition particle* even before synthesis is complete. These proteins pass through the endoplasmic

Sites of protein deployment

- Intracellular:
  - nuclear
  - cytoplasmic
  - membrane-related
  - mitochondria
  - chloroplast
  - endoplasmic reticulum, Golgi stack (eukaryotes)
  - periplasmic (prokaryotes)
  - viral
- Extracellular:
  - secreted

reticulum and the Golgi apparatus; secondary signals then direct them to lysosomes, to the plasma membrane, or to discharge from the cell (**exocytosis**). Proteins targeted to the nucleus, or to intracellular organelles such as the mitochondria, chloroplasts, or peroxisomes, are translocated after they are synthesized. Proteins lacking targeting sequences remain in the cytoplasm.

Related to the intracellular protein distribution systems are mechanisms for the uptake of proteins from outside the cell. Some viruses take advantage of this machinery to gain entry.

Protein turnover

Cells contain, or secrete, many different types of proteinases. Many function to digest nutrients, or for defence; others are involved in **apoptosis** (programmed cell death). A complex macromolecular particle, the **proteasome**, coordinates *protein turnover*, the controlled degradation of proteins:

- Cells must destroy normal functional proteins that have reached the end of their natural lifetimes. Some proteins, for example glycolytic enzymes, are long-lived. Others, such as those that initiate DNA recognition or cell division, must have their activities turned off after their signals have been received. The N-terminal amino acid determines the half-life of a protein. Proteins beginning with Arg or Lys have short half-lives (2–3 minutes), and those beginning with Gly, Val, or Met have long half-lives (>1 day). Defective regulation of protein turnover is associated with several diseases. For example, the half-life of the protooncogene ornithine decarboxylase differs between normal tissues and tumours.
- Misfolded proteins must be removed. They are potentially dangerous—especially those present in high concentrations, which can form aggregates. Cells try to rescue misfolded proteins with chaperones (see Chapter 5), but if this is unsuccessful, degradation is the necessary alternative.
- The immune system uses proteolytically generated oligopeptides (~8–10 residues), presented by MHC proteins, to recognize foreign proteins (see Chapter 7).

A fundamental problem with an intracellular proteinase is keeping it away from the many proteins that the cell does not want to see degraded. The solution is to sequester the proteinase activity inside a large macromolecular assembly called the '**proteasome**', with a narrow channel leading to its interior.

Proteasomes or related structures appear in Archaea, Eukaryotes and some Bacteria. A typical human cell contains 30 000 of them, distributed between nucleus and cytoplasm.

The proteasome is a barrel-shaped structure, 15 nm high and 11 nm in diameter. The complete proteasome particle contains a 20S proteasome-core complex together with one or two copies of a regulatory complex of comparable size forming a cap at both ends. In Archaea, the core complex contains multiple copies of two homologous proteins α and β, assembled into a stack of four 7-fold symmetrical rings: α_7, β_7, β_7, α_7. The proteolytic activity resides on the inside of the central rings. The catalytic residue is the N-terminal threonine of the β-subunits. In Eukaryotes, the α- and β-subunits have diverged to form 14 independent proteins. Not all the β-subunits are proteolytically active. In higher organisms, the cap structure and some of the β-subunits can vary under the control of the cytokine interferon-γ during activation of the immune response.

To target a protein for degradation, the cell attaches a linear chain of ubiquitins to the condemned protein. Ubiquitin is a small, 76-residue protein. An enzyme attaches its C-terminus to a lysine of the original protein, or to the preceding ubiquitin in the chain. At the proteasome, a cap domain removes the ubiquitin adducts and unfolds the protein. The unfolded protein is threaded through the narrow channel into the central chamber of the proteasome, where hydrolysis occurs.

DNA microarrays

DNA microarrays analyse the mRNAs in a cell to reveal the expression patterns of proteins; or genomic DNAs to reveal absent or mutated genes (see Box).

- For an integrated characterization of cellular activity, we want to determine what proteins are present, where, and in what amounts.

  To determine the expression pattern of all a cell's genes, it is necessary to measure the relative amounts of many different mRNAs. Hybridization is an accurate and sensitive way to detect whether a particular sequence is present. The key to high-throughput analysis is to run many hybridization experiments in parallel.
- Knowing the human genome sequence can help to identify genes associated with diseases. Some diseases, such as cystic fibrosis, are associated with single genes. For these, isolating a region by classical genetic mapping usually leads to identification of the gene. Other diseases, such as asthma, depend on interactions between many genes, with environmental factors as complications. To understand the aetiology of these multifactorial diseases requires the ability to determine and analyse patterns of genes, which may be distributed around different chromosomes.

DNA **microarrays**, or DNA chips, are devices for *simultaneously* checking a sample for the presence of many sequences. DNA microarrays can be used: (1) to determine the

Applications of DNA microarrays

- **Identifying genetic individuality in tissues or organisms.** In humans and animals, this permits the correlation of genotype with susceptibility to disease. In bacteria, this permits the mechanisms of development of drug resistance by pathogens to be identified.
- **Investigating cellular states and processes.** Patterns of expression that change with cellular state or growth conditions can give clues to the mechanisms of processes such as sporulation, or the change from aerobic to anaerobic metabolism.
- **Diagnosis of disease.** Testing for the presence of mutations can confirm the diagnosis of a suspected genetic disease, including detection of a late-onset condition such as Huntington disease. Detection of carriers of genetic diseases can help in counselling prospective parents.
- **Genetic warning signs.** Some diseases are not determined entirely and irrevocably from genotype, but the probability of their development is correlated with genes or their expression patterns. A person aware of an enhanced risk of developing a condition can, in some cases, improve his or her prospects by adjustments in lifestyle, or in some cases even prophylactic surgery.
- **Drug selection.** Allows detection of genetic factors that govern responses to drugs, that in some patients render treatment ineffective and in others cause unusual serious adverse reactions.
- **Specialized diagnosis of disease.** Different types of leukaemia can be identified from different patterns of gene expression. Knowing the exact type of the disease is important for prognosis, and for selecting the optimal treatment.
- **Target selection for drug design.** Proteins showing enhanced transcription in particular disease states might be candidates for attempts at pharmacological intervention.
- **Pathogen resistance.** Comparisons of genotypes or expression patterns, between bacterial strains susceptible and resistant to an antibiotic, point to the proteins involved in the mechanism of resistance.
- **Following temporal variations in protein expression** permits timing the course of: (1) responses to pathogen infection; (2) responses to environmental change; and (3) changes during the cell cycle.

expression patterns of different proteins by detection of mRNAs; or (2) for genotyping, by detection of different variant gene sequences, including, but not limited to, single-nucleotide polymorphisms (SNPs). It is possible to measure simple presence or absence, or to quantitate relative abundance.

The basic idea is this:

- to detect whether one oligonucleotide has a particular known sequence, test whether it can bind to an oligo with the complementary sequence;
- to detect the presence or absence of a query oligo in a mixture, spread the mixture out, and test each component of the mixture for binding to the oligo complementary to the query. (This is a Northern or Southern blot.)

- to detect the presence or absence of *many* oligos in a mixture, synthesize a set of oligos, one complementary to each sequence of the query list, and test each component of the mixture for binding to each member of the set of complementary oligos.

Microarrays provide an efficient, high-throughput way of carrying out these tests in parallel.

To achieve parallel hybridization analysis, a large number of DNA oligomers are affixed to known locations on a rigid support, in a regular two-dimensional array. The mixture to be analysed is prepared with fluorescent tags, to permit detection of the hybrids. After the array is exposed to the mixture, each element of the array to which some component of the mixture has become attached bears the tag. Because we know the sequence of the oligomeric probe in each spot in the array, measurement of the *positions* of the probes identifies their sequences. This analyses the components present in the sample.

A DNA microarray is based on a small wafer (chip) of glass or nylon, typically 2-cm square. Oligonucleotides are attached to the chip in a square array, at densities between 10 000 and 250 000 positions per cm^2. The spot size may be as small as ~150 μm in diameter. The grid is typically a few centimetres across. A *yeast chip* contains over 6000 oligos, covering all known genes of *Saccharomyces cerevisiae*. A DNA array, or DNA chip, may contain 100 000 probe oligomers. Note that this is larger than the total number of genes even in higher organisms (excluding immunoglobulin genes).

To analyse a mixture, expose it to the microarray under conditions that promote hybridization, then wash away any loose probe. The oligos in the target mixture are prepared with fluorescent reporter molecules, making detection easy. To compare two sets of oligos, tag the samples with differently coloured fluorophores (Fig. 2.26). Scanning the array collects the data in a computer-readable form.

Different types of chips designed for different investigations differ in the types of DNA immobilized:

- In an **expression chip**, the immobilized oligos are cDNA samples, typically 20–80 base pairs long, derived from mRNA of known genes. The target sample is a mixture of mRNA from normal or diseased tissue.
- In **genomic hybridization**, one looks for gains or losses of genes or changes in copy number. The target sequences, fixed on the chip, are large pieces of genomic DNA, from known chromosomal locations, typically 500–5000 base pairs long. The probe mixture is genomic DNA from a normal/disease state. For instance, some types of cancer arise from chromosome deletions, which can be identified by microarrays.
- In **mutation microarray analysis** one looks for patterns of single-nucleotide polymorphisms (SNPs).

Microarrays are, in principle, capable of comparing concentrations of probe oligos. However, mRNA levels, detected by the array, do not always reflect protein levels quantitatively. Indeed, mRNAs are usually reverse transcribed into more stable cDNA for microarray analysis; the yields in this step may also be non-uniform.

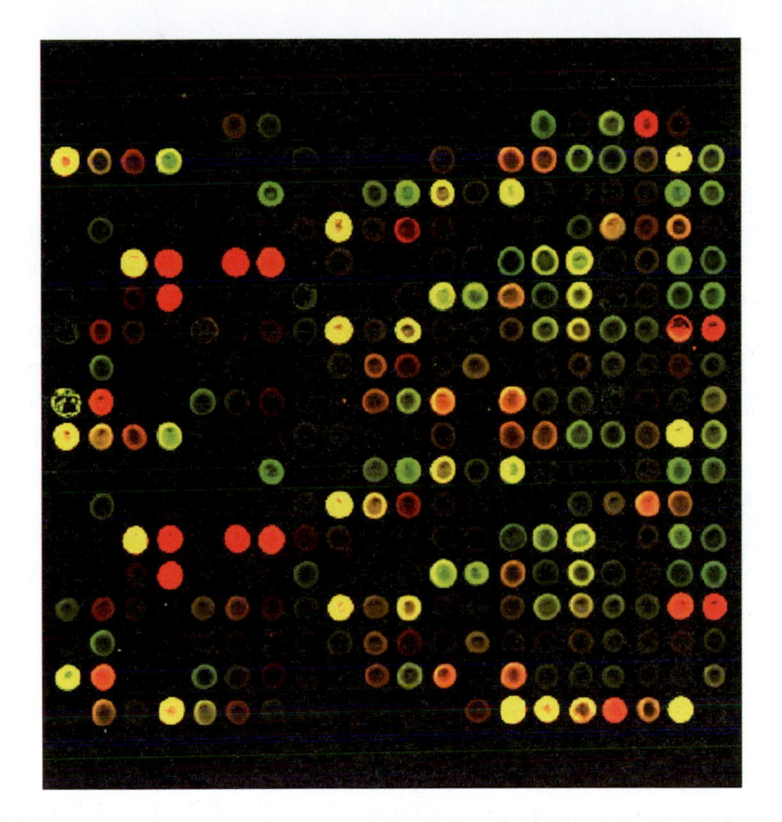

Fig. 2.26 Comparison of gene expression patterns in liver (red) and brain (green). The liver RNA is tagged with a red fluorophore, the brain RNA with a green one, then both are exposed to the array. Red spots correspond to genes active in the liver but not in the brain. Green spots correspond to genes active in the brain but not in the liver. Yellow spots correspond to genes active in both brain and liver. (Courtesy Dr P.A. Lyons.)

Mass spectrometry

Mass spectrometry is a physical technique that characterizes molecules by measuring the masses of their ions. Applications to molecular biology include:

- rapid identification of the components of a complex mixture of proteins;
- sequencing of proteins and nucleic acids;
- analysis of post-translational modifications, or substitutions relative to an expected sequence;
- measuring the extents of hydrogen–deuterium exchange, to reveal the solvent exposure of individual sites; this provides information about static conformation, dynamics, and interactions.

Investigations of the large-scale expression patterns of proteins—**proteomics**—require the use of methods that give high-throughput rates as well as accuracy and precision. Mass spectrometry, like microarrays, achieves this, which has stimulated its development into a mature technology in widespread use.

Identification of components of a complex mixture (see Fig. 2.27)

First the components are separated by electrophoresis, then the isolated proteins are digested by trypsin to produce peptide fragments with an r.m.m. of about 800–4000. Trypsin cleaves proteins after Lys and Arg residues. Given a typical amino acid composition, a protein of 500 residues yields about 50 tryptic fragments. The spectrometer measures the masses of the fragments with very high accuracy. The list of fragment masses, called the *peptide mass fingerprint*, characterizes the protein (Fig. 2.28). Searching a database of fragment masses identifies the unknown sample.

Construction of a database of fragment masses is a simple calculation from the amino acid sequences of known proteins, translations of open reading frames (ORFs) in genomes, or (in a pinch) of segments from EST (expressed sequence tag) libraries. The fragments correspond to segments cut by trypsin at lysine and arginine residues, and the masses of the amino acids are known. (Note that trypsin doesn't cleave Lys–Pro peptide bonds, and may also fail to cleave Arg–Pro peptide bonds.)

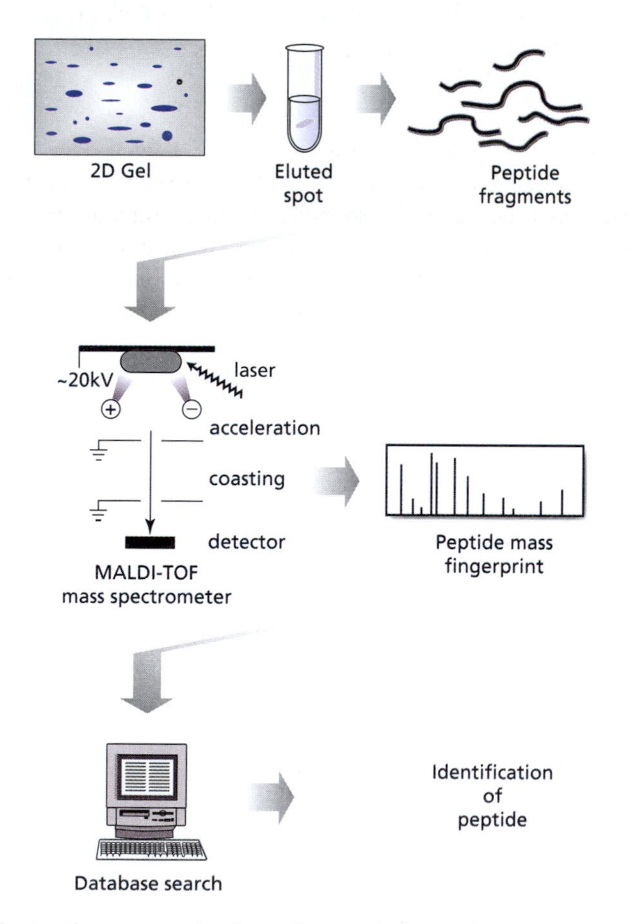

Fig. 2.27 Identification of components of a mixture of proteins by elution of individual spots, digestion, and fingerprinting of the peptide fragments by MALDI–TOF (matrix-assisted laser desorption ionization–time-of-flight) mass spectrometry, followed by looking up the set of fragment masses in a database.

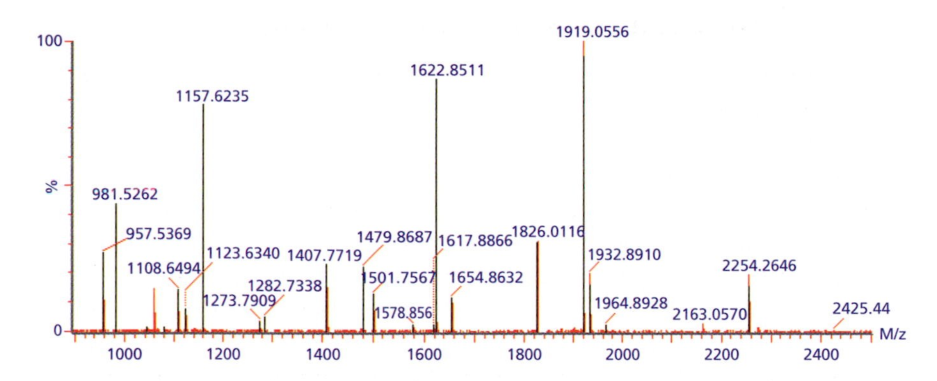

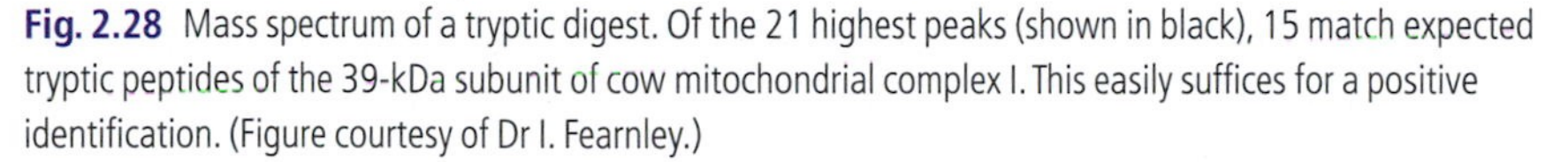

Fig. 2.28 Mass spectrum of a tryptic digest. Of the 21 highest peaks (shown in black), 15 match expected tryptic peptides of the 39-kDa subunit of cow mitochondrial complex I. This easily suffices for a positive identification. (Figure courtesy of Dr I. Fearnley.)

Two commonly used databases used for the identification of proteins from peptide mass fingerprints are: **www.matrixscience.com** and **prospector.ucsf.edu/ucsf.html3.4/msfit.htm**.

Mass spectrometry is sensitive and fast. Peptide mass fingerprinting can identify proteins in sub-picomole quantities. Measurement of fragment masses to better than 0.1 mass units is quite good enough to resolve isotopic mixtures. It is a high-throughput method, capable of processing 100 spots/day (though sample preparation time is longer). However, there are limitations. Only proteins of known sequence can be identified from peptide mass fingerprints, because only their predicted fragment masses are included in the databases. Also, post-translational modifications interfere with the method because they alter the masses of the fragments.

As with other fingerprinting methods, it would be possible to show that two proteins from different samples are likely to be the same, even if no identification is possible.

The results shown in Fig. 2.28 are from an experiment in which the molecular masses of the ions were determined from their **time-of-flight** (TOF) over a known distance, as illustrated in Fig. 2.27.

The operation of the spectrometer involves the following steps:

(1) production of the sample in an ionized form in the vapour phase;

(2) acceleration of the ions in an electric field, each ion emerging with a velocity proportional to its charge/mass ratio;

(3) passage of the ions into a field-free region, where they 'coast';

(4) detection of the times of arrival of the ions, the 'time-of-flight' (TOF) indicating the mass-to-charge ratio of the ions;

(5) the result of the measurements is a trace showing the flux as a function of the mass-to-charge ratio of the ions detected.

Proteins being fairly delicate objects, it has been challenging to vapourize and ionize them without damage. Two 'soft-ionization' methods that solve this problem are:

1. In **matrix-assisted laser desorption ionization (MALDI)**, the sample is introduced into the spectrometer in dry form, and mixed with a substrate or matrix that moderates the delivery of energy. A laser pulse, absorbed initially by the matrix, vaporizes and

ionizes the protein. The MALDI–TOF combination, which produced the results shown in Fig. 2.28, is a common experimental configuration.

2. The **electrospray ionization (ESI)** method starts with the sample in liquid form. Spraying it through a small capillary with an electric field at its tip creates an aerosol of highly charged droplets. As the solvent evaporates, the droplets contract, bringing the charges closer together and increasing the repulsive forces between them. Eventually the droplets explode into smaller droplets, each with less total charge. This process repeats, creating ions, (which may be multiply charged) devoid of solvent. These ions are transferred into the high-vacuum region of the mass spectrometer.

Because the sample is initially in liquid form, ESI lends itself to automation in which a mixture of tryptic peptides passes through a high-performance liquid chromatograph (HPLC) directly into the mass spectrometer.

Protein sequencing by mass spectrometry

Fragmentation of a peptide produces a mixture of ions. Conditions under which cleavage primarily occurs at peptide bonds yield series of ions differing by the masses of single residues (Fig. 2.29). The amino acid sequence of the peptide can therefore be deduced from analysis of the mass spectrum (Fig. 2.30), subject to ambiguities: for instance, Leu and Ile have the same mass and cannot be distinguished in fragments cleaved only at peptide bonds. Discrepancies from the masses of standard residues signal post-translational modifications. In practice, a sequence of about 5–10 amino acids can be determined from a peptide up to about <20–30 residues long.

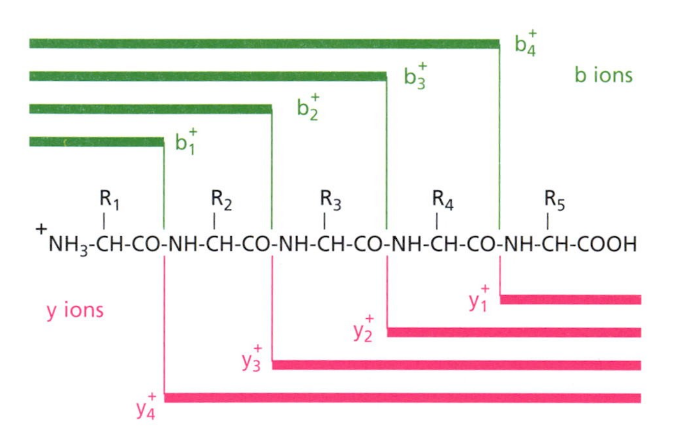

Fig. 2.29 Fragments produced by peptide bond cleavage of a short peptide. b-ions contain the N-terminus; y-ions contain the C-terminus. The difference in mass between successive b-ions or successive y-ions is the mass of a single residue, from which the peptide sequence can be determined. Two ambiguities remain: Leu and Ile have the same mass and cannot be distinguished, and Lys and Gln have almost the same mass and usually cannot be distinguished. In collision-induced dissociation, bond breakage can be largely limited to peptide linkages by keeping to low-energy impacts. Higher energy collisions can fragment side chains, occasionally useful to distinguish Leu/Ile and Lys/Gln.

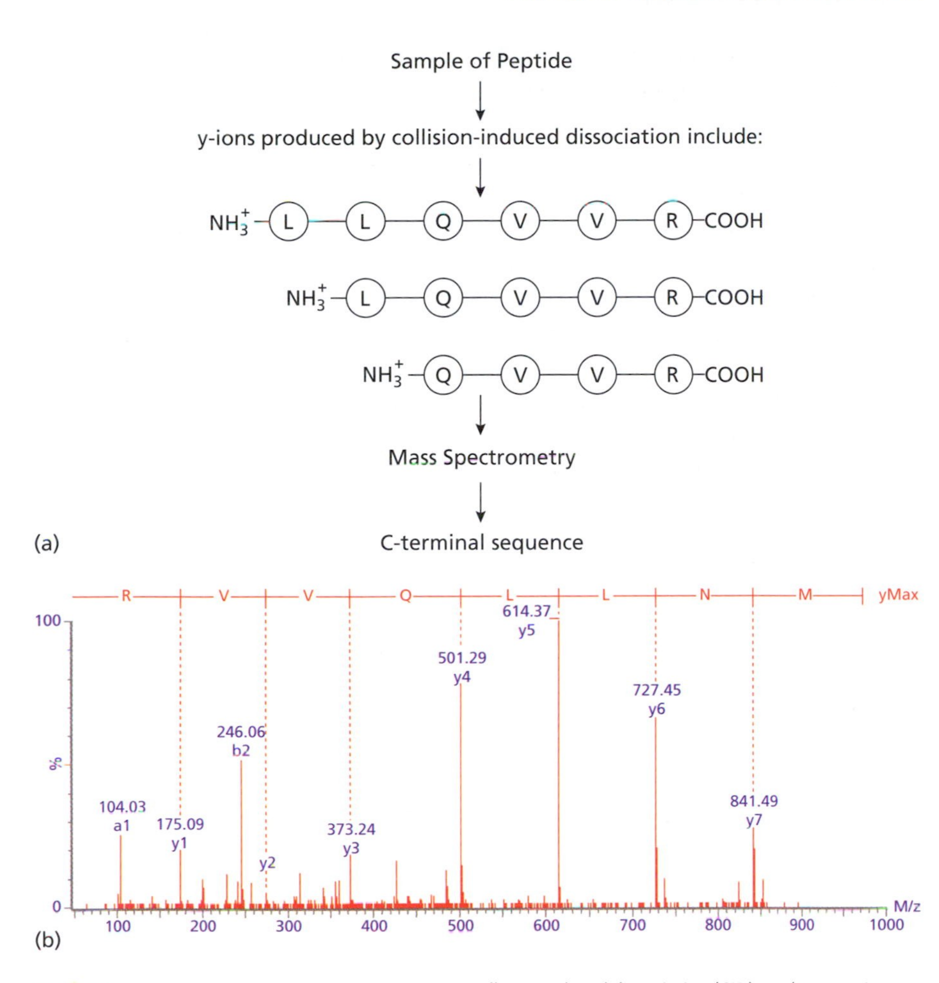

Fig. 2.30 Peptide sequencing by mass spectrometry. Collision-induced dissociation (CID) produces a mixture of ions. (a) The mixture contains a series of ions, differing by the masses of successive residues in the sequence. In CID the ions are *not produced* in sequence as suggested by this list, but the mass-spectral measurement automatically sorts them in order of their mass/charge ratio. (b) Mass spectrum of fragments suitable for C-terminal sequence determination. The greater stability of y-ions over b-ions in fragments produced from tryptic digests simplifies the interpretation of the spectrum. The mass differences between successive y-ion peaks are equal to the individual residue masses of successive residues in the sequence. Because y-ions contain the C-terminus, the y-ion peak of smallest mass contains the C-terminal residue, etc., and therefore the sequence comes out 'in reverse'. The two leucine residues in this sequence could not be distinguished from isoleucine in this experiment. (From Carroll, J., Fearnley, I.M., Shannon, R.J., Hirst, J., and Walker, J.E. (2003). Analysis of the subunit composition of complex I from bovine heart mitochondria. *Mol. Cell. Proteomics*, 2, 117–26 (Supplementary figure S138).)

In earlier techniques, vaporization *followed* sequential digestion of the peptide in solution. Methods of digestion included modifications of the classical Edman degradation, or treatment with non-specific exopeptidases such as leucine aminopeptidase or carboxypeptidase. In current practice, the fragments are produced *in situ*: first the peptide is vaporized, then it is fragmented by collision-induced dissociation (CID) with argon gas.

Masses of amino acid residues, standard isotopes

Gly	57.02146	Ala	71.03711	Ser	87.03203
Pro	97.05276	Val	099.06841	Thr	101.04768
Cys	103.00919	Leu	113.08406	Asn	114.04293
Asp	115.02694	Gln	128.05858	Lys	128.09496
Glu	129.04259	Met	131.04049	His	137.05891
Phe	147.06841	Arg	156.10111	Tyr	163.06333
Arg	156.10111	Trp	186.07931		

This approach requires the use of two mass analysers, operating in tandem in the same instrument (called MS/MS). The vaporized sample first passes through one mass analyser, to separate an ion of interest. The selected ion passes into the collision cell where impact with argon atoms excite and fragment it. By keeping the energy of impact low, the fragmentation can be largely limited to peptide bond breakage (Fig. 2.29). The second mass analyser determines the masses of the fragments.

Measuring deuterium exchange in proteins

If a protein is exposed to D_2O (deuterium oxide, heavy water), mobile hydrogen atoms will exchange with deuterium at rates dependent on their exposure in the protein conformation. By exposing proteins to D_2O for variable times, mass spectrometry can give a conformational map of the protein. Applied to native proteins, the results give information about the structure. Using pulsed exposure, the method can give information about intermediates in folding.

Computing in protein science

Computing has made major contributions to protein science, including:

- bioinformatics—the creation, coordination, and applications of large-scale databases;
- molecular graphics;
- simulations, including molecular dynamics; and
- computer-based instrumentation.

Let consider these in more detail—in reverse order.

Computer-based instrumentation

One theme of our discussions has been the trend towards using high-throughput experimental methods, in molecular biology in general and protein science in particular.

Computers are essential for the control, capture, and reduction of the data. Large-scale sequencing projects such as the human genome project are entirely computer controlled. Even studies of individual molecules often generate enough data to require the use of computer data-acquisition methods. X-ray crystallography is a good example; in a protein structure determination the intensities of hundreds of thousands of reflections may be measured.

Simulations, including molecular dynamics

Proteins obey the laws of physics. Why not apply these laws to predict the structures and properties of proteins? (After all, this worked fine for the solar system.) *Molecular dynamics* is just such an attempt. The atoms of the protein, and of the solvent in which it is embedded, are the particles. Newton's laws allow their trajectories to be predicted, knowing the forces the particles exert on one another. One problem is that the forces are known only approximately, and therefore the results can, at best, be only approximate. Another is that even the very powerful computers now available cannot simulate the motion of a protein long enough to see it fold up from a denatured state. However, with guidance from experimental data—in X-ray diffraction or nuclear magnetic resonance (NMR) spectroscopy—molecular dynamics does a fine job of finding a suitable model consistent with the data. Such simulations now contribute to protein structure determinations by X-ray crystallography and NMR spectroscopy.

Not all simulations treat full-atom representations by Newtonian mechanics. Other computational approaches to understanding and predicting protein structure permit: (1) other ways of applying experimental information; and (2) testing ideas about protein folding using simplified models and force fields.

Knowledge-based methods of protein structure prediction

Databases contain very large amounts of information about amino acid sequences and protein structures. Should it not be possible to apply them to predicting protein structure from an amino acid sequence, even if the target structure is an entirely novel fold? One crucial idea is the extent to which *local* sequence patterns determine *local* structural patterns—for instance, is there a signal in a stretch of residues that implies that this region will fold into an α-helix? It is possible to do a fairly good job of predicting the structures of local regions, or at least to narrow the possibilities. Of course, the problem is then to put the regions together into a complete structure. Such methods are under active development, and have achieved some impressive successes.

Protein structure prediction is discussed in Chapter 5.

Simplified simulations

A *lattice model* represents the structure of a protein as a connected set of points distributed at discrete and regular positions in space, with simplified interaction rules for calculating the energies of different conformations. For small lattice polymers, it is possible to enumerate every possible configuration. There is then no chance that a computer program failed to find the optimum because the simulation could not be carried out for long enough times.

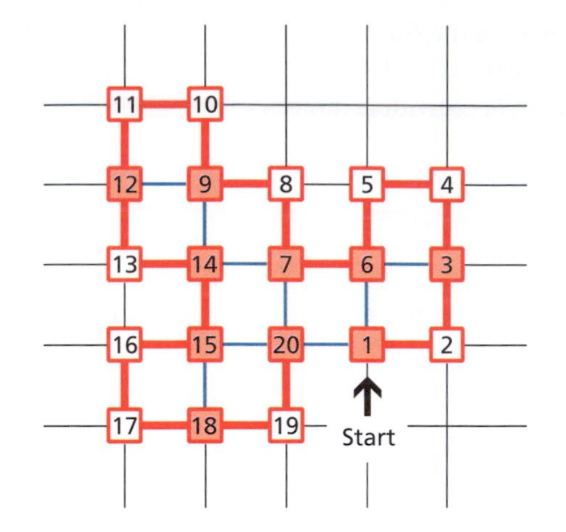

Fig. 2.31 A two-dimensional lattice model of a protein. There are two types of residues: hydrophobic (H) shown in filled red squares, and polar (P) shown in open squares. The 'protein' contains 20 residues, with the 'sequence' HPHPPHHPHPPHPHHPPHPH. Light blue lines show H–H contacts between residues not consecutive in the sequence. In this conformation there are 9 H–H contacts. Can the reader find a lower energy conformation of this system?

The simplest lattice model is the two-dimensional hydrophobic–polar (HP) model, with the assumption that:

1. The positions of residues are restricted to points on a square lattice in a plane.
2. Each residue can be of one or two types for hydrophobic or polar—H or P
3. The interaction energy is measured only between residues in contact: a neighbouring H–H pair contributes a constant attractive energy—E, and all other neighbours make no contribution to the conformational energy.

Figure 2.31 shows a two-dimensional lattice 'protein' in a low-energy conformation.

Another intensively studied lattice model represents a 27-residue polymer on a three-dimensional $3 \times 3 \times 3$ lattice. There are exactly 57 704 compact structures. It is very comforting to do statistical-mechanical calculations on systems with small, discrete sets of states, for which it is possible to consider every conformation explicitly! It's just too bad that they're not real proteins.

The late Irving Geis produced superb pictures, but of course of only a few structures. Among museum artists, works of Salvador Dali and Ben Shahn include representations of molecules.

Molecular graphics

There is nothing like vision to appreciate a three-dimensional structure. With so many known protein structures, it is impossible to build physical models, or to draw pictures of all of them by hand. Computers can generate many different representations of proteins, from fully detailed ones to simplified 'cartoons'. There is an active cottage industry designing different representations. A skilled molecular illustrator will combine them to show different parts of a structure in varying and finely tuned degrees of detail. At a computer screen it is possible to rotate the picture under manual control, a great advantage for perceiving three-dimensional relationships.

A crucial advantage of computer graphics over physical models is that it is possible to superpose structures. Only in this way can structural differences be easily seen.

Fig. 2.4 showed a structural superposition.

Bioinformatics

Bioinformatics is a hybrid science, including computer science and molecular biology. Computing is absolutely essential to managing the flood of data that the high-throughput methods are generating. A major goal of bioinformatics is to map out the world of sequences, structures, and functions of biological molecules, and the relationships among them. Using these maps, molecular biologists travel in time—back, to deduce events in evolutionary history; and onwards to achieve greater controlled modifications of biological systems. The methods of bioinformatics support applications to medicine, agriculture, and technology.

At the basis of the enterprise are the large-scale databases available on the World Wide Web. Readers will have used the Web for reference material, for access to databases in molecular biology, for checking out personal information about individuals—friends or colleagues or celebrities—or just for browsing. The Web provides a complete global village, containing the equivalent of library, postoffice, shops, and schools. It is not merely the number of entries in the Web but the density of the connections between them—their reticulation—that makes it such a powerful tool. The Web has already transformed the world. Its professional and social implications continue to be revolutionary.

Databanks for protein science

The World Wide Web contains many compilations of molecular biology information. Some are general and comprehensive databanks of sequence, structure, function, or bibliography; others are specialized 'boutique' collections. Still others are primarily indices of other Web sites. Indeed, given the crucial role of the reticulation of information sources, the utility of a site depends on the quality of the links it contains as well as the data it presents. Every January, issues of the journal *Nucleic Acids Research* contain annual contributions from most of the important database projects. A companion issue collects descriptions of Web-based software.

Major primary data-curation projects in molecular biology archive and annotate nucleic acid sequences, protein sequences, and macromolecular structures. Many individuals or groups select and recombine data focused on particular topics, and provide links affording steamlined access to information about the topic of interest. For instance, the protein kinase resource is a compilation that includes sequences, structures, functional information, laboratory procedures, lists of interested scientists, tools for analysis, a bulletin board, and links (**http://pkr.sdsc.edu/html/index.shtml**). The HIV protease database archives structures of human immunodeficiency virus-1 (HIV-1) proteases, human immunodeficiency virus-2 (HIV-2) proteinases, and simian immunodeficiency virus (SIV) proteases and their complexes, and provides tools for their analysis and links to other sites with AIDS-related information (**http://mcl1.ncifcrf.gov/hivdb/**).

Archival databanks of amino acid sequences The archiving of amino acid sequences of proteins, now determinined almost exclusively by translation of gene sequences, is being carried out by the United Protein Database (UniProt): a merger of the databases SWISS-PROT, the Protein Identification Resource (PIR), and Translated EMBL (TrEMBL). Many derived databanks attempt to classify proteins into families based on the similarities of their sequences.

Archival databanks of protein structures The Protein Data Bank (PDB) is a collection of the publicly available structures of proteins, nucleic acids, protein–nucleic acid complexes, and other biological macromolecules. Related archives include: the Nucleic Acid Data Bank, containing oligo and polynucleotide structures, and a database of protein–DNA complexes; and the BioMagRes Data Bank, containing NMR structures. The Cambridge (UK) Crystallographic Data Centre archives the structures of small organic molecules.

Currently (1 January 2004), the PDB contains 23,792 sets of coordinates. However, there is a certain amount of duplication. In some cases there have been several determinations of the structure of a protein, crystallized under different conditions, or re-solved at higher resolution. In some cases the structure of a protein has been determined in different states of ligation; for example, an enzyme with no ligand, and the same protein binding an inhibitor.

In other cases the structures of very closely related molecules have been determined. This makes it difficult to state precisely how many unique protein structures the databank contains, but this number may be estimated at 10 000. Several sites present classifications of protein structures (see Chapter 5).

Submitted coordinates are subjected to stereochemical checks, and translated into a standard entry format. Entries contain technical information about the structure determination, some structural analysis such as assignments of helices and sheets, references to papers describing the structure, and the atomic coordinates.

The PDB Web Site www.rcsb.org and its 'mirrors' around the world permit the retrieval of entries in computer-readable form. The Macromolecular Structure Database www.ebi.ac.uk/msd/ is a European counterpart. These and other sites provide search tools for identifying entries that match stated specifications, including molecule name, species of origin, depositor name, experimental technique, etc. Protein data banks in the USA, UK and Japan are collaborating to form an umbrella organization called the worldwide PDB.

Information-retrieval tools

One function of archival databanks is conservation; protecting information against loss or corruption. However, conservation is not enough: databanks without effective access become data graveyards. Effective access requires computer programs that allow users to identify and retrieve data according to versatile sets of criteria. It is the responsibility of the database designers to organize the data with an adequate internal logical structure that makes writing such programs possible.

In addition to scanning databanks for items of interest, facilities available on the Web provide a wide range of computational tools for data analysis. They take the form of Web Servers—sites that allow the user to select input and launch calculations. Some of these are straightforward operations, such as the calculation of the r.m.m of a protein. Some, at the other extreme, are at the cutting edge of research—examples are the servers that attempt to predict the three-dimensional structure of a protein from its amino acid sequence.

If the computation is a fast one, the results may be returned by the browser 'on the fly'. If the computation is slow, the program may send the results by e-mail. The old model of 'install programs on your computer and download the data on which to run them' is giving way to world-wide distributed computer facilities. Only connect.

Web access to the scientific literature

We are in an era of transition to paper-free publishing.

The US National Library of Medicine supports a bibliographical database of the biomedical literature www.ncbi.nlm.nih.gov/PubMed. More and more scientific publications are appearing on the Web. A journal may post its table of contents, a table of contents together with abstracts of articles, or even full articles.

Although most commercial publishers are moving towards Web access, they and their customers are still feeling their way while a new economic *modus vivendi* evolves. Some journals place their issues on the Web immediately upon publication. Some place their issues on the Web after a delay—to encourage the purchase of paper copies. Others make only their tables of contents, or only summaries of their articles available electronically. Moreover, although the medium of distribution has changed, the underlying 'news-stand' economic model—payment at the point of access to the information—has remained the same.

A radical alternative, PubMed Central, has created a central repository for literature in the life sciences. Negotiations with publishers, to allay their fears of loss of subscription income, involved agreements that the publishers would retain their customary copyright assignments for commercial use of the material published, and that there would be a delay between the publication of a journal on paper and free electronic access to its contents.

A new organization, the Public Library of Science, has the goal of making the scientific (including medical) literature publicly and freely accessible. A non-profit organization, The Public Library of Science has received support from various foundations, for its efforts in distributing literature published by others, and to start its own publications, which will permit exploration of different relationships—including, but not limited to, economic ones—between authors, publishers, and readers.

USEFUL WEB SITES

The Protein Data Bank: www.rcsb.org
The Macromolecular Structure Database: www.ebi.ac.uk/msd/
The worldwide Protein Data Bank: www.wwpdb.org
Home page of ReLiBase, a system for analysing receptor–ligand complexes in the Protein Data Bank: www.relibase.ebi.ac.uk/
Collection of on-line analysis tools, including database searches: www.biol.univ-mrs.fr/english/logligne.html
Index to web sites in molecular biology, including specialized databases: www.cbs.dtu.dk/biolink.html

BCM search launcher—various database searches and associated tools:
http://kiwi.imgen. bcm.tmc.edu:8088/search-launcher/launcher.html
Collection of protein analysis tools:
http://www.bioscience.org/urllists/protanal.htm
http://alpha2.bmc.uu.se/embo/structdb/links.html
World Index of Molecular Visualization Resources: www.molvisindex.org
Site about electronic scholarly publishing, with emphasis on genetics: www.esp.org/

RECOMMENDED READING

A TRENDS guide to proteomics. (2002). *Trends Biotechnol.*, **20**, Supplement.

Berman, H.M., Goodsell, D.R. and Bourne, P.E. (2002). Protein structures: from famine to feast. *Am. Sci.*, **90**, 350–9.

Parry, D.A. and Steinert, P.M. (1999). Intermediate filaments: molecular architecture, assembly, dynamics and polymorphism. *Q. Rev. Biophys.*, **32**, 99–187.

Ponting, C.P., Schultz, J., Copley, R.R., Andrade, M.A. and Bork, P. (2000). Evolution of domain families. *Adv. Prot. Chem.*, **54**, 185–244.

Richards, F.M. (1991). The protein folding problem. *Sci. Am.*, **264**(1), 54–7; 60–3.

Ureta-Vidal, A., Ettwiller, L. and Birney, E. (2003). Comparative genomics: genomewide analysis in metazoan eukaryotes. *Nat. Rev. Genet.*, **4**, 251–62.

Vale, R.D. (1999). Millenial musings on molecular motors. *Trends Cell Biol.*, **9**, M38–M42.

EXERCISES, PROBLEMS, AND WEBLEMS

Exercises

1. On photocopies of the drawing of the helix, on the left in the box on page 27, indicate residue positions, as in the drawing on the right, for a helix containing (a) 2 residues/turn, (b) 1.5 residues per turn.
2. If a helix has a pitch of 10 Å and a radius of 4 Å, what is the steepness, that is, the rise per residue?
3. Collagen chains contain three helices. How many different combinations of parallel and antiparallel combinations are geometrically possible?
4. On the helix of GCN4 (Fig. 2.6(a)), which positions of the heptad repeat have the highest proportion of charged residues? For each of these positions, do all the charged residues appearing have charges of the same sign?
5. What other protein illustrated in this chapter resembles the human growth hormone receptor (green and orange chains of Fig. 2.17) in structure?
6. The orientation of sperm-whale myoglobin in Fig. 2.22(a) is close (within about 45°) to that of one subunit of human haemoglobin in Fig. 2.22(c). What is the colour of this subunit of human haemoglobin?

7. On a photocopy of Fig. 2.18, circle the following groups within the GTP: (a) the sugar, (b) the central phosphate.

8. On a phototocopy of Fig. 2.21, circle a region containing an α-helix.

9. **(a)** If the direction of one strand in acylphosphatase were reversed, the sheet topology would be a pure antiparallel sheet (that is, the direction of every pair of adjacent strands would be opposite). On a photocopy of Fig. 2.20(a), indicate which strand would have to be reversed.

   (b) Could a β-sheet containing an odd number of strands form a β-barrel with a pure antiparallel β-sheet? Explain your answer.

10. Draw a topology diagram, similar to those on page 42, of a 5-stranded sheet made of four hairpins.

11. On a photocopy of Fig. 2.20(b), indicate where the fungal toxin has the following structural pattern: a *strand* followed by a *helix* followed by a *strand positioned two strands to the left* (in the orientation shown).

12. On photocopies of the diagram on page 49, indicate points:

    (a) where a pair of highly diverged homologous proteins with similar structure but without obvious sequence similarities might lie;

    (b) where a pair of non-homologous proteins with similar structure might lie;

    (c) where a pair of enzymes that share a function but not a structure, such as chymotrypsin and subtilisin, might lie.

13. **(a)** What is the sequence of the fragment y_6 in Fig. 2.30(b)?

    (b) To which peak in Fig. 2.30(b) does the fragment NH_3^+LQVVR correspond?

14. In the two-dimensional HP lattice model of a protein (Fig. 2.31):

    (a) what is the lowest-energy conformation of the peptide PHHP?

    (b) why would the peptide PHP tend to form aggregates?

15. Hen egg-white lysozyme has a relative molecular mass of about 14 300. If mass spectroscopy can measure mass to within 0.01%, could the following be confidently distinguished from the unmodified protein: (a) N-terminal acetylation? (b) phosphorylation of a single serine residue? (c) a single Lys → Gln substitution?

16. On photocopies of Fig. 2.30(b), indicate the positions of the peaks if the sequence were: (a) MNLVQVR, (b) GNLQVVR, (c) MNLQVVG.

Problems

1. The sequences of the leucine zipper domains of jun and fos are:

   fos: RRELTDTLQAETDQLEDEKSALQTEIANLLKEKEK

   jun: RIARLEEKVKTLKAQNSELASTANMLREQVAQL

   (a) Draw helical wheels for each, and identify the zipper positions. Use an angular difference of 103°, corresponding to the value appropriate for a supercoil, rather than the 100° appropriate for a straight α-helix. This will improve the register (compare Fig. 2.6).

(b) Photocopy and cut out the two helical wheels, and assemble them in the relative orientation expected for formation of the jun–fos dimer.

2. The geometry of binding of human growth hormone to human growth hormone receptor has interesting effects on the dose–response curve. Recall that in the active complex of hormone with dimeric receptor, two molecules of receptor use the *same* active site to interact with *different* sites on the hormone.

Call the two sites on the hormone 1 and 2 and let * signify 'is ligated to'. Then possible species are:

monomeric receptor

dimer: monomeric receptor*site 1–hormone

dimer: monomeric receptor*site 2–hormone

trimer: site 1–hormone–site 2

```
        *                    *
     receptor     *      receptor (dimeric receptor)
```

It is believed that there is little receptor dimerization in the absence of ligand. Because one of the sites on the hormone forms a more stable complex with monomeric receptor than the other, it is likely that binding is sequential:

hormone $\rightarrow$ receptor*site 1–hormone

$\rightarrow$ receptor*site 1–hormone–site 2*receptor.

In the presence of a fixed amount of receptor, the dose–response curve shows an initial increase in dimer formation as the concentration of hormone is raised, followed by decrease. Explain this result. First sketch the species present during the assumed sequential binding. What species is likely to predominate at high ratios of hormone:receptor concentration? Why does formation of this species prevent the formation of active dimer?

3. On the aligned sequences of the histidine-containing phosphocarrier proteins of *E. coli* and *S. faecalis,* indicate with highlighter the worst-fitting regions. Is there any correlation with the parts of the sequences with the fewest local matches and the regions of insertions and deletions?

4. Draw a topology diagram similar to those on page 42 of the histidine-containing phosphocarrier proteins in Fig. 2.4.

5. The mass per unit length of human hair is quite variable, but 2 mg m^{-1} is a typical value. Calculate a rough, order-of-magnitude estimate of the number of coiled coils that pack side-by-side in a human hair.

6. What is the general structural difference between an oligomeric protein (such as aspartate carbamoyltransferase) and a modular protein (such as Factor VIIa)? Give an example of a protein that is *both* modular and oligomeric, other than an immunoglobulin, which was illustrated in Fig. 2.23.

7. Consider a process of two irreversible steps:

$$A \rightarrow B \rightarrow C.$$

Suppose you start with a very large amount of A and no B or C. What will be the ratio of the amount of C to the amount of B at subsequent times, before A is exhausted, if:

(a) the rate of conversion of one molecule of A to one molecule of B is twice the rate of conversion of one molecule of B to one molecule of C?

(b) the rate of conversion of one molecule of A to one molecule of B is half the rate of conversion of one molecule of B to one molecule of C?

8. **(a)** What is the mass of the octapeptide $^{+}H_3N$–$(Ala)_4Lys^{+}(Ala)_4COO^{-}$?

(b) What kinetic energy would be imparted to this ion by passage through a potential drop of 20 kV?

(c) To what velocity would this kinetic energy correspond?

(d) What would be the time-of-flight of this ion through a field-free path of 1.2 m?

(e) To what precision would the detector have to measure the time of arrival to separate the parent ion from one in which one ^{12}C atom was replaced by ^{13}C?

Hint for (b) and (c): $eV = \frac{1}{2} mv^2$

9. The mass of the peptide TAPGEQGTTTTR is 1201.5573.

(a) What is the mass of the peptide identical in sequence except for the substitution of Lys for the Gln?

(b) How precise must the measurement of mass be, in parts per million, so that these two peptides are not be confused when interpreting a mass spectrum?

(c) How precise must the measurement of mass be, in parts per million, so that it would be possible to distinguish a molecule of this peptide containing only ^{12}C carbons from one containing a single ^{13}C atom and the rest ^{12}C? The masses of isotopic carbon atoms are:

^{12}C	^{13}C	^{14}C
12.0000(0)	13.003 354 8378(10)	14.003 241 988(4)

The numbers in parentheses are uncertainties; the uncertainty in the mass value for ^{12}C is 0 because the mass of this isotope defines the atomic weight scale

10. The mass spectrum of a peptide from a tryptic digest of a spot from a two-dimensional gel contains y-ion peaks at the following masses: 377.2, 524.2, 637.3, 752.3, 899.3, 1012.4, 1125.5, 1212.5. What is the C-terminal amino acid? What are the possible amino acid sequences of the peptide?

11. In the two-dimensional HP lattice model of a protein:

(a) what are the closest possible analogues of a helix (not necessarily an α-helix) and a sheet?

(b) Do you see any examples of either of these in Fig. 2.31? On a photocopy of Fig. 2.31 indicate any examples that you see.

(c) Write a 10-residue sequence of Hs and Ps that in its lowest-energy state would form the analogue of a β-hairpin.

(d) Write a 10-residue sequence of Hs and Ps that in its lowest-energy state would form the analogue of a helix.

12. In the two-dimensional HP lattice model of a protein:

(a) how many conformations of a tripeptide are possible? Draw them.

(b) How many conformations of a tetrapeptide are possible? Draw them. (Consider two conformations that are superposable with the chain running in opposite directions to be *different*.)

13. (a) How many positions in all are there in the microarray in Fig. 2.26?

(b) How many are complementary to RNAs from liver?

(c) How many are complementary to RNAs from brain?

(d) How many are complementary to RNAs from liver and brain?

(e) How many are complementary to neither?

Weblems

1. Find the amino acid sequences of:

(a) baboon α-lactalbumin,

(b) cytochrome *b* from the duckbill platypus.

2. Find an example of a protein containing one or more SH2 domains that is: (a) a protein kinase, (b) a phosphatase, (c) involved in Ras signalling, (d) involved in ubiquitination. For each, provide a diagram showing the complete domain structure of the protein.

3. What are the Protein Data Bank entry codes of:

(a) three structures containing SH3 domains?

(b) three structures containing human thrombin-binding inhibitors (not including [1ABI] which was shown in Fig. 2.10)? what inhibitors are bound?

(c) three structures containing antibodies binding hen egg-white lysozyme.

4. From what protein or proteins could the fragment MNLLQVVR (Fig. 2.30) be derived?

5. (a) Find the citations and PubMed identification numbers of three papers published in the last 3 years by C. Chothia.

(b) (Harder) Find the citations and PubMed Identification numbers of three papers published in the last 3 years by J.D. Yeast.

Chapter 3

THE CHEMICAL STRUCTURE AND ACTIVITY OF PROTEINS

LEARNING GOALS

1. **To understand the basic chemistry of proteins**: including the distinction between the common repetitive backbone or main chain and the variable sequence of side chains.
2. **To know the 20 natural side chains**: structure, properties—size, polarity, charge, titratable groups, atom nomenclature.
3. **To be able to define the degrees of freedom that specify different conformations of the polypeptide chain**: including definitions of conformational angles, allowed values of main chain conformational angles (the Sasisekharan–Ramakrishnan–Ramachandran plot), rotamers, and rotamer libraries.
4. **To appreciate the energetic factors stabilizing native states of proteins**: covalent bonds, including disulphide bridges; hydrogen bonding; the hydrophobic effect and accessible surface area; the coordination of metal ions; van der Waals forces and the dense packing of protein interiors. The net effect is a relatively small stabilization energy of about 20–60 kJ mol^{-1}.
5. **To understand the results available from spectroscopic measurements on proteins**: absorbance, circular dichroism (CD) and its relationship to secondary structure content, fluorescence and fluorescence resonance energy transfer (FRET).
6. **To become familiar with methods of protein structure determination**: X-ray crystallography, nuclear magnetic resonance (NMR) spectroscopy, cryoelectron microscopy.
7. **To understand basic features of protein-ligand interactions** and the Michaelis–Menten model of enzyme kinetics.
8. **To appreciate that some proteins undergo conformational changes in response to ligands**, and know some of the roles of conformational changes in function.
9. **To know some basic general modes of control over protein activity**: including, but not limited to, the mechanism of action of simple inhibitors, and allosteric conformational changes.

The polypeptide chain and protein conformation

The polypeptide chain links amino acids through the **peptide bond**:

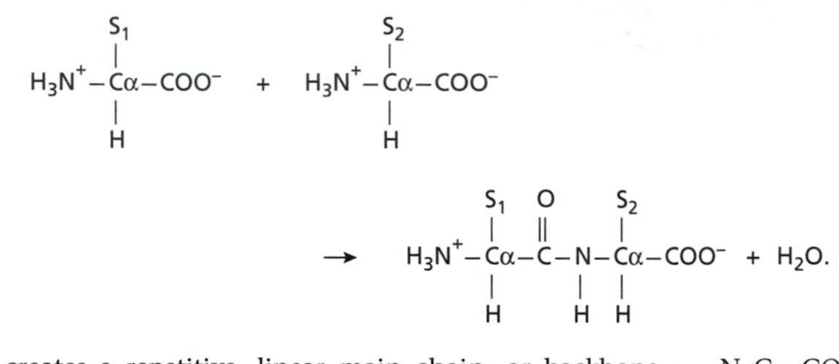

This creates a repetitive, linear main chain, or backbone— ... N–Cα–CO–N–Cα–CO– ... A side chain S is attached to the Cα of each unit.

Because the backbone is flexible, it can adopt an infinite number of spatial conformations, just as a piece of string thrown repeatedly on the floor will form random and unrelated patterns. In natural proteins, interactions among the side chains select one three-dimensional structure that, under physiological conditions of solvent and temperature, has much greater thermodynamic stability than any other. Each protein thereby takes up a unique *native*, biologically active conformation.

The amino acids

Collagen was discussed in Chapter 2.

Natural proteins contain a basic repertoire of 20 amino acids (see Box and Fig. 3.1). Some proteins do contain other amino acids outside the usual set of 20. In collagen we met hydroxyproline and hydroxylysine. These are produced by enzymatic modification after the protein is synthesized. The unusual amino acid selenocysteine is introduced

Naming the atoms

Atoms in the main chain of each residue are denoted N, Cα, C, and O. The side chain is attached to the Cα. Side chain atoms are identified by their chemical symbol, and by successive letters from the Greek alphabet, proceeding out from the Cα. Thus the side chain of methionine has atoms Cβ, Cγ, Sδ, Cε:

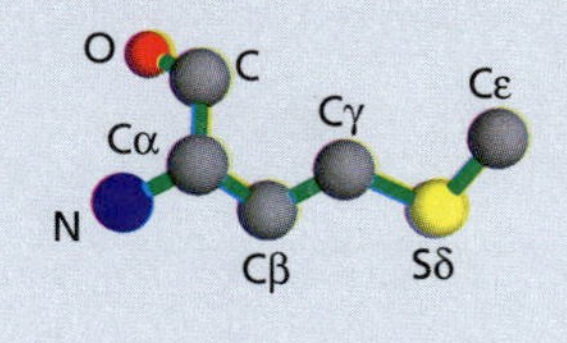

during translation. In addition to the amino acids, we know that ions, small organic ligands and even water molecules are integral parts of many protein structures. For instance, haem groups are an essential component of haemoglobin and cytochrome *c*.

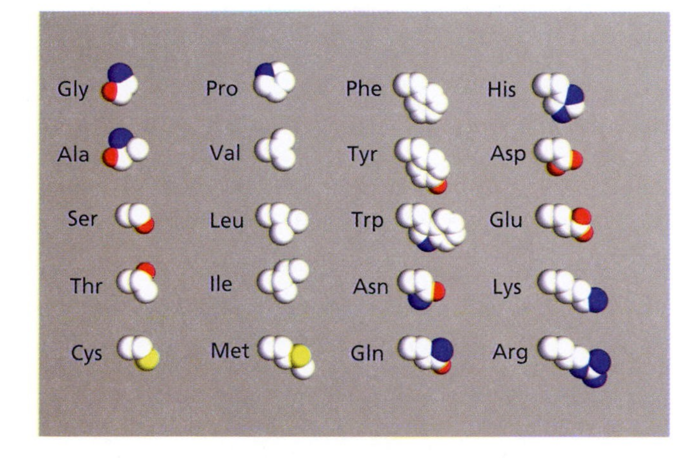

Fig. 3.1 The 20 natural amino acids. Grey or black = carbon, red = nitrogen, blue = oxygen, yellow = sulphur. This is the cast of characters, that play all the different roles in different proteins.
The side chains vary in their physicochemical properties: size, hydrogen bonding potential, and charge (see Box).

- Some side chains are electrically neutral. Because of the thermodynamically unfavourable interaction of hydrocarbons with water, they are called 'hydrophobic' residues.
- Other side chains are polar. Asparagine and glutamine contain amide groups; serine, threonine, and tyrosine hydroxyl groups. Polar side chains, like main chain peptides, can participate in hydrogen bonding.
- Other side chains are charged. Aspartic acid and glutamic acid are negatively charged; lysine and arginine are positively charged. The charged atoms occur at or near the ends of the relatively long and flexible side chains. The atoms nearest to the backbone are non-polar. Two side chains with positive and negative charge can approach each other in space to form a 'salt bridge'.

Acid/base properties of proteins

Amino acid	p*K* of side chain
Aspartic acid	3.90
Glutamic acid	4.07
Histidine	6.04
Cysteine	8.37
Tyrosine	10.46
Lysine	10.54
Arginine	12.48
Terminal amino group	~8.0
Terminal carboxyl group	~2.4

pK_a values of ionizable groups vary by about 0.5–1.0 unit, depending on their environments within the protein.

The chemical structures of the amino acids

Glycine	$^{+}NH_3$–Cα–COO^-
Alanine	–Cα–CH_3
Serine	–Cα–CH_2–OH
Cysteine	–Cα–CH_2–SH
Threonine	–Cα–CH(OH)–CH_3
Proline	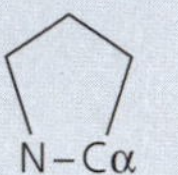 N–Cα
Valine	–Cα–CH–$(CH_3)_2$
Leucine	–Cα–CH_2–CH–$(CH_3)_2$
Isoleucine	–Cα–CH(CH_3)–CH_2–CH_3
Methonine	–Cα–CH_2–CH_2–S–CH_3
Phenylalanine	–Cα–CH_2–(phenyl ring)
Tyrosine	–Cα–CH_2–(phenyl ring)–OH
Aspartic acid	–Cα–CH_2–COO^-
Glutamic acid	–Cα–CH_2–CH_2–COO^-
Histidine	–Cα–CH_2–C (ring: N=CH, NH, HC)
Asparagine	–Cα–CH_2–$CONH_2$
Glutamine	–Cα–CH_2–CH_2–$CONH_2$
Lysine	–Cα–CH_2–CH_2–CH_2–CH_2–NH_3^+
Arginine	–Cα–CH_2–CH_2–CH_2–NH–C(NH_2)=NH_2^+
Tryptophan	–Cα–CH_2–(indole ring, N)

Protein main chain conformation

By the **conformation** of a polypeptide chain we mean the space curve that the backbone traces out. The main chain is flexible, and the number of possible spatial patterns very large. Indeed, in the denatured state, the conformation of different molecules in the sample are all different, and the chain can be extended over a large volume. The native states of proteins, in contrast, are special. Under physiological conditions of solvent and temperature, all molecules with the same amino acid sequence adopt the *same* conformation, the **native state**.

The native state conformation of a protein consists of a compact assembly of residues. Features consistent with stability of the native state include: satisfaction of the hydrogen-bonding potential of polar groups, burying of non-polar atoms, and dense packing of residues in the interior. The optimal solution of this thermodynamic jigsaw puzzle is a special conformation in which the backbone describes a curve traversing the space occupied by the molecule. The shape of this curve, called the protein's '*folding pattern*', usually includes standard elements of secondary structure—helices and sheets—and makes use of a common repertoire of ways in which helices and sheets pack against one another.

Protein architecture is the study of how protein folding patterns may be classified, how to understand the relationships between the local interactions and the overall folding pattern of the chain, and how the entire structure is determined by the amino acid sequence.

Chapter 4 discusses protein architecture.

The conformation of a polypeptide chain can be described quantitatively in terms of angles of internal rotation around the bonds in the main chain (Fig. 3.2). The bonds between the N and Cα, and between the Cα and C, are single bonds. Internal rotation around these bonds is not restricted by the electronic structure of the bond, but only by possible steric collisions in the conformations produced. (See Box: The Sasisekharan–Ramakrishnan–Ramachandran diagram.) The peptide bond has partial double-bond character, and has two possible conformations: *trans* (by far the more common) and *cis* (rare). Proline is an exception: the side chain is linked back to the N of the main chain to form a pyrrolidine ring. This restricts the main chain conformation of proline residues. It disqualifies the N atom as a hydrogen-bond donor, for instance in helices or sheets.

Also, the energy difference between *cis* and *trans* conformations is less for proline residues than for others. Most *cis* peptides in proteins appear before prolines.

The entire conformation of the protein can be described by these angles of internal rotation. Each set of four successive atoms in the main chain defines a dihedral angle. In each residue i (except for the N- and C-termini) the angle ϕ_i is the angle defined by atoms C(of residue $i-1$)–N–Cα–C, and the angle ψ_i is the angle defined by atoms N–Cα–C–N(of residue $i+1$). ω_i is the angle around the peptide bond itself, defined by the atoms Cα–C–N(of residue $i+1$)–Cα(of residue $i+1$). ω is restricted to be close to 180° (*trans*) or 0° (*cis*).

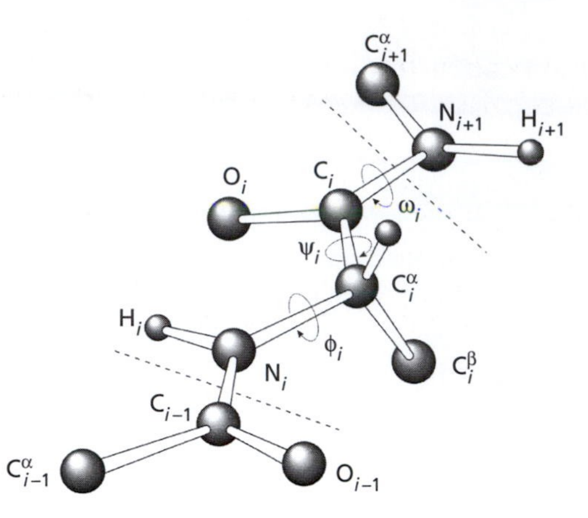

Fig. 3.2 Conformational angles describing the folding of the polypeptide chain. The main chain of each residue (except the C-terminal residue) contains three chemical bonds: N–Cα, Cα–C, and the peptide bond C–N linking the residue to its successor. The conformation of the main chain is described by the angles of rotation around these three bonds:

Rotation around:	N–Cα bond	Cα–C bond	peptide bond (C–N)
Name of angle:	ϕ	ψ	ω

Geometrically, the main chain of a protein is a succession of points in space: N–Cα–C–N–Cα–C ... (the carbonyl oxygens are not in the main chain itself). To a good approximation, the bond lengths and angles—the distances between every two successive points, and the angles between every three successive points—are fixed. The degrees of freedom of the chain then involve *four* successive atoms, and consist of rotations in which the first three atoms are held fixed, and the fourth atom is rotated around the bond linking the second and third. (An ordinary paperclip easily unwinds to give a chain with four vertices, and is a useful object with which to practise.) By convention, the cis conformation is 0°, and the positive sense of each angle corresponds to looking down the bond that serves as the axis of rotation, and turning the distant atom in a clockwise direction.

The Sasisekharan–Ramakrishnan–Ramachandran diagram

The main chain conformation of each residue is determined primarily by the two angles ϕ and ψ.

The angle of rotation around the peptide bond, ω, usually has the value $\omega = 180°$ (*trans*) and occasionally (most often before a proline residue) $\omega = 0°$ (*cis*). Taking the *trans* conformation of the peptide bond, $\omega = 180°$, for some combinations of ϕ and ψ atoms would collide, a physical impossibility. V. Sasisekharan, C. Ramakrishnan, and G.N. Ramachandran first plotted the sterically allowed regions (Fig. 3.3). There are two main allowed regions, one around $\phi = -57°$, $\psi = -47°$—denoted α_R—and the other around $\phi = -125°$, $\psi = +125°$—denoted β—with a 'neck' between them. The mirror image of the α_R conformation, denoted α_L, is allowed for glycine residues only. (Because glycine is **achiral**—that is, identical with its mirror image—a Ramachandran plot specialized to glycine must be right–left symmetrical. For non-glycine residues, collisions of the Cβ atom forbid the α_L conformation.)

continues...

The two major allowed conformations of the main chain, α_R and β, correspond to the two major types of secondary structure: α-helix and β-sheet. The α-helix is right-handed, like the threads of an ordinary bolt. In the β-region, the chain is nearly fully extended.

A graph showing the ϕ, ψ angles for the residues of a protein, against the background of the allowed regions, a Sasisekharan–Ramakrishnan–Ramachandran plot, is often called a Ramachandran plot for short (Fig. 3.3).

It is no coincidence that the same conformations that correspond to low-energy states of individual residues also permit the formation of structures with extensive main chain hydrogen bonding. The two effects thereby cooperate to lower the energy of the native state.

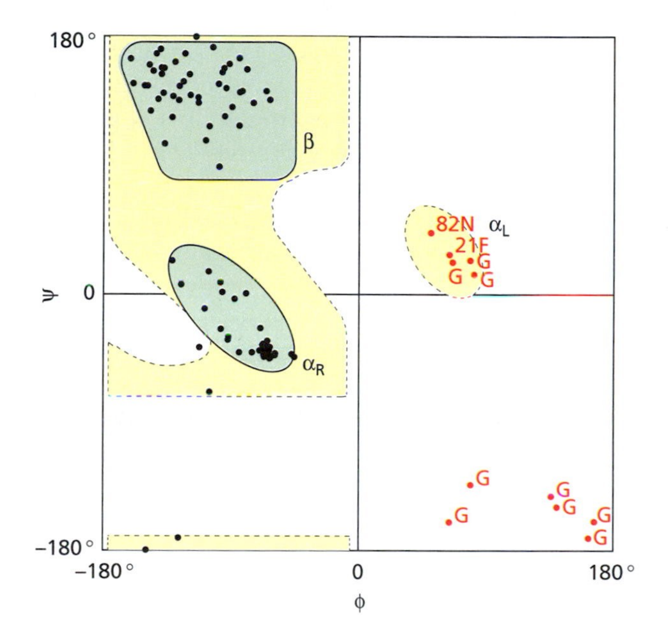

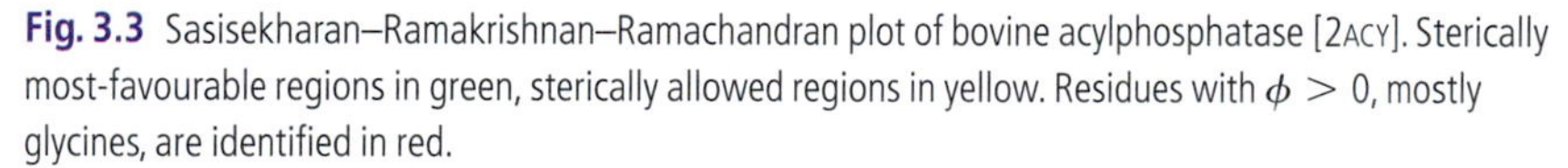

Fig. 3.3 Sasisekharan–Ramakrishnan–Ramachandran plot of bovine acylphosphatase [2ACY]. Sterically most-favourable regions in green, sterically allowed regions in yellow. Residues with $\phi > 0$, mostly glycines, are identified in red.

Side chain conformation

Side chain conformations are also described by angles of internal rotation. Different side chains have different numbers of degrees of freedom. An Arg side chain has five angles of internal rotation, and Gly has none at all. The side chain conformational angles are denoted $\chi_1, \chi_2, \ldots$. For example, for a lysine side chain χ_1 is the angle determined by the atoms N, Cα, Cβ, Cγ.

Different conformations of any side chain are called '*rotamers*'. Are there preferred conformations of side chains? For many side chains, the rotation angle around the Cα–Cβ bond tends to cluster around $\chi_1 = 180°$, $+60°$, and $-60°$, avoiding eclipsed states. For some side chains, the distributions of observed conformational angles may be very highly skewed indeed. For valine, the conformation $\chi_1 \sim 180°$ is populated almost exclusively.

Rotamer libraries

Most side chains have relatively small repertoires of preferred conformations. **Rotamer libraries** are collections of preferred side chain conformations. The local backbone structure influences side chain conformation, because placement of the backbone and Cβ atoms creates specific loci for potential steric collision. Secondary structure can thereby bias the side chain rotamer distribution. For example, the $\chi_1 = 60°$ conformation around the Cα–Cβ bond is hardly ever observed for any residue in a helix, except for serine, which can form a hydrogen bond to the C=O of the preceding residue. *Backbone-dependent rotamer libraries* specify the side chain conformational states preferred for different backbone conformations.

Rotamer libraries are useful for modelling, because only a small, discrete set of possible conformations needs to be considered. Also useful for modelling is the observation that as proteins evolve, side chain conformation tends to be conserved. That is, homologous residues in related proteins tend to have similar side chain conformations, even when they are mutated. The reason is that side chains pack against their neighbours. Each side chain fits into a 'cage' created by neighbouring side chains. When a residue is changed by mutation, the new side chain, in order to fit into the old cage, tends to adopt a conformation similar to that of the side chain it replaces.

Stabilization of the native state

Observations of protein structures have revealed general principles by which nature stabilizes native states of proteins:

- **Covalent and coordinate chemical bonds**. Many proteins contain covalent chemical bonds in addition to those of the polypeptide backbone and the side chains. Disulphide bridges, between cysteine residues, are quite common. Figure 3.4 shows the small protein crambin, which contains three disulphide bridges. Disulphide bridges can also link different polypeptide chains, as in insulin (Fig. 3.5) and in immunoglobulins.

  Metal ions are integral parts of the structures of many proteins. The 4-zinc form of the pig insulin hexamer illustrates both disulphide bridges and metal-ion binding directly to side chains (Fig. 3.5). In other cases, the metal is not bound directly to the

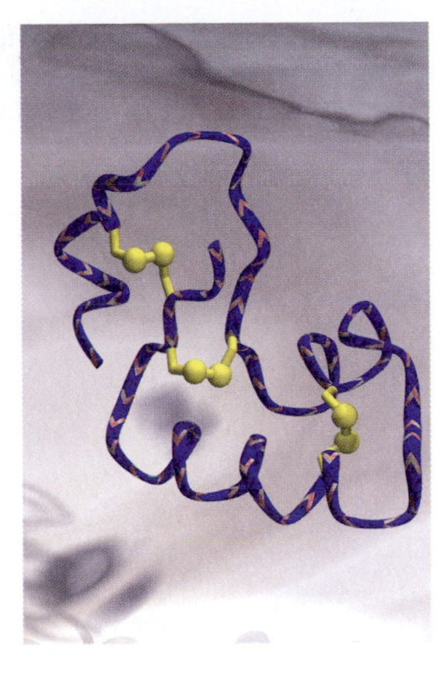

Fig. 3.4 Crambin [1CRN].

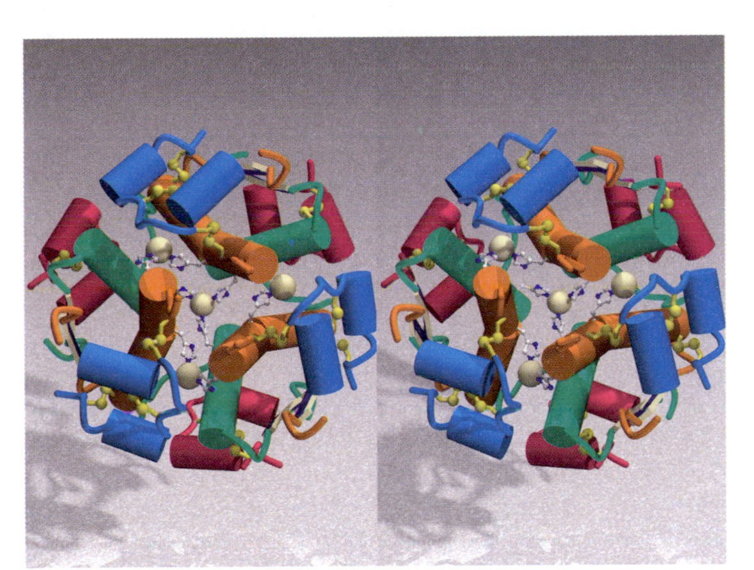

Fig. 3.5 Pig insulin hexamer, 4-zinc form [4ZNI].

protein, but is part of a larger ligand. Cytochrome *c* includes an iron-containing haem group (Fig. 3.6).

- **Hydrogen bonding**. Polar atoms in proteins make hydrogen bonds to water in the unfolded state. In the folded state, the hydrogen-bonding potential of atoms buried in the interior of the protein must somehow be satisfied. The main chain, containing peptide groups, *must* pass through the interior. Some polar side chains are also buried. They thereby lose their interactions with water. To recover the energy,

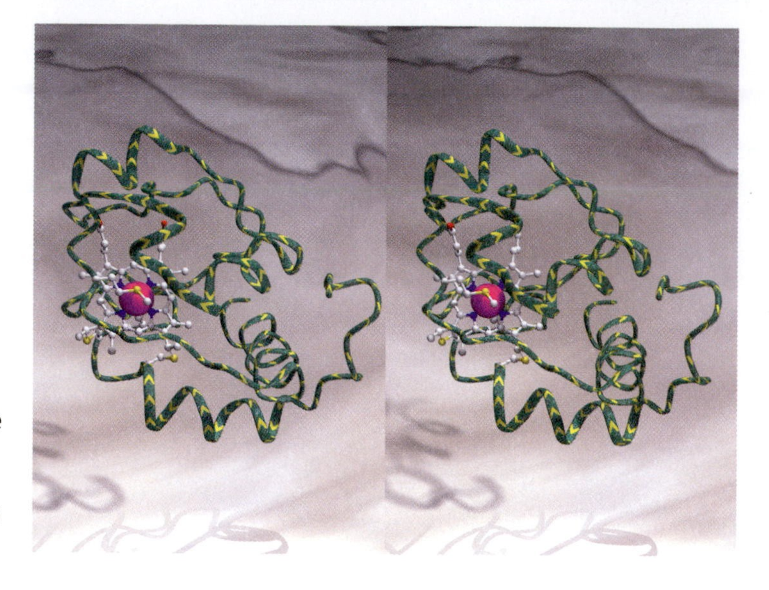

Fig. 3.6 Rice cytochrome c [1CCR]. The haem group is in ball-and-stick represention, and the purple sphere is the iron.

buried polar atoms form protein–protein hydrogen bonds. The standard secondary structures—helices and sheets—achieve the formation of hydrogen bonds by the main chain atoms. Because the hydrogen bonds of secondary structures involve main chain atoms only, they can potentially form from any amino acid sequence—except that the backbone N atom of proline cannot form a hydrogen bond.

Hydrogen bonds also contribute to the binding of cofactors and substrates.

- **The hydrophobic effect**. The **hydrophobicity** of an amino acid is a measure of the interaction between the side chain and water. Hydrocarbon side chains such as those of leucine and phenylalanine interact unfavourably with water, as does the oil in a salad dressing. Just as oil–water mixtures separate spontaneously into two phases, there is a tendency for the hydrophobic side chains to sequester themselves in the interior of a protein, away from contact with water. This effect provides an important driving force for protein folding. Different residues buried inside proteins contribute to protein stability according to a **hydrophobicity scale** (see Box; many other hydrophobicity scales have been proposed). The **accessible surface area** of the protein, calculatable from a set of atomic coordinates, measures the thermodynamic interaction between the protein and water (see Box). Surface-area calculations identify *which* residues are buried, contributing to the hydrophobic stabilization of a protein structure.
- **Van der Waals forces and dense packing of protein interiors**. The packing of atoms in protein interiors contributes to the stability of the structure in two ways. One is the exclusion of non-polar atoms from contact with water—the hydrophobic effect. The other is the force of attraction between the protein atoms.

The observed cohesion of ordinary substances shows that there are *attractive* forces between atoms and molecules. Conversely, the fact that substances do not collapse

Amino acid hydrophobicity scale*

2.25	Trp
1.80	Ile
1.79	Phe
1.70	Leu
1.54	Cys
1.23	Met
1.22	Val
0.96	Tyr
0.72	Pro
0.31	Ala
0.26	Thr
0.13	His
0.00	Gly
−0.04	Ser
−0.22	Gln
−0.60	Asn
−0.64	Glu
−0.77	Asp
−0.99	Lys
−1.01	Arg

*Fauchère, J. and Pliška, V. (1983). Hydrophobic parameters π of amino-acid side chains from the partitioning of *N*-acetyl-amino-acid amides. *Eur. J. Med. Chem.*, **18**, 369–75.

Accessible and buried surface area

The accessible surface area of a protein or protein complex is the area swept out by a water molecule (modelled as a sphere 1.4 Å in radius) rolling around the outside the protein. The accessible surface will include nooks and crannies in the protein surface that are larger than 2.8 Å wide, but smooth over finer wrinkles. Accessible surface area calculations rationalize the hydrophobic contribution to the thermodynamics of protein folding and interactions. Observed regularities include:

1. A basic calibration: each $Å^2$ of buried surface area contributes 6 J of free energy of stabilization.
2. The accessible surface area (ASA) of monomeric proteins of up to about 300 residues varies as the 2/3 power of the molecular weight M: ASA = $11.1\ M^{2/3}$.
3. The formation of oligomeric proteins from monomers buries an additional 1000–5000 $Å^2$ of surface. Lower values characterize proteins for which the monomer structure is stable in isolation; higher values characterize proteins in which association must stabilize the structure of the monomers as well as the complex.
4. Nature of the accessible and buried area in native protein structures: the average solvent-accessible surface of monomeric proteins—the protein *exterior*—is ~58% non-polar (hydrophobic), ~29% polar, and ~13% charged. The average buried surface of monomeric proteins—the protein *interior*—is ~60% non-polar (hydrophobic), ~33% polar, and ~7% charged. Many people expect the large *buried* hydrophobic surface but are surprised at how large the *exterior* hydrophobic area is. In fact, the main 'take-home message' about the difference between the surface and the interior is that proteins *almost never* bury charged groups.

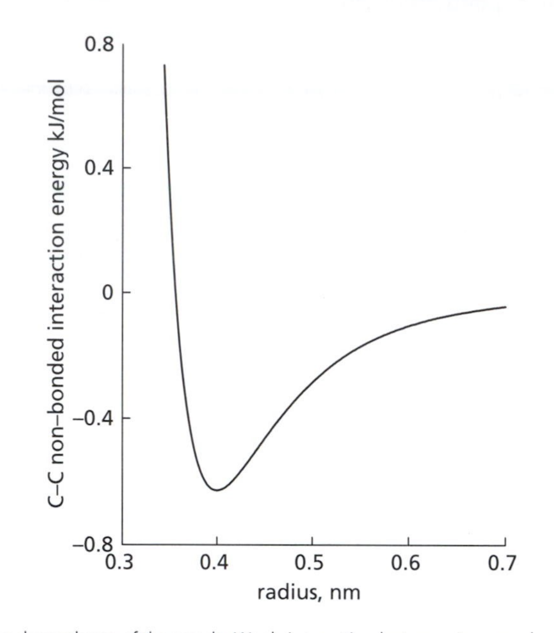

Fig. 3.7 The distance dependence of the van der Waals interaction between two non-bonded carbon atoms. This curve has a single minimum, rises very steeply at distances below the minimum, and rises more slowly above the minimum. If two carbon atoms were separated by the distance of minimum energy, attempts to push them closer together would be resisted by an increasingly strong force. Attempts to pull them apart would be resisted by an increasingly weak force.

completely, and indeed that there are limits to how far they can be compressed, shows that at short range these forces must be *repulsive*. The most general type of interatomic force, the van der Waals force, reflects this principle: the nearer the atoms, the stronger the force, until the atoms are actually 'in contact', at which point the forces become repulsive and strong (Fig. 3.7).

To maximize the total cohesive force, therefore, as many atoms as possible must be brought as close together as possible. It is the requirement for a dense packing that imposes a requirement for *structure* on the protein interior. It produces a jigsaw-puzzle-like fit of interfaces between elements of secondary structure packed together in protein interiors (see Fig. 3.8). Such a complementarity of opposing surfaces is also responsible for the specificity with which many enzymes bind their substrates, as in Fischer's 'lock and key' analogy.

- **A physical effect that opposes the formation of the native state**: it is unnatural for all the molecules in a sample with a flexible chain to adopt the same conformation.

The overall result is that the stabilization of the native states of protein structures is relatively small, typically 20–60 kJ mol^{-1}. This is the equivalent of no more than about two or three hydrogen bonds.

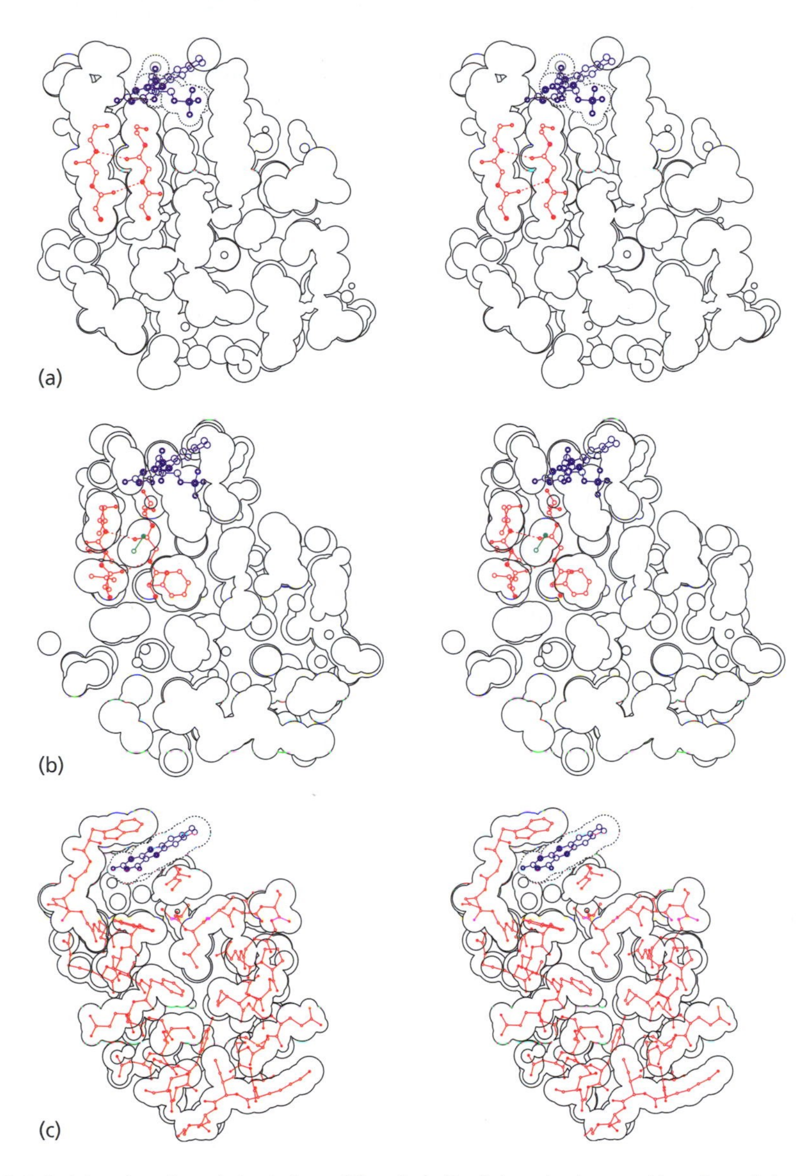

Fig. 3.8 Serial sections through the β-sheet of flavodoxin [5NLL] show the dense packing of protein interiors. Flavodoxin contains a five-stranded β-sheet with two helices packed against each side, and the prosthetic group flavin mononucleotide (FMN). Each drawing shows three serial sections through Van der Waals envelopes of the atom. The sections are cut 1 Å apart. The FMN is shown in blue, at the top of the picture, with broken contours. (a) Sections passing through the mean plane of the sheet. The main chains of three residues from each of two adjacent strands of sheet are also shown in red. (b) Sections passing through the side chains above the sheet (in this orientation). The main chains *and side chains* of three residues from each of two adjacent strands of sheet are also shown in red. The distal atoms of a methionine, from a helix packing against the sheet (shown in green), insert between side chains of residues on the two adjacent strands. (c) Sections passing through two helices packed against the sheet (below the sheet in this orientation). In this drawing only the flavin ring of the FMN is shown. Atoms of the polypeptide are shown in red.

Spectroscopic methods of characterizing proteins in solution

Key points about light and molecules are given in the Box.

The interaction of molecules with light provides information about molecular structure and dynamics. Absorbance, fluorescence, and circular dichroism are useful probes of protein conformation. Fluorescent tags, including fluorescent antibodies, are useful in mapping the intracellular locations of molecules. Fluorescent *in-situ* hybridization (FISH) locates gene sequences on chromosomes. Fluorescence recovery after photobleaching (FRAP) measures the regain of fluorescence at a particular cellular site after localized destruction of the tags by a laser beam. The motion of replacement molecules into the bleached area reveals their mobility.

* For a forceful statement of this point of view, see W.B. Gratzer (1970). Those woolly cotton effects. *Nature*, 227, 94–5.

For many purposes, spectroscopic methods have been superseded by atomic-resolution structure determinations, which contain much more detailed information.* However, spectroscopic measurements survive, at least as 'niche' methods, because they can be performed in real time, permitting observation of the kinetics of reactions or folding; and they provide sensitive information about molecular structure, dynamics, and interactions. Circular dichroism permits the determination of the secondary-structure content of proteins in solution, and is also in common use to monitor protein unfolding and refolding transitions. Fluorescence resonance energy transfer (FRET) gives information about the structures of macromolecular complexes too large to be easily treated by conventional crystallographic methods.

All spectroscopic experiments involve the temporary exchange of energy between light and matter; excitation involves only a 'short-term loan' of energy. An excited molecule will quickly release its energy and return to the ground state. Possible paths of de-excitation include:

- **Conversion of excitation energy to heat**. This is an efficient and fast process, occurring within about 10^{-12} s. As a result, some of the energy of a molecule that has absorbed visible or ultraviolet light is lost to warm up the surroundings.
- **Fluorescence**. This is the *efficient* loss of excitation energy by the re-emission of radiation. Typically, fluorescence occurs within about 10^{-8}–10^{-6} s. Because of the loss of some excitation energy to heat, the re-emitted light must be at a lower energy—that is, higher frequency—than the exciting radiation. This displacement of the wavelengths of absorption and fluorescence is useful, experimentally. It makes it possible to filter out the exciting light, facilitating the detection of a relatively weak fluorescent signal. The *quantum yield* is a measure of the fluorescence intensity: it is the ratio of the number of photons emitted to the number of photons absorbed.
- **Phosphorescence**. This is the *inefficient* loss of excitation energy by the re-emission of radiation. Phosphorescence may last for as long as 10^2–10^4 s. The reason for the delay is an intrinsically slow process involving a change in the orientation of an electron spin.
- **Energy transfer**. Under certain circumstances, an excited molecule can transfer its energy to another molecule (which then in turn has to decide what to do with it). If energy transfer leads to radiationless de-excitation, we say that the fluorescence of the originally excited chromophore has been **quenched**. Alternatively, fluorescence

Light and molecules: key points

- Atoms and molecules have energy levels. In dilute gases these energy levels are discrete and well separated. In condensed phases, interactions between molecules perturb the states, giving a continuous spectrum of energies.
- Absorption of light raises a molecule from its lowest energy level, or **ground state** to a higher energy level, an **excited state**.
- For absorption of light of frequency ν, the excitation energy is $\Delta E = h\nu$, in which Planck's constant $h = 6.626176 \times 10^{-34}$ J s^{-1}. A packet of light energy is called a *photon*. Light of frequency ν can be regarded as a beam of particles each of energy $h\nu$.
- The frequency ν and wavelength λ of *any* wave are related by the formula:

  $$\nu \times \lambda = \text{the speed of propagation of the wave.}$$

  The speed of light in a vacuum is a constant, independent of the frequency: $c = 2.99792458 \times 10^8$ m s^{-1}. For light in vacuum, $\nu\lambda = c$.
- Frequency and wavelength are therefore equivalent variables with which to describe light. Most spectroscopic measurements in protein science are recorded in terms of wavelength rather than frequency.
- Different wavelengths of visible light correspond to different colours. Light of a wavelength of about 510 nm appears green. This corresponds to a frequency of 5.878×10^{14} Hz, a photon energy of 3.89×10^{-19} J, equivalent to 234.6 kJ mol^{-1}.
- Many substances appear coloured, implying that their interaction with light varies with the frequency. If illuminated with white light (a mixture of different wavelengths), they differentially absorb some of the wavelengths, transmitting or reflecting light enriched in some colour. In many cases, a specific group within a molecule is responsible for interaction with light. This group is called a **chromophore**. For instance, the chromophore in haemoglobin is the haem-complexed iron. Groups that absorb and fluoresce only outside the visible region do not appear coloured but are called chromophores nevertheless.
- The *visible region* of the electromagnetic spectrum comprises wavelengths in the range 780–380 nm, corresponding to photon energies in the range 2.55×10^{-19}–5.23×10^{-19} J, equivalent to 153.37–314.81 kJ mol^{-1}. (Note that the *higher* the frequency the *lower* the energy, because of the relationship $E = h\nu = hc/\lambda$.) This energy is below the energy of typical chemical bonds, which is why ordinary visible light rarely produces photochemical reactions, including not causing severe sunburn. The *near-ultraviolet* comprises wavelengths in the range 380–200 nm, corresponding to photon energies of 5.22×10^{-19}–9.93×10^{-19} J, equivalent to 314.81–598.13 kJ mol^{-1}. The *far-ultraviolet* lies from 100 to 200 nm, corresponding to photon energies in the range 9.93×10^{-19}–1.99×10^{-18} J, equivalent to 598.13–1196.27 kJ mol^{-1}.

may be observed from the molecule to which the excitation energy was transferred (Fig. 3.9). One requirement for efficient energy transfer is that the donor and acceptor molecules be nearby—within about 20–70 Å. Another requirement is that the donor and acceptor have the correct relative excitation energies. If the acceptor molecule can *absorb* light at the wavelength at which the donor molecule would *fluoresce*, then the transfer would balance in energy, and be more probable. The spectroscopist's criterion is that there should be good overlap between the fluorescence spectrum of the donor and the absorption spectrum of the acceptor.

- **Fluorescence resonance energy transfer (FRET).** FRET takes advantage of the requirement for proximity to map out neighbouring groups in large macromolecular complexes; either statically or to detect conformational changes. The reasoning is that if excitation of a donor molecule is followed by fluorescence from an acceptor molecule, the two chromophores must be nearby in space.
- **Photochemical reaction.** Excited states are intrinsically more reactive than ground states, if only because they have more energy to spare. Some photochemical reactions are irreversible, and can damage proteins and other substances. (A molecule that loses its fluorescence after photochemical damage is said to be '*bleached*'.) In other cases—notably photosynthesis and vision—systems have evolved to capture excitation energy safely, channel it, and store it as chemical energy (in photosynthesis) or to trigger a nerve impulse (in vision), and then to restore the initial state.

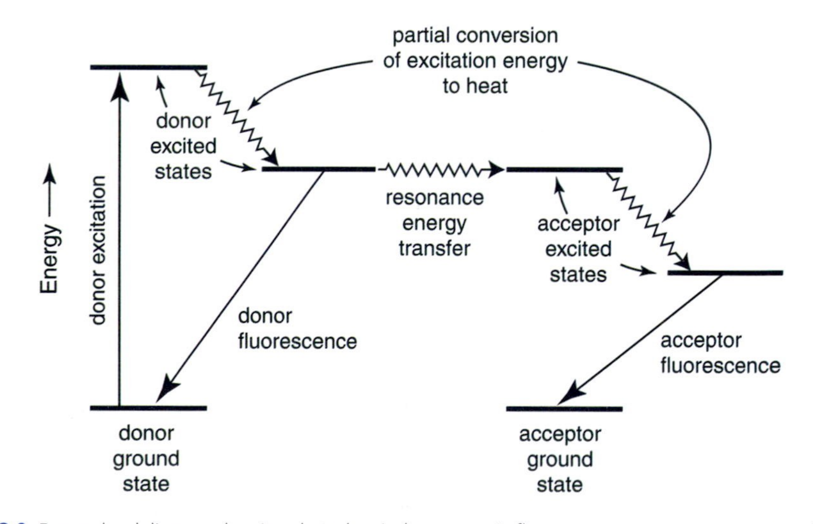

Fig. 3.9 Energy-level diagram showing photochemical processes in fluorescence resonance energy transfer (FRET). Initial excitation of a donor molecule is followed by a fast radiationless de-excitation with the release of energy as heat, leaving the donor in a low-lying excited state. Next, the donor may fluoresce to its own ground state, or transfer the excitation energy to the acceptor, which then converts some of *its* excitation energy to heat, leading to the observation of fluorescence from the acceptor. The excitation energy transfer is most probable between states of equal energy. 'Molecular excitation is like skiing: going up is much less complicated than getting back down'.

Natural FRET in vision and photosynthesis

- Insect visual pigments contain a chromophore, **3-hydroxyretinal**, bound to a protein, **opsin**. Excitation of the chromophore triggers a conformational change in the protein. In flies, a second, auxiliary, chromophore, 3-hydroxyretinol, is hydrogen-bonded to the opsin, but not mechanically linked to trigger the protein conformational change directly. The absorption maximum of 3-hydroxyretin**ol** is ~350 nm. The absorption maximum of 3-hydroxyretin**al** is 500 nm. The auxiliary pigment acts as a sensitizer, transferring its excitation energy to the main chromophore. As a result, the spectral response of the fly photoreceptor is broadened and its efficiency enhanced.
- Chloroplasts contain complex arrangements of accessory pigments that channel excitation energy to special sites of photochemical reaction. Some of these accessory pigments are chlorophylls, some are bound to proteins similar to globins, and some are similar to visual pigments.
- Conversely, and very unusually, the deep-sea dragon fish (*Malacosteus niger*) uses a molecule related to chlorophyll as a sensitizing pigment for vision.

Absorbance and fluorescence of proteins

Absorption of light by proteins involves energy levels of:

- **The peptide bond**, which has an intense transition at 190 nm, in the far-UV, and weak absorption at 210–220 nm. Side chain carboxyl and amide groups of Asp, Glu, Asn, Gln; and Arg and His side chains; also absorb in this region.
- **Side chains of aromatic amino acids** absorb at longer wavelengths, in the near-UV. Phe has an absorption band at 257 nm, Tyr at 274 nm, and Trp at 280 nm. The wavelengths of these transitions vary with the environment of the side chain, including pH and solvent polarity. Ionization shifts the tyrosine transition to 295 nm. Shifts in absorption bands arising from changes in the polarity of the environment are sensitive probes of conformation. They can be applied to distinguish exposed from buried aromatic side chains.
- **Ligands** that absorb in the visible region are responsible for most proteins that appear coloured. Examples include the red haem group of globins, and the blue complexed copper ion of plastocyanin. Nucleotides and related compounds such as NADH and NAD absorb strongly in the near-UV at 260 nm. In addition, NADH, but not NAD, absorbs at 340 nm; this provides the classic method for following NAD-linked enzymatic reactions.
- The **green fluorescent protein (GFP)** from the jellyfish *Aequorea victoria*, and its homologues, are a special case. GFP absorbs in the blue (λ_{max} = 395 nm and 475 nm), and fluoresces in the green (λ_{max} = 509 nm) (Fig. 3.10). The chromophore of GFP is formed as a post-translational modification of a Ser–Tyr–Gly tripeptide within the protein. The transformation is autocatalysed, requiring no enzyme to effect it.

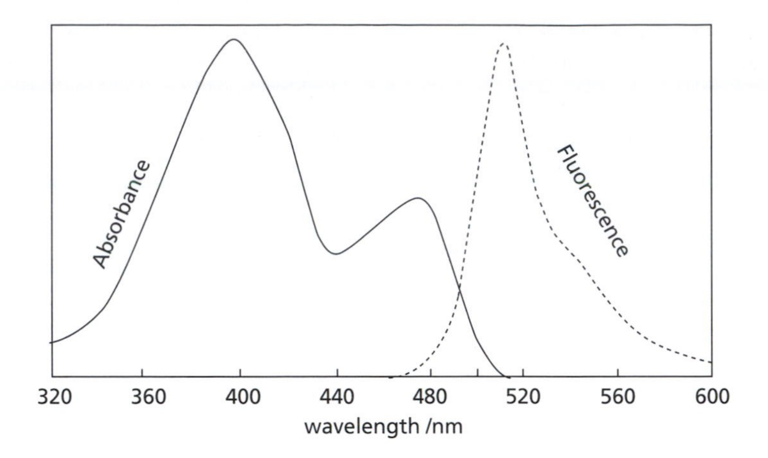

Fig. 3.10 Absorption (solid line) and fluorescence (broken line) spectra of green fluorescent protein (GFP) from the jellyfish *Aequorea victoria*. Note the shift of the fluorescence spectrum to a higher wavelength (lower energy) relative to absorbance.

Some chromophores in protein structures

Chromophore	Wavelength of absorption maximum	Colour
Cu^{2+}	787 nm	blue
Haem (deoxy)	550 nm	blue–purple
Haem (oxy)	940 nm	red
Bacteriorhodopsin	498 nm	purple
NAD	260 nm	none
NADH	260 nm and 340 nm	none

However, Ser-Tyr-Gly sequences in many other proteins do not form chromophores. Moreover, denatured GFP is not fluorescent. The special spectral properties depend on *both* the chemical modification and the environment of the altered amino acids within the three-dimensional structure of the protein. The sensitivity of the spectral properties to the environment of the chromophore has led to the exploration of natural variants and engineered mutants that fluoresce at different wavelengths.

Fluorescence is sensitive to the environment and dynamics of the chromophore

The intensity of fluorescence depends on the competition between fluorescence and alternative paths of de-excitation, such as excitation energy transfer. Acceptor molecules in the vicinity of a chromophore can quench its fluorescence. This provides a technique for determining solvent exposure of aromatic side chains or ligands.

Interaction of a molecule with light may depend on the relative orientation of the molecule and the plane of polarization of the light. If such a chromophore is fixed in orientation, there will be a correlation between the polarization of its fluorescence and that of the exciting light. If the chromophore is mobile within a structure, the degree of polarization of the fluorescence depends on how much time the chromophore has available to reorient itself during the lifetime of its excited state. Therefore, measurements of **depolarization of fluorescence** reveal the mobility of the chromophore. (Rates of overall molecular tumbling, which also reorient chromophores and can lead to fluorescence depolarization, are usually slower than fluorescent lifetimes.)

Spectroscopic measurements take advantage of speed and time resolution with which data can be collected. For example, it has been possible to determine the rate of loss of mobility of tryptophan side chains, during protein folding.

Fluorescence resonance energy transfer (FRET)

Fluorescence resonance energy transfer (FRET) is a method for probing protein structure, based on the observation that excitation energy transfer only occurs between chromophores nearby in the structure. It has been used to study the conformation and dynamics of proteins and nucleic acids, including static structure, conformational changes, folding pathways, and macromolecular interactions.

Two chromophores are required: one to receive the initial excitation, and the other as the destination of the energy transfer. The observable effect is a decrease of the fluorescence from the donor and an increase in the fluorescence from the acceptor. There is a strong distance dependence, varying as $1/[1 + (R/R_0)^6]$, where R is the distance between the chromophores, and R_0 is the distance at which 50% of the energy is transferred. For this reason, FRET is often called a molecular ruler. There is also an orientation effect on the efficiency of energy transfer.

Circular dichroism

Circular dichroism (CD) measures the asymmetry of a chromophore or its environment. When polarized light passes through an asymmetric medium, it emerges with the plane of polarization reoriented. This is called *optical rotation*. In addition, there is some loss of polarization, called *ellipticity*, the effect measured as circular dichroism.

The near-UV circular dichroism of a native protein is much stronger in the native state than in the denatured state. In the native state the peptide groups are in fixed and asymmetric surroundings, whereas in the denatured state the environment is random and fluctuating.

The most common applications of CD spectra of proteins are: (1) monitoring unfolding and refolding transitions (discussed in Chapter 5), and (2) determining the amounts of helix and sheet in a protein structure.

Placing peptide groups in an asymmetric environment, for instance in an α-helix, gives rise to a CD signal sensitive to the secondary structure. This effect arises from the interaction of the peptide groups. A β-sheet, although intrinsically less asymmetric than an α-helix, also gives rise to a CD signal.

The CD spectrum of a protein is the sum of contributions from its individual secondary structures. By measuring CD spectra of a set of proteins with known secondary structure contents, it is possible to solve for the spectra of α-helices and β-sheets (Fig. 3.11). The CD spectrum of a protein of unknown structure can then be fitted to a sum of contributions from helix, sheet, and a residual signal from other parts of the structure (see Box).

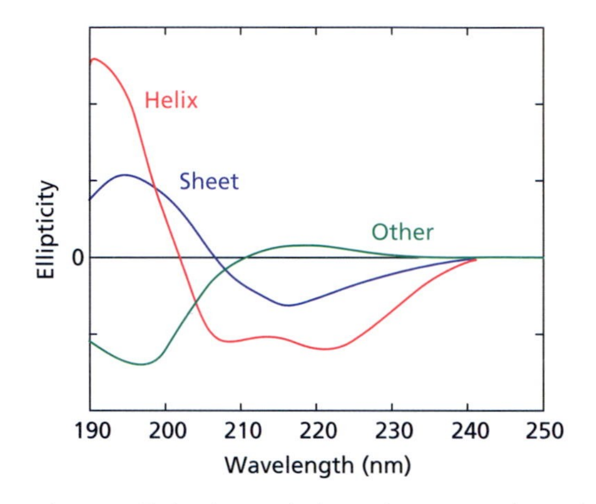

Fig. 3.11 Contributions of regions of helix, sheet, and other conformations to the circular dichroism (CD) of a globular protein. A measured spectrum of a protein of unknown secondary structure content could be analysed as a weighted sum of these curves.

Conformational changes in the prion protein

Transmissible **spongiform encephalopathies** are a group of diseases causing central nervous system degeneration in humans and animals. They include **kuru** and **Creutzfeldt–Jakob disease (CJD)** in humans, **scrapie** in sheep and goats, and **bovine spongiform encephalopathy** (BSE) (colloquially, '**mad-cow disease**'). These diseases tend to be species-specific, and to have long incubation periods. A variant CJD (vCJD) with a much shorter incubation period in humans is believed to have arisen by a jumping of BSE across the species barrier.

The infectious agent of these diseases appears to be a **prion** protein that changes from a normal state (PrP^{C} = prion protein—cellular) to a disease one (PrP^{Sc} = prion protein—scrapie). The disease form catalyses the $PrP^{C} \rightarrow PrP^{Sc}$ transformation, converting additional protein to PrP^{Sc}, leading to fibrillar aggregates of PrP^{Sc}.

Circular dichroism measurements show that the prion protein changes from a predominantly α-helical conformation in the PrP^{C} state, to a predominantly β-structure in the PrP^{Sc} state.

continues...

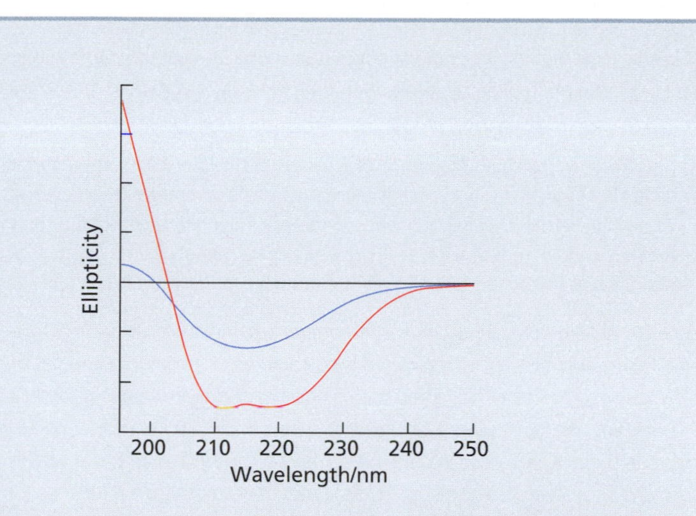

The near-UV CD spectra of the normal form of prion protein, PrP^{C} (blue) and the disease form, PrP^{Sc} (red). From Fig. 3.11 a change can be recognized in secondary structure in the prion protein from predominantly α-helical to a predominantly β-structure. (See: Jackson, G.S., *et al.* (1999). Reversible conversion of monomeric human prion protein between native and fibrilogenic conformations. *Science*, **283**, 1935–7.).

Protein structure determination

X-ray crystallography, and nuclear magnetic resonance (NMR) spectroscopy, and **cryoelectron microscopy** (**cryoEM**) are the major techniques for determining the structures of proteins.

Some history

Pasteur was one of the first scientists to think in three-dimensional terms about the structure of molecules in general and biological molecules in particular. Readers will know of his classic experiment in separating racemic tartaric acid by the manual selection of crystals of different shape. Pasteur recognized that the different crystal forms reflect a difference in underlying chemical constitution. Later, in studying the fermentation of tartaric acid, he observed that enzymes were also selecting only one form, discriminating between the two kinds of tartaric acid molecules on the basis of their three-dimensional structure.

Protein crystals had been known since at least 1840, when haemoglobin crystals were observed. The determination of detailed molecular structures from crystallography became possible in 1912, with the discovery of X-ray diffraction. In 1913, while the Braggs were solving the structure of simple minerals such as rock salt, Nishikawa and Ono took X-ray diffraction photographs of silk, and concluded, qualitatively, that the material must contain some ordered structure at the molecular level.

continues...

Bernal took the first X-ray photograph of a crystalline globular protein, pepsin, in late April 1934. He was excited by the implication that someday a complete atomic model of a protein would be revealed!

> ... The wet crystals gave individual X-ray reflections, which were rather blurred owing to the large size of the crystal unit cell, but which extended all over the films to spacings of about 2 Å. That night, Bernal, full of excitement, wandered around the streets of Cambridge, thinking of the future and of how much it might be possible to know about the structure of proteins if the photographs he had just taken could be interpreted in every detail. (Hodgkin and Riley, 1968).

From 1934 it took 25 years to determine a complete protein structure—J.C. Kendrew, M.F. Perutz, and colleagues published the structures of myoglobin and haemoglobin in 1959/60—followed by first a trickle and then a flood of crystal structures of other proteins.

For many years, X-ray crystallography was the only source of detailed macromolecular structures. A rival appeared in the 1980s, when K. Wüthrich and R.R. Ernst developed methods for solving protein structures by nuclear magnetic resonance (NMR) spectroscopy. Protein structure solution by NMR is now a thriving field.

With a third technique, cryoelectron microscopy (cryoEM), it is possible to solve structures of large aggregates, including viral capsids, large protein complexes, and intact ribosomes.

Solving the phase problem of X-ray crystallography

The **phase problem** of X-ray crystallography arises because the experimental data are incomplete. It is possible to measure the absolute value of each diffraction peak, but, even in the simplest case, not its sign.

- The earliest successes in solving the phase problem for proteins used the method of **isomorphous replacement**. Phases are determined by combining diffraction data from a native crystal with data from other crystals containing the same protein packed in the same way, but modified by the addition of a heavy atom.

- Many new proteins are similar to proteins of known structure. Calculation of the complete diffraction pattern—intensities and phases—expected from the related structure provides approximate phases with which to solve the new one. This method is called **molecular replacement**.

- Certain atoms absorb as well as scatter X-rays, with effects on the diffraction pattern that contain phase information. The solution of crystal structures from measurements of the variation of the intensity distribution in the diffraction pattern over a range of wavelengths is called the method of **multiwavelength anomalous dispersion**, or MAD. It requires a tunable X-ray source, available at a synchrotron.

  Most newly determined structures are solved using molecular replacement or multiwavelength anomalous dispersion methods.

- Knowledge of the general features of electron-density distributions in crystals—for instance, that they must always be positive, and have certain statistical properties—permits calculation of phases directly from the experimental data. **Direct methods** have routinely solved the structures of small molecules for many years, and are now applicable in protein crystallography.

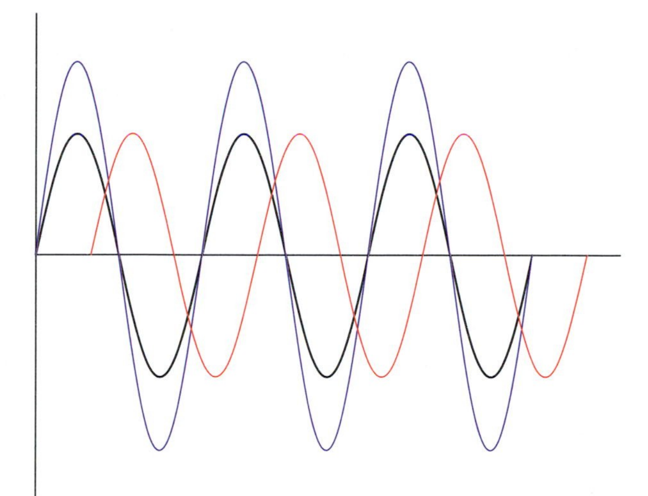

Fig. 3.12 What is the phase of a wave? Relative to the black curve as a reference, the blue one differs in amplitude, but matches the reference curve in the positions of its crests and troughs. The black and blue curves have the same phase. In contrast, the peaks and troughs of the red curve are displaced from those of the other two. The red curve has the same amplitude as the reference curve, but a different phase.

X-ray detectors record only the integrated intensity of a wave impingent on them. The difference in amplitude of the black and blue curves would be detectable. The phase difference between the black and red curves cannot be measured, but it is essential information that must somehow be reconstructed to solve the structure.

X-ray crystallography

The X-ray structure determination of a protein begins with its isolation, purification, and crystallization. If a suitable crystal is placed in an X-ray beam, diffraction will be observed, arising from the regular microscopic arrangement of the molecules in the crystal. But the experimental data are incomplete—we can only measure the *intensities* of the diffracted rays. To compute the electron density we need more information, about the *phases* of the diffracted rays (see Fig. 3.12 and Box).

Crystallographers next derive a model of the protein by fitting atoms into the electron-density map. This can be done either manually, using computer displays, or by computer programs. Once a model is available the process of *refinement* adjusts it to optimize both the fit to the experimental data and the stereochemical quality of the structure.

Protein crystallography is now a mature technique. The equipment and procedures for data collection, and software for data reduction and structure solution, are now integrated into an effective high-throughput technology (see Box). The rate-limiting steps are now in the preparation—expression and purification, and getting good crystals.

How accurate are the structures?

Most protein scientists are consumers rather than providers of crystal structures. It is essential to appreciate what details of the coordinates can be trusted.

Structural genomics

In analogy with full-genome sequencing projects, **structural genomics** has the commitment to deliver structures of the complete protein repertoire. X-ray crystallographers and NMR spectroscopists will solve enough structures to make it possible to build models for all proteins from known experimental structures.

The theory and practice of modelling suggests that at least a 30% sequence identity between the target and some experimental structure is necessary. This means that experimental structure determinations will be required for an exemplar of every sequence family, including many that share the same basic folding pattern. Experiments will have to deliver the structures of something like 10 000–30 000 domains. In the year 2003, 4673 structures were deposited in the PDB, so the throughput rate is not far from what is required.

Methods of bioinformatics can help select targets for experimental structure determination that offer the highest payoff in terms of useful information. Goals of target selection include:

- elimination of redundant targets—proteins too similar to known structures;
- identification of sequences with undetectable similarity to proteins of known structure;
- identification of sequences with similarity only to proteins of unknown function, or
- proteins of unknown structure with 'interesting' functions—for example, human proteins implicated in disease, or bacterial proteins implicated in antibiotic resistance; and
- proteins with properties favourable for structure dermination—likely to be soluble and contain methionine (which facilitates solving the phase problem of X-ray crystallography by the MAD method) (see previous Box).

The machinery for carrying out the modelling is already up and running. MOD-BASE collects homology models of proteins of known sequence but without experimentally determined structures:

http//alto.compbio.ucsf.edu/modbase-cgi/index.cgi

There are a number of fairly reliable indicators of the accuracy of the atomic coordinates in published protein structures. Some are derived during the process of the structure determination and reflect experimental observations; others are derived from the atomic coordinates themselves.

The *resolution* of an X-ray structure determination is a measure of how many data were collected. The more data, the greater the ratio of the number of observations to the number of atomic coordinates to be determined, and, in principle, the more accurate the results. Resolution is expressed in Å, with a *lower* number signifying a higher resolution. Experience has shown that the confident determination of different structural features is dependent on different thresholds of resolution (see Box).

A report of an X-ray structure determination will include a statistic called the '**R-factor**', a measure of how well the coordinates reproduce the experimental data. Other things being equal, the lower the R-factor the better the structure. A typical, well-refined protein structure, based on 2.0-Å resolution data, will have an R-factor of less

Confidence in the structural features of proteins determined by X-ray crystallography at different resolutions

Structural feature	Resolution				
	5 Å	3 Å	2.5 Å	2.0 Å	1.5 Å
Chain tracing	—	Fair	Good	Good	Good
Secondary structure	Helices fair	Fair	Good	Good	Good
Side chain conformations	—	—	Fair	Good	Good
Orientation of peptide planes	—	—	Fair	Good	Good
Protein hydrogen atoms visible	—	—	—	—	Good

These are *rough* estimates, and depend strongly on the quality of the data.

than 20%. A related quantity, the *free R-factor*, measures the agreement between the model and a subset of the experimental data withheld during the refinement. The free R-factor gives an unbiased measure of the agreement.

Now that we have seen enough well-determined protein structures to know what they should look like, it is possible to subject atomic coordinate sets to scrutiny independent of the experimental data. Good protein structures: (1) are compact, as measured by their surface area and packing density; (2) have hydrogen bonds of reasonable geometry, with few hydrogen bonds 'missing' in places where they would be expected, e.g. secondary structures; and (3) the residues have a distribution of backbone conformation angles confined almost entirely to the allowed regions of the Sasisekharan–Ramakrishnan–Ramachandran plot (see Fig. 3.3). Programs are available to carry out this analysis, and the results are available on the Web.

Nuclear magnetic resonance spectroscopy (NMR)

NMR spectra measure the energy levels of the magnetic nuclei in atoms. These energy levels are sensitive to the chemical environment of the atom. From effects transmitted between atoms bonded to each other, which affect the precise frequency of the signal from an atom (the **chemical shift**), NMR can determine the values of conformational angles. In particular, chemical shifts can define secondary structures. From interactions through space between non-bonded atoms <5 Å apart (the Nuclear Overhauser Effect or NOE), NMR can identify pairs of atoms close together in the structure, including those *not* close together in the *sequence*. The information about atoms distant in the sequence but close together in space is crucial to being able to assemble individual regions into the correct overall structure.

In protein structure determination by NMR, the experimental data provide a set of distance constraints: that is, a list of atoms that are close together in space. These specify the secondary structure, and give indications of other interactions. Computations that optimize the fit of the coordinates to the experimental data, together with force fields to enforce proper stereochemistry, produce sets of atomic coordinates.

Proteins in solution are not subject to the constraints of crystal packing, and are expected to flounce around somewhat. Indeed, typically the result of an NMR experiment is a set of similar, but not identical, structures (often ~15–20 of them), all of which are comparably consistent with the combination of experimental data and stereochemical restraints.

X-ray crystallography and NMR spectroscopy each have advantages and disadvantages. The main advantage of NMR is that it is not necessary to produce crystals; sometimes a severe impediment to X-ray structure determination. NMR gives us a window into protein dynamics, on a time scale of about 10^{-9}–10^{-6} seconds. A disadvantage of NMR is the limit on the size of a protein that can be studied. Roughly speaking, a protein with fewer than 150 residues is a 'piece of cake'. A 150–300-residue protein is solvable without problems by using suitable strategies of isotopic labelling (e.g., uniform double and triple labelling—with ^{2}H, ^{13}C, and ^{15}N—or selective labelling of different amino acids).

Low-temperature electron microscopy (cryoEM)

What are the prospects for extending structural studies to very large aggregates, which may be difficult to prepare as single crystals suitable for X-ray diffraction, and are beyond the size limit of applicability of NMR?

Electron microscopy of specimens at liquid-nitrogen temperatures has revealed the structures of assemblies in the range r.m.m. = 5×10^5 to 4×10^8, 100–1500 Å in diameter, such as the hepatitis B core shell or the clathrin coat. These results do not achieve atomic resolution—'blobs' are seen rather than individual atoms. However, at 3–4 Å resolution, attainable in some cases, features such as helices and sheets can begin to be recognized.

An exciting idea is the combination of electron microscopy with separate high-resolution X-ray crystal structures of the component proteins. By determining the positions of the individual proteins within the aggregate, the high-resolution structures can be assembled into a full atomic model of the entire complex.

In cryoEM a sample under native conditions is fast-frozen. The solvent is not given time to form the regular hexagonal equilibrium crystal form of ice that we see in snowflakes, but rather remains in a snapshot of the liquid state. The solid sample is then sectioned without melting.

There are two main approaches to deriving a three-dimensional structure:

1. **Diffraction from two-dimensional crystals**. Some proteins form two-dimensional crystals in nature, including bacteriorhodopsin, and ribulose-1,5-bisphosphate carboxylase/oxygenase. Others can be persuaded to form artificial two-dimensional crystals, e.g., aquaporin. Two-dimensional crystals *diffract* electrons. By measuring the diffraction pattern from tilted samples, it is possible to collect an almost complete three-dimensional diffraction pattern. The structure of bacteriorhodopsin (Fig. 2.14) was solved this way. This is crystallography without the phase problem, as phases can be measured experimentally.
2. **Image reconstruction**. Many samples do not form crystals. Electron micrographs show images of individual particles. Each image gives a different *projection* of the structure. From multiple images of the same object, in different orientations, a three-dimensional structure can be assembled (Fig. 3.13).

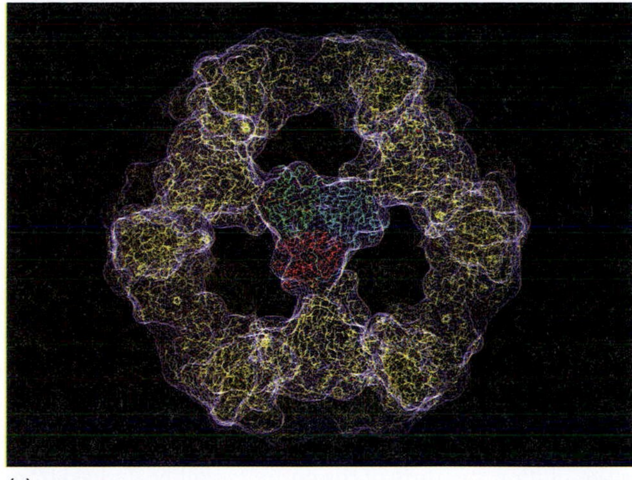

(a)

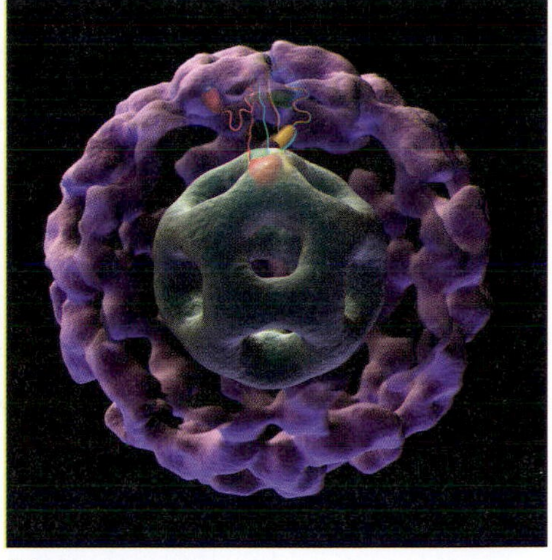

(b)

Fig. 3.13 Pyruvate dehydrogenase catalyses the transfer of an acetyl group from pyruvate, via thiamine pyrophosphate and dihydroxylipoamide, to coenzyme A. The structure of the complete enzyme comprises three component enzymes, assembled in a complex containing over 100 subunits, arranged in concentric shells, about 300 Å in diameter. Mobile domains on the inner shell can access several active sites on the inner surface of the outer shell.

(a) The core of the complex is an icosahedral array of 60 copies of one of the enzymes, E2. This figure shows a contoured electron-density map for the icosahedral E2 catalytic core of *Bacillus stearothermophilus* pyruvate dehydrogenase, determined by averaging images of the particles obtained by cryoelectron microscopy. Copies of X-ray structures of the subunits (yellow, cyan, red, and green) are superimposed on the density. The red, cyan, and green subunits are related by a threefold axis of symmetry. (Figure courtesy of Dr P.B. Rosenthal.)

(b) Representation of the structure and mechanism of pyruvate dehydrogenase. The figure shows the inner icosahedral core domain (green)—the structure shown in (a)—containing the E2 subunits, and approximately half the outer shell (purple), containing the E1 subunits. The outer shell is complete in the structure but cut away here to expose the inside of the particle.

The outer diameter of the outer shell is 460 Å. The outer diameter of the inner shell is 225 Å. The width of the gap between the shells is 90 Å.

Mobile regions of the E2 core extend into the space between the shells, and can make contact with several active sites of *both* E1 and E2 subunits.(Figure courtesy of Drs J. Milne, R. Perham, and S. Subramaniam. See: Milne, J.L. *et al*., (2002). Molecular architecture and mechanism of an icosahedral pyruvate dehydrogenase complex: a multifunctional catalytic machine. *EMBO J.*, **21**, 5587–98.)

The relationship between structure determinations of isolated proteins, and protein structure and function *in vivo*

A question that has been asked since the earliest days of protein structure determination is: How do we know that the results obtained from purified samples derived by X-ray crystallography and NMR spectroscopy are relevant to the structure of the molecule in the cell?

Some of the evidence is direct. For a few proteins, it has been possible to diffuse substrate into the crystal and observe enzymatic activity. The implication is that the structure in the crystal is close enough to the native state for the protein to be active. Certainly it justifies interpreting the mechanism of function on the basis of the crystal structure.

In most cases, the interactions within the crystal are only small irritations rather than large perturbations. In comparisons of the structures of proteins prepared in different crystal forms—with different packing patterns against their neighbours in the crystal—the atomic positions differ, on average, by only a few tenths of an Å. There can be local conformational change at the sites of crystal-packing contacts. Even crystal structures and NMR structures of the same protein are reassuringly similar.

For some macromolecular aggregates, including the ribosome, the structures of several component proteins have been solved in isolation. Their structures can be fitted into the envelope of a lower-resolution structure of the entire assembly, derived from electron microscopy (see Fig. 3.13a).

On the other hand, current methods of structure determination are reticent about conformational changes. Many proteins undergo structural changes as part of their function. We can see snapshots of different states, but cannot visualize the pathway between them. In particular, it is difficult to capture short-lived intermediates or transition states. Two attempts to overcome these limitations are: (1) slowing down reactions by lowering the temperature; and (2) speeding up the data collection. These specialized methods have been applied to a few systems.

It may also be pointed out that the concentration of protein inside a cell is much higher than in a classical enzymologist's assay tube. Intracellular concentrations of macromolecules in prokaryotes are estimated at 300–400 mg ml^{-1}; somewhat lower in eukaryotes. Perhaps the question, historically directed at crystallographers, should be turned around—why should we think that measurements on purified proteins in dilute solutions are relevant to the living medium?

Protein–ligand interactions

The reversible binding of ligands to proteins involves equilibria of the form:

$$\text{Protein} + \text{Ligand} = \text{Protein} \bullet \text{Ligand}$$
$$P + L = P \bullet L$$

for a one-to-one complex, or:

$$\text{Protein} + n\text{Ligand} = \text{Protein} \bullet \text{Ligand}_n$$

for the binding of n identical ligands to a single protein. These do not exhaust the possibilities. Many proteins bind two or more different ligands at the same time: enzymes binding a substrate and a cofactor provide many examples (see Fig. 2.8). A common index of the affinity of a complex is the **dissociation constant**, K_d, the equilibrium constant for the *reverse* of the binding reaction:

$$\text{Protein} \bullet \text{Ligand} = \text{Protein} + \text{Ligand} \qquad K_d = \frac{[P]\,[L]}{[PL]};$$

where: $[P]$, $[L]$, and $[PL]$ denote the numerical values of the concentrations of protein, ligand, and protein–ligand complex, respectively, expressed in mol l^{-1}. The lower the K_d, the tighter the binding. K_d corresponds to the concentration of free ligand at which half the proteins bind ligand and half are free: $[P] = [PL]$.

The K_d is related to the Gibbs free energy change of dissociation by the relationship:

See Dictionary of thermodynamic quantities (p. 198)

$$P \bullet L = P + L \quad \Delta G^{\circ} = \Delta H^{\circ} - T\Delta S^{\circ} = -RT \ln K_d$$

where: G = Gibbs free energy, H = enthalpy, I = absolute temperature, S = entropy, R = gas constant, and the 'London underground' symbol superscript indicates standard states of reactants and products.

The Michaelis constant of an enzyme is the dissociation constant of the enzyme–substrate complex, assumed in the Michaelis–Menten model to be at equilibrium with respect to enzyme + substrate (see next section).

Assuming no structural change on ligation, the entropy term will always favour dissociation, because two objects will have greater conformational freedom if they are kinetically independent than if they are tethered. Therefore, to achieve a stable complex, the enthalpy term must provide attractive forces adequate to overcome the intrinsic entropic penalty. Raising the temperature, which gives more importance to the entropy term, will promote dissociation.

To get a feel for the numbers, at 300 K the purely kinetic entropy gain upon dissociation, $T\Delta S^{\circ}$, is about 20 kJ mol^{-1}. This is equivalent, in terms of attractive interactions, to about a hydrogen bond, or burial of about 200 Å^2 of hydrophobic surface. A value of ΔG° for a dissociation reaction of 50 kJ mol^{-1} corresponds at 300 K to a dissociation constant $K_d \sim 2 \times 10^{-9}$.

Dissociation constants of protein–ligand complexes span a very wide range, as shown in the table.

Biological context	**Ligand**	**Typical K_d**	**ΔG° at 298 K kJ mol^{-1}**
Allosteric activator	Monovalent ion	10^{-4}–10^{-2}	11–23
Coenzyme binding	NAD, for instance	10^{-7}–10^{-4}	23–40
Antigen–antibody complexes	Various	10^{-4}–10^{-16}	23–91
Thrombin inhibitor	Hirudin	5×10^{-14}	76
Trypsin inhibitor	Bovine pancreatic trypsin inhibitor	10^{-14}	80
Streptavidin	Biotin	10^{-15}	85.6

Catalysis by enzymes

Enzymes are examples of protein–ligand complexes. They bind substrates and cofactors *selectively* and *in specific geometric orientations*. In this way, they can ensure that two molecules approach each other in the correct orientation for favourable reaction. If the same two molecules collide in random orientation, the probability of reaction would be much lower.

Some enzyme-catalysed reactions follow the same pathway as the uncatalysed reactions, but with lower activation barriers. Other enzymes substitute different reaction pathways, with intermediates very different from those of the uncatalysed reaction. To understand rate enhancement by activation-barrier lowering, compare the affinities of the initial enzyme-substrate complex and the enzyme–transition state complex (S = substrate, $S^\ddagger$ = transition state, E = enzyme, ES = enzyme–substrate complex, $ES^\ddagger$ = enzyme–transition state complex):

$$\text{Free energy of activation in the presence of an enzyme} = G(ES^\ddagger) - G(ES)$$

$$\text{Free energy of activation in the absence of an enzyme} = G(S^\ddagger) - G(S)$$

$$\begin{aligned}\text{Subtracting: } \Delta\Delta G^\ddagger &= [G(ES^\ddagger) - G(S^\ddagger)] - [G(ES) - G(S)] \\ &= [G(ES^\ddagger) - G(S^\ddagger) - G(E)] \\ &\quad - [G(ES) - G(S) - G(E)] \\ &= \text{binding affinity of the transition state} \\ &\quad -\text{binding affinity of the substrate}\end{aligned}$$

The rate enhancement is directly related to the lowering of the activation energy, $\Delta\Delta G^\ddagger$. The effect of the enzyme on $\Delta G^\ddagger$ is the *difference* between the affinity of the enzyme for the transition state $S^\ddagger$ and the binding affinity of the enzyme for the substrate S. (Here $\Delta G = G(ES) - G(S) - G(E)$ is the Gibbs free energy change of the *association* reaction E + S = ES.) An efficient enzyme will bind its substrate adequately to get the process started, but bind the transition state more tightly. Some enzymes are rigid, and have better complementarity to the transition state than to the substrate. Others undergo conformational changes upon binding substrate, from a form adapted to bind the substrate to one adapted to bind the transition state.

The Michaelis–Menten equation describes the velocity of enzymatic reactions as a function of substrate concentration

For many enzymatic reactions, the dependence of the initial steady-state velocity v on substrate concentration $[S]$ looks like this:

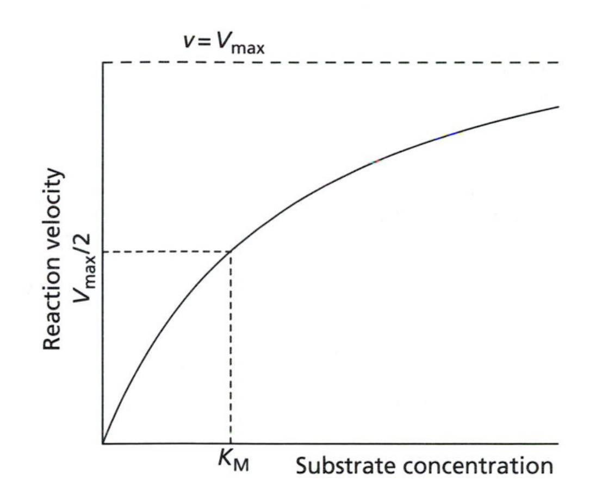

Michaelis and Menten proposed a simple model to account for the shape of this curve: (1) equilibrium formation of an enzyme–substrate complex; and (2) slow, irreversible, bleeding of product from the complex:

$$E + S \rightleftarrows [ES] \xrightarrow{k_{cat}} E + P.$$

Here E stands for enzyme, S for substrate, and P for product, and the *turnover number* k_{cat} is an apparent first-order rate constant for the generation of a product from enzyme–substrate complex. This model implies the equation:

$$v = \frac{V_{max}[S]}{K_M + [S]};$$

in which V_{max} is the limiting velocity for high substrate concentrations, and $[E]$ and $[S]$ are the concentrations of enzyme and substrate. (More precisely, $[E]$ is the concentration of active sites; each enzyme molecule may contain more than one active site.) K_M, called the *Michaelis constant*, is the equilibrium constant for the dissociation of substrate:

The values of V_{max} and K_M characterize the enzyme, the substrate, and the conditions of reaction, such as temperature, pH, and ionic strength.

$$K_M = \frac{[E][S]}{[ES]}$$

From the model we can interpret the variation of rate in molecular terms. For low values of $[S]$, v and $[S]$ are proportional. In this region, the enzyme is accommodating all substrate molecules equally well. The rate-limiting step is the encounter of substrate and free enzyme. As $[S]$ increases, v rises less and less steeply and approaches a limiting value V_{max}. This is attributable to saturation of the enzyme. Increasing $[S]$ will convert all the enzyme to complex; that is, for large $[S]$, $[ES] \sim [E]$ and $v = k_{cat}[ES] \sim k_{cat}[E] = V_{max}$ (*independent* of substrate concentration). The observed rate at the plateau corresponds to the rate of some step of the reaction that occurs *after* binding of the substrate. The substrate concentration at which $v = \frac{1}{2}V_{max}$ is equal to K_M.

Conformational change

Many proteins undergo conformational changes as part of their activities. In many, but not all, cases, binding of ligands triggers the change. For example, the enzyme citrate synthase binds substrate and cofactor in a cleft between two domains. In the unligated form, the two domains swing open to receive the ligands. In the ligated form, the domains close over the ligands, excluding water from the active site. Allosteric proteins show another class of conformational changes induced by ligand binding.

Structurally, various types of conformational changes have been observed.

- **Localized conformational changes in loop regions.** Loops are generally the most flexible regions in proteins. In enzymes such as triose phosphate isomerase, and in p21 Ras, a change in the state of ligation affects the conformation of local mobile regions (see Figs 2.18 and 2.19).
- **Hinge motions reorientating two domains.** Lactoferrin shows a hinge motion in response to binding iron. Two domains remain individually rigid but change their relative orientation, by means of structural changes in only a few residues in the regions linking the domains (Fig. 3.14). Hinge motion in myosin is responsible for the impulse in muscle contraction (see Fig. 3.16).
- **Cumulative shifts of secondary structures packed at interfaces.** As Fig. 3.14 shows, hinge motion is only possible if the domains are free to move as rigid bodies. In some proteins, a well-packed and extensive interface constrains the relative motion of the two domains that are to move with respect to each other. A rule of thumb states that two packed helices can move by only about 2 Å with respect to each other, by small readjustments of side chain conformations. Think of this 2-Å threshold as the limit of plastic deformation of a helix interface. In order to achieve a large structural change, relative motions of 2 Å in individual pairs of helices can be coupled to give large cumulative effects. The ligand-induced conformational change in citrate synthase is an example of a large motion that is the resultant of several small motions at individual helix–helix interfaces.
- **Changes in the relative geometry of subunits.** Many changes, including allosteric changes, involve changes in **quaternary structure**. Haemoglobin shows a relatively small change—a rotation of about 15° of one $\alpha\beta$ dimer with respect to the other (see pp. 109–116). F_1-ATPase shows a very large change—it is a rotary motor in which the γ-subunit turns with respect to the $\alpha_3\beta_3$ hexamer, successively inducing conformational changes in the subunits of the hexamer (see Chapter 6, ATPase section 6.9). The GroEL–GroES complex is a large reciprocating motor that depends for its activity on a crucial conformational change in the GroEL protein.
- **A few proteins appear to have more than one possible folding pattern.** These include the prion protein, and the serpins (see Chapter 7).
- **Amyloid aggregates.** Many proteins can form fibrillar aggregates with structures different from their free native state. In at least some cases the structure adopted has a 'crossed-β' form (see Chapter 7).

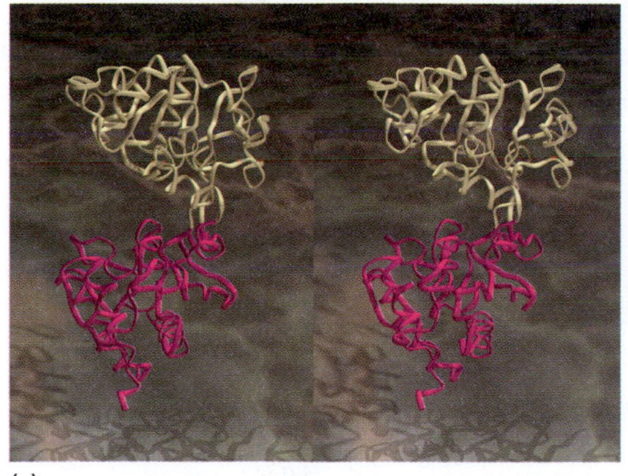
(a)

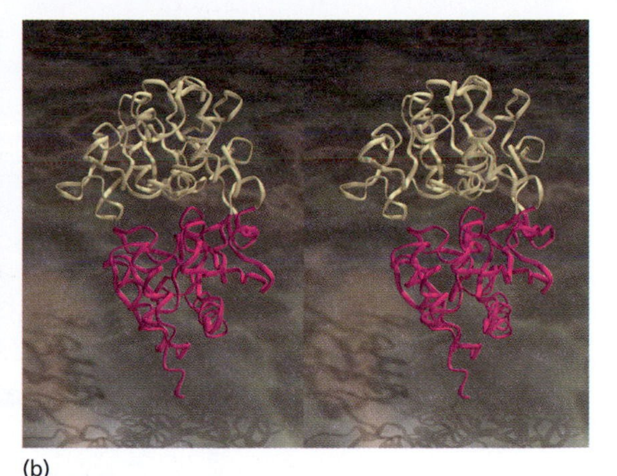
(b)

Fig. 3.14 Lactoferrin is an iron-binding protein found in secretions such as milk and tears. It contains a single polypeptide chain of approximately 700 residues, folded into two lobes of related sequence and similar structure. Each lobe contains two domains. The binding of iron causes the N-terminal lobe to change from an 'open' to a 'closed' conformation. This figure shows the two conformations of the N-terminal lobe[1FH, 1LFG]. The individual domains stay rigid during the conformational change. The only residues that alter their main chain conformation are those at the bridge regions between the two domains.

The sliding filament mechanism of muscle contraction

The structural and mechanical unit of vertebrate skeletal muscle is an intracellular organelle called the **sarcomere**. Sarcomeres contain interdigitating filaments of actin and myosin (Fig. 3.15). The actin filaments are fixed to structures called the *Z-disks* at the ends of the sarcomere. The motor protein myosin pulls the actin filaments inwards towards the centre of the sarcomere. During contraction, the actin and myosin filaments do not themselves shorten, but slide past one another, shortening the sarcomere by increasing the region of overlap. Think of the shortening of a bicycle pump during its compression stroke.

A large muscle might contain ~10^4–10^5 sarcomeres, laid end-to-end. Each sarcomere has a resting length of ~2.5 μm, and can contract by ~0.3 μm. Therefore the entire muscle can contract by about ~1–2 cm.

Individual myosin molecules are large fibrous proteins of r.m.m. ~5×10^5. They contain a long fibrous section ~1.6–1.7 μm long, and a globular head. Each thick filament contains ~200–300 myosin molecules. The mechanical coupling between the actin and myosin occurs through the myosin head, as shown in Fig. 3.15.

During the power stroke, the myosin head undergoes a cycle of attachment-detachment, and conformational change. From left to right in the inset in Fig. 3.15, attachment of the myosin head is followed by conformational change that propels the actin towards the centre of the sarcomere. Detachment is followed by restoration of the initial conformation of the myosin. The myosin heads are like oars that 'row' the actin filaments towards the centre of the sarcomere. The displacement of the actin is

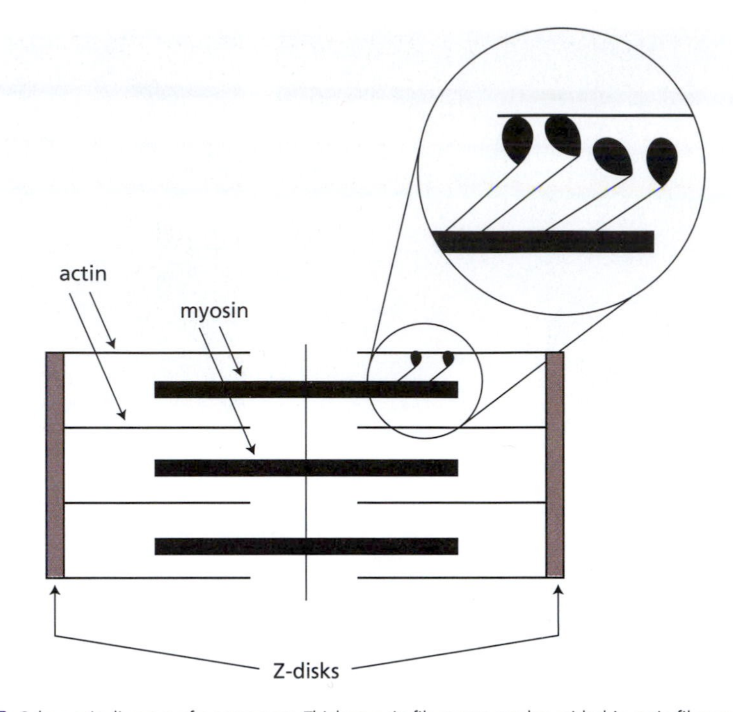

Fig. 3.15 Schematic diagram of a sarcomere. Thick myosin filaments overlap with thin actin filaments. In the main diagram, it is cursorily indicated that multiple myosin molecules from thick filaments interact with adjacent thin filaments. In fact, each thick filament contains several hundred myosin molecules. The inset shows different stages of the power stroke. From left to right: attachment, conformational change propelling the thin filament inwards by ~10 nm, detachment, recovery of original conformation of the myosin head.

~10 nm per myosin molecule per cycle. Hydrolysis of one molecule of ATP during each cycle of each myosin molecule provides the energy.

Structures of myosin fragments containing the globular head have defined the mechanism of the conformational change (Fig. 3.16).

Control of protein activity

Regulation of enzyme activity is essential to control the flow of matter and energy through metabolic pathways. Many mechanisms are available, including:

- **Feedback inhibition**: given a sequence of reactions:

$$\begin{array}{c} A \rightarrow B \rightarrow C \rightarrow D, \\ \uparrow \\ D \end{array}$$

  if D inhibits the enzyme catalysing the conversion of A to B, then flow through this pathway will be downregulated when product D is in adequate supply.

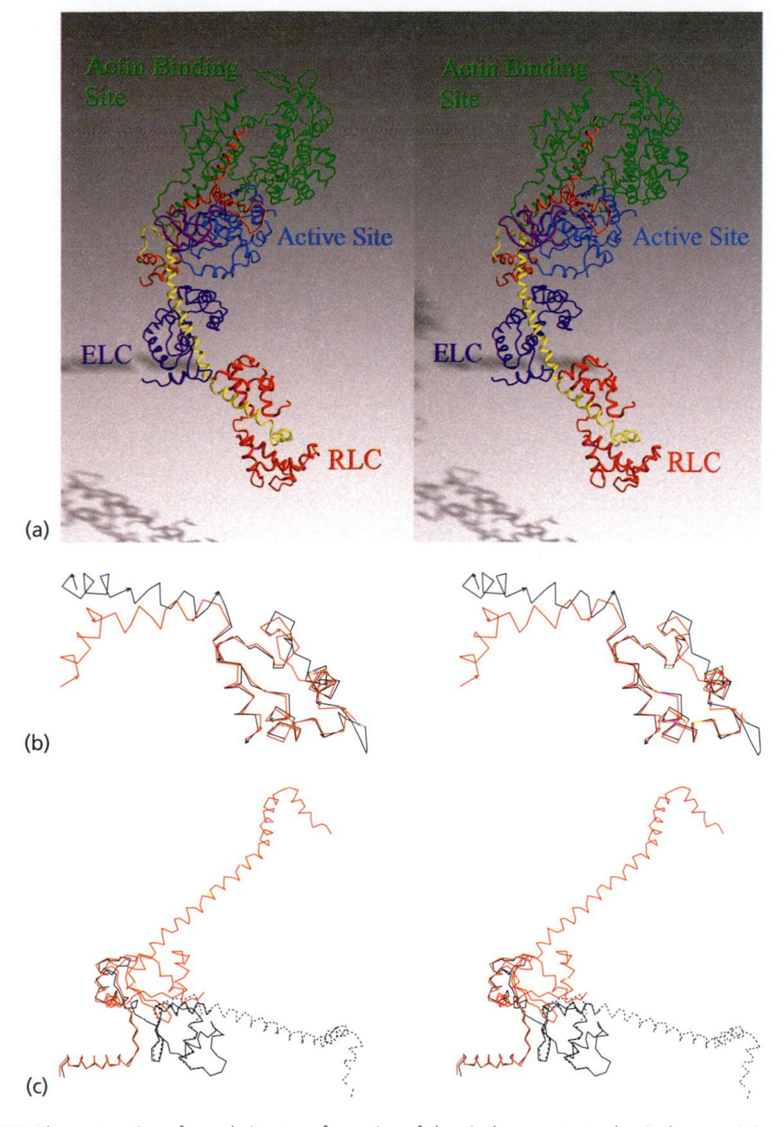

Fig. 3.16 The contraction of muscle is a transformation of chemical energy to mechanical energy. It is carried out at the molecular level by a hinge motion in myosin, while myosin is attached to an actin filament. The cycle of *attach to actin–change conformation–release from actin,* in a large number of individual myosin molecules, creates a macroscopic force within the muscle fibre. (a) The structure of myosin subfragment-1 from chicken. The active site binds and hydrolyses ATP·ELC and RLC are the essential and regulatory light chains [2MYS]. (b) Hinge motion in myosin. Comparison of parts of chicken myosin open form [2MYS] (no nucleotide bound) and closed form binding the ATP analogue ADP·AlF_4^- [1BR2]. This shows the segments of the structure that surround the hinge region. (c) Model of the swinging of the long helical region in myosin as a result of the hinge motion. The dashed line shows a model of the position that the complete long helix would occupy in the closed form [2MYS] and [1BR1]. This conformational change is coupled to the hydrolysis of ATP. It takes place while myosin is bound to actin, providing the power stroke for muscle contraction. In the context of the assembly and mechanism of function of a muscle filament, it is arguable that one should regard the helix as fixed and the head as swinging. But that would not show the magnitude of the conformational change as dramatically.

- **Control of the concentrations of inhibitors** is a mechanism of regulating the activity of their protein targets. For instance, **α_1-antitrypsin** in the lungs inhibits the activity of elastase. People lacking active α_1-antitrypsin are prone to emphysema, caused by damage to lung tissue by elastase.
- **Proteolytic cleavage** is a mechanism for activation of digestive enzymes. Clearly the body doesn't want rogue proteinases active in inappropriate locations. For example, trypsin is synthesized in the pancreas in an inactive form, trypsinogen. In the gut, cleavage of trypsin between residues 15 and 16 produces a charged amino terminus at residue 16, which can form a salt bridge. This induces a conformational change to the active form. Similar mechanisms participate in the very complex control network regulating the enzymes involved in blood clotting.

Discussion of allostery is expanded in the next section.

- **Allosteric regulation** gives proteins an alternative to the simple Michaelis–Menten relationship between ligand concentration and enzyme activity or simple binding. In haemoglobin, positive cooperativity allows for the efficient uptake of oxygen in the lungs, where it is plentiful, and the efficient release to tissues where it is not. Oxygen requirements during periods of intense physical activity are critical, indicating a clear selective advantage of effective transport to the tissues. It is the release of oxygen by haemoglobin that is responsible for the efficiency of transfer: knock-out mice lacking myoglobin appear normal.

  Allosteric proteins are also subject to feedback inhibition or stimulation. Aspartate carbamoyltransferase is an allosteric enzyme at the first step of the pathway of pyrimidine biosynthesis. Cytidine triphosphate, a product of the pathway, is an allosteric effector that downregulates the activity of aspartate carbamoyltransferase. Adenosine triphosphate—a *purine* triphosphate—upregulates it.

A 'knock-out' mouse is an animal in which the expression of a selected gene has been suppressed.

These examples illustrate the effects of endogenous molecules directly affecting the activity of proteins to which they themselves bind. Protein activity can also be regulated by signals external to the metabolic pathway in which the protein participates, and also even external to the cell.

- **Control of activity through reversible dimerization**. Some enzymes are active only in dimeric form. These can provide a mechanism for transmission of a signal from the cell surface, as illustrated in Figs 2.16 and 2.17. Other proteins are active only when *dissociated*; for instance, the heterodimeric G-proteins (see page 38) and protein kinase A (see below).
- **Covalent modification** can affect activity. Phosphorylation at a specific Ser, Thr, or Tyr residue is a common mechanism for the reversible inactivation of proteins in signal transduction cascades.

These mechanisms of regulation operate within seconds or minutes. Control over gene expression also controls protein activity, on longer time scales.

Regulation of tyrosine hydroxylase illustrates several control mechanisms

Tyrosine hydroxylase catalyses the conversion of L-tyrosine to L-3,4-dihydroxyphenylalanine (L-dopa) in neurons, a step in the synthesis of the neurotransmitters dopamine and adrenaline (epinephrine). Tyrosine hydroxylase is the focus of many diverse forms of regulation, including control over transcription and RNA processing and turnover. One regulatory pathway is triggered by the arrival of a neurotransmitter at the external cell surface:

- External binding to a receptor activates adenylate cyclase inside the cell.
- The cyclic AMP produced activates protein kinase A by a mechanism involving subunit dissociation. The resting, inactive form of protein kinase A is a tetramer of two catalytic subunits, C, and two regulatory subunits, R. In the resting state the regulatory subunits inhibit the activity of the catalytic subunits. Binding of cyclic AMP to protein kinase A dissociates the tetramer, releasing the catalytic subunits in active form.
- Active protein kinase A phosphorylates tyrosine hydroxylase at Ser40, upregulating its activity.
- The mechanism resetting this stimulation is the specific dephosphorylation of Ser40 by phosphatase 2A.

L-Dopa is used to treat Parkinson disease.

Control of protein function: allosteric regulation

Allosteric proteins show 'action at a distance': ligand binding at one site affects activity at another. An impulse at the first site must cause a conformational change affecting the second. This fundamental mechanism permeates the regulatory networks of molecular biology.

Some allosteric changes produce **cooperativity**. Haemoglobin is a classic case: the binding of oxygen at any site enhances the affinity of other sites. Figure 3.17 compares non-cooperative and cooperative binding curves. Haemoglobin shows *positive* cooperativity: binding of oxygen *increases* the affinity for additional oxygen. Some other proteins show negative cooperativity: binding of ligand *reduces* the affinity for additional ligand.

In contrast to a ligand that induces a cooperative effect on activity towards the *same* ligand, an **effector** is a ligand that alters the activity of a protein towards a *different* ligand. Diphosphoglycerate (DPG) is an effector for haemoglobin. It binds preferentially to the deoxy form, decreasing the oxygen affinity and enhancing O_2 release.

People and animals living at high altitude have higher concentrations of DPG than their sea-level relatives.

GTP-activated p21 Ras (Figs 2.18 and 2.19) also shows ligand-induced regulation of activity. In p21 Ras, the region of conformational change is in direct contact with the binding site. More challenging to explain are the properties of haemoglobin, in which the shortest distance between binding sites is over 20 Å.

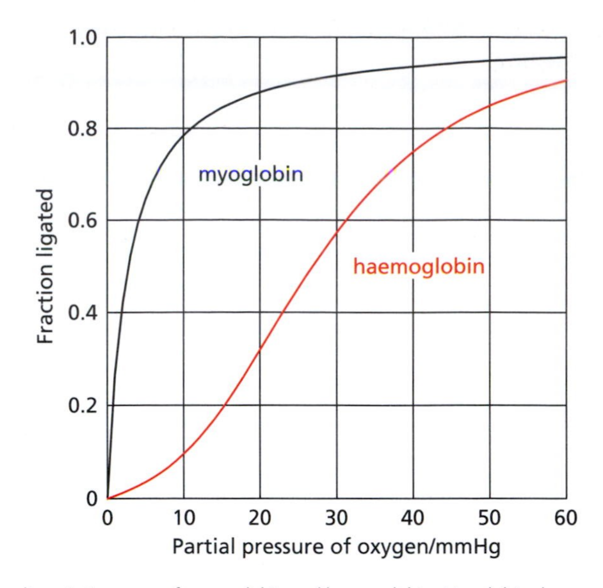

Fig. 3.17 Oxygen-dissociation curves for myoglobin and haemoglobin. Myoglobin shows a simple equilibrium, with a binding constant independent of oxygen concentration. Haemoglobin shows positive cooperativity, the binding constant for the first oxygen being several orders of magnitude smaller than the binding constant for the fourth oxygen. The units of mmHg for partial pressure are traditional in the literature about this topic. 760 mmHg = 1 atmosphere = 101 325 Pa.

A general model for cooperativity, explaining influence extending over long distances between binding sites (as in haemoglobin) is that the subunits of a protein are in equilibrium between two forms, with different binding constants, and that ligands shift the equilibrium between them. One form, called 'Tense' (T) is structurally constrained in a way that inhibits binding. The other, 'Relaxed' (R) form is unconstrained. This model rationalizes the properties of haemoglobin:

1. At low partial pressures of O_2, all the subunits of haemoglobin are in the T form. The binding constant for oxygen is low, because binding to the T state is inhibited (K_1, the dissociation constant for the first O_2 to be bound, is 26 mmHg).
2. At high partial pressures of O_2, all the subunits of haemoglobin are in the R form. The binding constant for oxygen is high, comparable to that of simple globins such as myoglobin, or dissociated α- or β-chains of haemoglobin, because binding of O_2 to the R state is unconstrained (K_4, the dissociation constant of the fourth and last O_2 to be bound, is 0.17 mmHg).
3. Binding of between two and three oxygen molecules shifts the subunits *concertedly* from all T state to all R state.

Writing the equilibrium between T and R states in the free and fully ligated forms:

$$\mathrm{T} \rightleftharpoons \mathrm{R}$$
$$\mathrm{TL}_n \rightleftharpoons \mathrm{RL}_n.$$

A ligand-induced switch between T and R states is consistent with the following free-energy relationships:

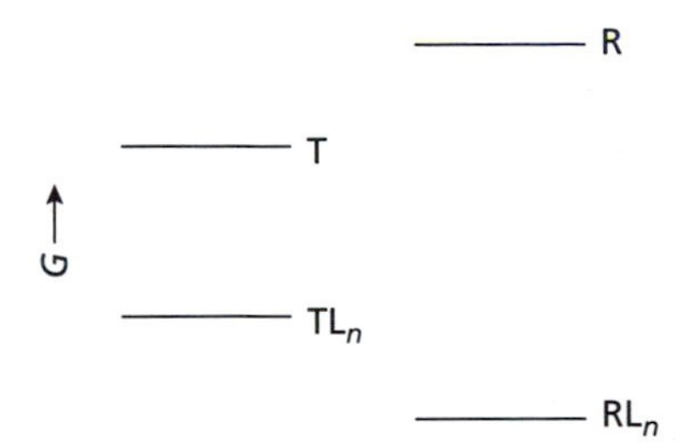

implying that in the absence of ligand the T state is more stable than the R, but in the ligated state RL_n is more stable than TL_n. Although not all allosteric proteins fit this picture, the terminology of T and R states continues to be used.

The interpretation of this scheme in structural terms was one of early triumphs of protein crystallography. It involves coupled changes in tertiary and quaternary structure.

The allosteric change of haemoglobin

To play its physiological role in oxygen distribution effectively, haemoglobin must capture oxygen in the lungs as efficiently as possible, and release as much as possible to other tissues. To achieve this 'take from the rich, give to the poor' effect, haemoglobin has a high oxygen affinity at high oxygen partial pressures (pO_2) and a low affinity at low pO_2. The binding is *cooperative*, in that binding of some oxygen enhances the binding of additional oxygen.

The vertebrate haemoglobin molecule is a tetramer containing two α-chains and two β-chains. These can adopt either of two structural forms. One is characteristic of deoxy (unligated) haemoglobin, and has low oxygen affinity; the other is characteristic of oxyhaemoglobin (four oxygen molecules bound) and has high oxygen affinity. The difference in colour between arterial and venous blood reveals the different state of the iron.

In the **erythrocyte** (red blood cell), haemoglobin is an equilibrium mixture of deoxy and oxy forms; the concentration of partially ligated forms is tiny. (The structures of partially ligated haemoglobin can, however, be observed in crystal structures.) Binding of oxygen induces structural changes that alter the relative free energy of the two forms, shifting the equilibrium towards the high-affinity form. Starting from the deoxy state, partial ligation (between two and three oxygens) is enough to shift the equilibrium to the oxy state, which will then pick up the remaining oxygens with greater affinity. Conversely, starting with the fully ligated oxy structure, a partial loss of oxygen will shift the equilibrium in the opposite direction, stimulating the release of the remaining oxygen. Other molecules that modify oxygen affinity, such as diphosphoglycerate, a natural allosteric effector, operate in part by shifting this equilibrium, by preferentially stabilizing one of the two forms.

The oxygen affinity of the oxy form of haemoglobin is similar in magnitude to that of isolated α- and β-subunits, and to that of myoglobin. The oxygen affinity of the deoxy form is much less: the ratio of the binding constants for the first and fourth oxygens is 1:150–300, depending on conditions. Therefore it is the deoxy form which is special, which has had its oxygen affinity 'artificially' reduced. In the terminology of Monod, Wyman, and Changeux, the reduced oxygen affinity of the deoxy form of haemoglobin arises from structural constraints that hold the structure in a 'tense' (T), internally inhibited state; while the oxy form is in a 'relaxed' (R) state, and free to bind oxygen as the isolated monomer.

The binding of oxygen is accompanied by a change in the state of the iron. In the deoxy state, the iron has five ligands—the four pyrrole nitrogens of the haem group and the proximal histidine—and is in a high-spin Fe (II) state with an ionic radius of 2.06 Å. In the oxy state, the iron has six ligands—five are the same as in the deoxy state and the sixth is oxygen—and is in a low-spin Fe (II) state with an ionic radius of 1.94 Å. These radii are important because the distance from the pyrrole nitrogens to the centre of the haem is 2.03 Å; this implies that the iron will fit in the plane of the pyrrole nitrogens in the oxy state but not in the deoxy state.

Oxy and deoxy forms of human haemoglobin

Property	Oxy	Deoxy
Oxygen affinity	High (K_4 = 0.17 mm/Hg)	Low (K_1 = 26 mm/Hg)
Spin state of iron	Low spin	High spin
Fe^{2+} radius	1.94 Å	2.06 Å
Monod, Wyman, and Changeux state	Relaxed (R)	Tense (T)

Structural differences between deoxy- and oxyhaemoglobin

The structures of haemoglobin in different states of ligation have been studied with intense interest, partly because of their physiological and medical importance, and partly because they were thought to offer a paradigm of the mechanism of allosteric change. (As the structures of other proteins showing allosteric changes have been determined, it is becoming apparent that different systems achieve cooperativity by different types of mechanisms.)

The two crucial questions to ask of the haemoglobin structures are:

1. What is the mechanism by which the oxygen affinity of the deoxy form is reduced?
2. How is the equilibrium between low- and high-affinity states altered by oxygen binding and release?

Comparison of the oxy and deoxy structures has defined the changes in the tertiary structures of individual subunits; and in the quaternary structure, or the relative geometry of the subunits and the interactions at their interfaces. Here is a simple description of how these changes are coupled: the quaternary structure is determined by the way the subunits fit together. This fit depends on the shapes of their surfaces. The tertiary-structural changes alter the shapes of the surfaces of the subunits, changing the way they fit together.

Tertiary-structural changes between deoxy- and oxyhaemoglobin

The tertiary-structural changes in α- and β-subunits are similar but not identical.

At the haem group itself, in the deoxy form the iron atom is out of the plane of the four pyrrole nitrogens of the haem group. There are two reasons for this: the larger radius of the iron in its high-spin state, and steric repulsions between the Nϵ of the proximal histidine and the pyrrole nitrogens. The haem group is 'domed'; that is, the iron-bound nitrogens of the pyrrole rings are out of the plane of the carbon atoms of the porphyrin ring of the haem, by 0.16 Å in the α-subunit and 0.10 Å in the β-subunit.

Forming the link between iron and oxygen would, in the absence of tertiary structural change, create strain in the structure. Without constraint the haem would become planar, and the iron would move into this plane. These changes are resisted by the steric interactions between the proximal histidine and the haem group, and the packing of the FG-corner against the haem group (the FG-corner is the region between the C-terminus of the F-helix and the N-terminus of the G-helix). In the β-subunit, an additional barrier to the binding of oxygen without tertiary structural change is the position of Val E11 in the region of space to be occupied by the oxygen itself.

E11 = eleventh residue of the E-helix

Helices are designated by letters in order of their appearances in the sequence: A, B, ... H (see Fig. 4.13). The F-helix contains the proximal histidine, bound to the iron.

In the oxy form, these impediments are relieved by changes in tertiary structure. Describing these changes locally, relative to the haem group, in both subunits there is a shift of the F-helix across the haem plane by about 1 Å, and a rotation relative to the haem plane (Fig. 3.18). The effect is to permit a reorientation of the proximal histidine so that the iron atom can enter the haem plane. Associated with this shift in the F-helix are conformational changes in the FG-corner.

To make the connection between tertiary-and quaternary-structural changes, we must describe the tertiary structural changes in terms of their effects on the shape of the entire subunit. The purely local description of what happens around the haem group is important for rationalizing the energetics of ligation, but is the wrong frame of reference to account for the change in quaternary structure.

Intersubunit interactions in haemoglobin

The haemoglobin tetramer can be thought of as a pair of dimers: $\alpha_1\beta_1$ and $\alpha_2\beta_2$. In the allosteric change, the $\alpha_1\beta_1$- and $\alpha_2\beta_2$-interfaces retain their structure, as does a portion of the molecule adjoining these interfaces, including the B,C,G,H regions of both subunits and the D-helix of the β-subunit. The overall allosteric change involves a rotation of 15° of the $\alpha_1\beta_1$-dimer with respect to the $\alpha_2\beta_2$, around an axis approximately perpendicular to their interface. (The motion is like that of a pair of shears with α_1 and α_2 = the blades, and β_1 and β_2 = the handles.)

Given that the structure of the $\alpha_1\beta_1$-interface is conserved, it provides the appropriate frame of reference for describing the tertiary-structural changes in a way that relates them to the changes in surface topography that account for the quaternary-structural change.

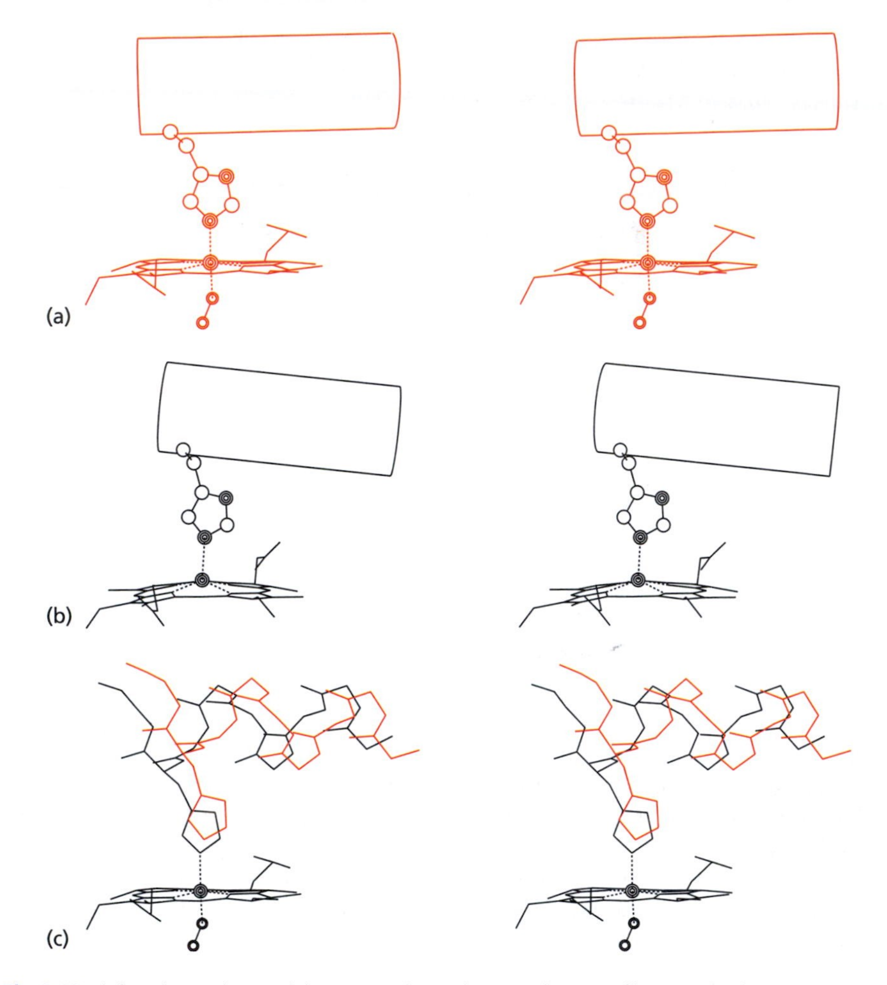

Fig. 3.18 Shifts in human haemoglobin as a result in a change in the state of ligation. This figure shows the F-helix, proximal histidine, and haem group of the α-chain in the (a) deoxy [2HHB] and (b) oxy [1HHO] forms of human haemoglobin [1HHO, 2HHB]. (c) Superposition of oxy (black) and deoxy (red); only the oxy haem is shown. The structures were superposed on the haem group.

With respect to the $\alpha_1\beta_1$-interface, the tertiary-structural changes appear as follows (Fig. 3.19a): In both subunits, in going from deoxy to oxy structures the haem groups move into the haem pockets, ending up 2 Å closer together in oxy than in deoxy structures. The backbone of the F-helix moves with the haem (it also moves relative to the haem as discussed above). The FG-corner also shifts. These tertiary structural changes of the F-helix and FG-corner do not extend beyond the EF-corner on the N-terminal side and the beginning of the G-helix on the C-terminal side. Note that this is consistent with the statement that the C- (and, in Fig. 3.19(b), the D) and G-regions are in the part of the structure that is conserved in the allosteric change.

To understand the quaternary-structural change, we must analyse how the interface between the $\alpha_1\beta_1$- and $\alpha_2\beta_2$-dimers changes. The most important intersubunit contacts are between the α_1–β_2 and α_2–β_1 subunits. (In the open-shears image, the

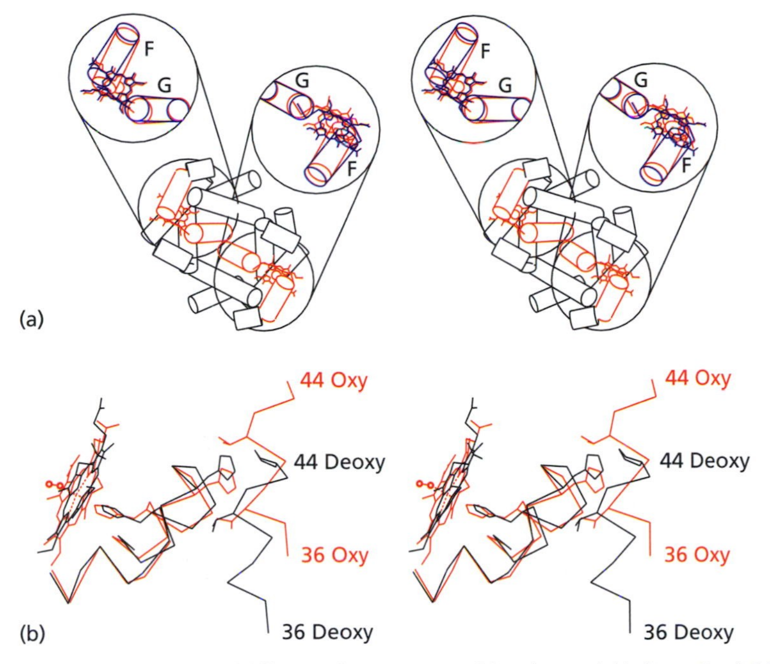

Fig. 3.19 Some important structural differences between oxy- and deoxyhaemoglobin [1HHO, 2HHB]. (a) The $\alpha_1\beta_1$ dimer in oxy (red) and deoxy (black, in blown-up regions only) forms. In the blown-up regions only the F-helix, FG-corner, G-helix and haem group are shown. The oxy and deoxy $\alpha_1\beta_1$-dimers have been superposed on their interface; in this frame of reference there is a small shift in the haem groups, and a shift and conformational change in the FG-corners. (b) Alternative packings of α_1- and β_1-subunits in oxyhaemoglobin (red) and deoxyhaemoglobin (black). The oxy and deoxy structures have been superposed on the F- and G-helices of the α_1 monomer. The helix oriented vertically is the C-helix, showing the large shift. Although for purposes of this illustration we have regarded the α_1-subunit as fixed and the β_2-subunit as mobile, only the relative motion is significant.

important variable contacts are between each blade and the opposite handle. The contacts between each blade and its own handle are—in haemoglobin as well as in shears—rigid.)

The interacting regions are the α_1 FG-corner—β_2 C-helix and β_2 FG-corner—α_1 C-helix. These are not identical. (However, these interactions *are* the same, by symmetry, as those of the α_2 FG-corner—β_1 C-helix and β_1 FG-corner—α_2 C-helix, respectively.)

The α_1 FG–β_2 C interaction is very similar in oxy and deoxy structures. Residues Arg92 (FG4), Asp94 (G1), and Pro95 (G2) of the α_1-subunit are in contact with Tyr37 (C3) and Arg40 (C6) in the β_2-subunit. In comparing oxy and deoxy structures there are small conformational changes in these residues but the pattern of interactions is retained.

FG4 = fourth residue in the interhelical region between F- and G-helices; G1 = first residue of the G-helix.

The other region of contact, β_2 FG-corner—α_1 C-helix, differs substantially between oxy and deoxy structures (Fig. 3.19(b)). In the deoxy structure, His β_2 97 (FG4) packs between Thr α_1 41 (C6) and Pro α_1 (44) (CD2), and there is a hydrogen bond between the side chains of Asp β_2 99 (G1) and Tyr α_1 42 (C7). In the oxy structure, His β_2 97 packs between Thr α_1 38 (C3) and Thr α_1 41 (C6). (Because the C-helix is a 3_{10} helix, this corresponds to a jump of one turn relative to the His β_2 97 against which it packs. The shift in β_2 FG relative to α_1 C is approximately 6 Å.) The Asp–Tyr hydrogen bond is not made in the oxy structure.

This explains the two discrete quaternary states: The β_2 FG-corner–α_1 C-helix contact has two possible states, depending on the subunit shapes presented by the tertiary-structural state. The other contact, α_1 FG-corner—β_2 C-helix changes only slightly.

The requirement for the quaternary-structural change arises from the tertiary-structural changes; in particular, from the shifts that bring the FG-corners of the α_1- and β_1-subunits in the $\alpha_1\beta_1$-dimer 2.5 Å closer together in the oxy structure relative to the deoxy structure. One tertiary-structural state of the $\alpha_1\beta_1$- and $\alpha_2\beta_2$-dimers can form a tetramer with one state of packing at the interface; the other tertiary-structural state is compatible with the alternative packing.

In conclusion, it may be useful to trace the logical connection between the structural changes: starting from the deoxy structure, the ligation of oxygen requires the relief of strain around the haem group by shifting the F-helix and the FG-corner. To accommodate these changes there must be a set of tertiary-structural changes which change the overall shape of the $\alpha_1\beta_1$- and $\alpha_2\beta_2$-dimers; notably the shifting of the relative positions of the FG-corners. In consequence, the deoxy quaternary structure is destabilized because the dimers no longer fit together properly (having changed their shape). Adopting the alternative quaternary structure requires the tertiary-structural changes to take place even in subunits not yet liganded. As a result of the quaternary-structural change, these unligated subunits have been brought to a state of enhanced oxygen affinity. It is important to emphasize that this is a sequence of steps in a logical process, and not a description of a temporal pathway of a conformational change.

USEFUL WEB SITES

Definition of polypeptide conformation:
www.chem.qmw.ac.uk/iupac/misc/biop.html
Side chain rotamer libraries:
www.cmpharm.ucsf.edu/dunbrack
http://duc.urbb.jussieu.fr/rotamer.html
www.fccc.edu/research/ labs/dunbrack/molecularmodeling.html

RECOMMENDED READING

Cornish-Bowden, A. (ed.) (1997). *New beer in an old bottle: Edward Buchner and the growth of biochemical knowledge*, València University, València.

Drenth, J. (1994). *Principles of protein X-ray crystallography*. Springer-Verlag, New York.

Fuller, S. (1999). Cryo-electron microscopy: taking back the knight. *Microbiol. Today*, **26**, 56–8.

Guss, J.M. and King, G.F. (2002). Macromolecular structure determination: Comparison of crystallography and NMR. In *Encyclopedia of the life sciences* (ed. S. Robertson), Vol. 11, pp. 290–5. Nature Publications Group, London.

Hodgkin, D.C. and Riley, D.P. (1968). Some ancient history of protein X-ray analysis. In *Structural chemistry and molecular biology* (ed.) A. Rich and N. Davidson), pp. 16–28. W.H. Freeman & Co., San Francisco.

Kay, L.E. (1997). NMR methods for the study of protein structure and dynamics. *Biochem. Cell. Biol.*, **75**, 1–15.

Kleywegt, G.J. and Jones, T.A. (2002). *Homo crystallographicus—Quo vadis? Structure*, **10**, 465–72.

Ladokhin, A.S. (2000). Fluorescence spectroscopy in peptide and protein analysis. In *Encyclopedia of analytical chemistry* (ed. R.A. Meyers), pp. 5762–79. Wiley, Chichester.

Longhi, S., Czjzek, M. and Cambillau, C. (1998). Messages from ultrahigh resolution crystal structures. *Curr. Opin. Struct. Biol.*, **8**, 730-7.

Mattevi, A., Rizzi, M. and Bolognesi, M. (1996). New structures of allosteric proteins revealing remarkable conformational changes. *Curr. Opin. Struct. Biol.*, **6**, 824–9.

Perutz, M.F. (2002). *I wish I'd made you angry earlier: essays on science, scientists and humanity*. Oxford University Press, Oxford.

Šali, A. (1998). 100,000 protein structures for the biologist. *Nat. Struct. Biol.*, **5**, 1029–32.

Selvin, P.R. (2000). The renaissance of fluorescent resonance energy transfer. *Nat. Struct. Biol.*, **7**, 730–4.

Wüthrich, K. (1995). NMR—this other method for protein and nucleic acid structure determination. *Acta Cryst.*, **D51**, 249–70.

EXERCISES, PROBLEMS, AND WEBLEMS

Exercises

1. Write the chemical formula for selenocystein.
2. On a photocopy of the box on page 76 circle those side chains that are capable of forming hydrogen bonds.
3. On a photocopy of the picture of methionine on page 74, indicate the bond that is the axis of rotation for the conformational angles (a) ϕ, (b) χ_1, (c) χ_3.
4. On a photocopy of the picture of methionine on page 74, sketch in additional atoms and bonds required to extend it to a picture of the dipeptide Met–Ile, in the extended conformation, with a *trans* peptide bond.
5. Draw the chemical structures of tripeptides (a) Ala–Leu–Phe, (b) Ser–Pro–Asn (assuming that the peptide preceding the Pro is in the *cis* conformation).
6. In the alignment of the two histidine-containing phosphocarrier proteins (page 24), a I in the the middle line indicates positions at which two sequences have identical residues. On a photocopy of this alignment, mark with a + on the middle line each position at which the two sequences have residues with physicochemically similar properties.
7. What is the amino acid sequence of a short peptide that the inhibitor of Fig. 2.11 resembles most closely? On a photocopy of the structural formula in Fig. 2.11(b), highlight the atoms that the inhibitor has in common with the peptide you proposed.
8. What is the sequence of the peptide shown this figure? Colour code: Main chain, magenta. Side chain: carbon, black; nitrogen, blue; oxygen, red; sulphur, green. Remember that the amino acid sequence is always stated in the direction from the N-terminus to the C-terminus.

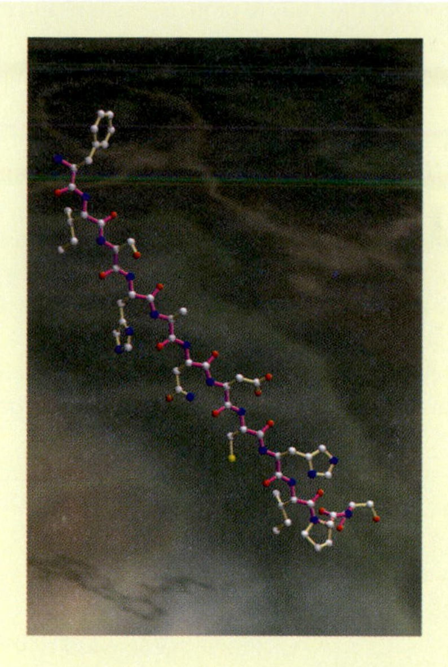

9. Which tripeptide would you expect to be more water soluble: (a) Ala–Thr–Ser or (b) Phe–Ile–Trp?

10. Write the chemical structure of the peptide W–V–E–Y–H–K, indicating the expected state of ionization, at: (a) pH 3, (b) pH 5, (c) pH 7, (d) pH 9, (e) pH 11. Ignore changes in pK arising from interactions between charged groups.

11. How many angles of internal rotation are there in a lysine side chain?

12. A Sasisekharan–Ramakrishnan–Ramachandran plot specialized to particular side chains can be constructed as follows: for any amino acid, e.g., Val, create a computer model of the tripeptide Ala–Val–Ala. For each combination of ϕ–ψ values of the central residue, find the conformation with lowest energy, and plot that energy as a function of ϕ and ψ. Regions of steric collision will appear as high energy, and can be marked as disallowed regions. Suppose such plots are constructed for Ala and Val. Which would contain the larger disallowed area?

13. The amino acids shown forming hydrogen bonds at the left of Fig. 3.8(a) form part of a β-sheet. By comparing the hydrogen bond pattern with those shown in Figs 2.3(b) and (c), are these two strands parallel or antiparallel?

14. The energy of a covalent bond is of the order 500 kJ mol^{-1}. To what frequency of electromagnetic radiation does this correspond?

15. The unit cell of a crystal of *E. coli* transaldolase is a rectangular solid with edges 68.9 Å, 91.3 Å, and 130.5 Å, that contains eight molecules of a protein of relative molecular mass 35 072. What is the concentration of protein in the crystal, in mg/ml? Compare with the estimates of 300–400 mg/ml for the intracellular concentration of

macromolecules in prokaryotic cells, and a typical protein concentration in an NMR structure determination, 30 mg/ml.

16. For an enzyme that follows the Michaelis–Menten model, calculate in terms of [E] and [*S*] the proportion of enzyme present in the form of enzyme–substrate complex ES.

17. Normal arterial pO_2 = 100 mmHg. Consider haemoglobin to be fully saturated under these conditions. Normal venous pO_2 = 40 mmHg. (a) From Fig. 3.17, estimate the fraction of O_2 bound to Hb that is delivered to the tissues. (b) The estimate based on Fig. 3.17 is in error because the increased CO_2 and lower pH of venous blood shifts the curve to the right. Is the estimate in part (a) based on Fig. 3.17 an underestimate or an overestimate?

Problems

1. One of the determinants of hydrophobicity is the size of the side chain. Figure 3.20 shows a graph of hydrophobicity and side chain volume for the 20 natural amino acids. (a) For which subset of side chains does there appear to be a fairly accurate linear relationship between hydrophobicity and side chain volume? (b) Which subset of amino acids deviates most from this relationship? (c) For the subset chosen in (a), determine the slope and intercept of a linear fit to the data, and give a formula for the dependence of hydrophobicity on side chain volume.

2. The crystal structure of pepsin (the subject of J.D. Bernal's original protein X-ray diffraction photograph) was solved in the late 1980s. The amino acid sequence, and the

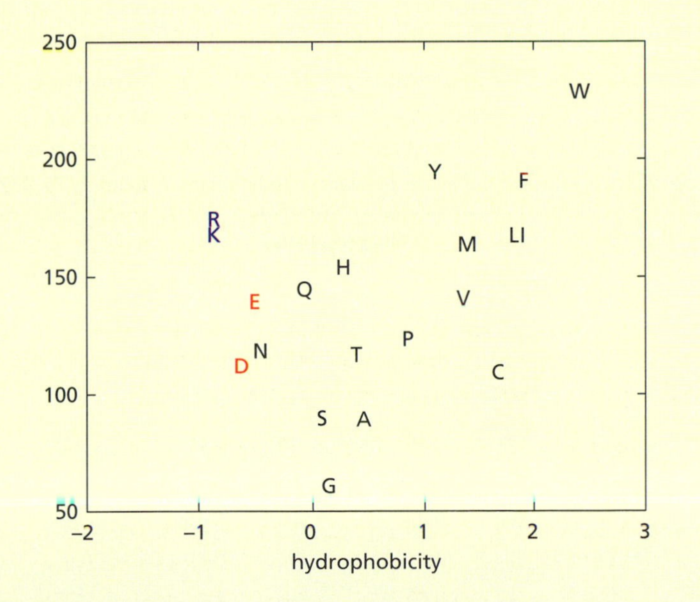

Fig. 3.20 The relationship between hydrophobicity and side chain volume.

residues on the surface (s) and buried in the interior (b) are:

```
         10        20        30        40        50        60
IGDEPLENYLDTEYFGTIGIGTPAQDFTVIFDTGSSNLWVPSVYCSSLACSDHNQFNPDD
sbsbsbsbsssssbsbsbbbbsssssbsbbbbbsbbbbbbbbssbsssbbsssssbsbss

         70        80        90       100       110       120
SSTFEATSQELSITYGTGSMTGILGYDTVQVGGISDTNQIFGLSETEPGSFLYYAPFDGI
sssssssssbssssssssbsbsbbsbsbsbsssssbssbbbbbssbssssssbsbbbs

        130       140       150       160       170       180
LGLAYPSISASGATPVFDNLWDQGLVSQDLFSVYLSSNDDSGSVVLLGGIDSSYYTGSLN
bbbbbsssbsssbsbbbssbsssssbsssbbbbsbbsssssssbsbbbbbsssssssssss

        190       200       210       220       230       240
WVPVSVEGYWQITLDSITMDGETIACSGGCQAIVDTGTSLLTGPTSAIANIQSDIGASEN
sbsssssbsbbbsbssbsssssssbssssssbbbbbssssbbbssssbssbbssbsbsss

        250       260       270       280       290       300
SDGEMVISCSSIDSLPDIVFTIDGVQYPLSPSAYILQDDDSCTSGFEGMDVPTSSGELWI
sssssbssssssssssbsbsbssssbsbsbsbbbssssssbsbbbssssssssssssbbb

        310       320
LGDVFIRQYYTVFDRANNKVGLAPVA
bbbbbbssbbbbbbbsssssbbbbbsss
```

(a) What percent of the residues are buried? (b) For each of the following physicochemical classes—(1) non-polar (GASTCVILPFYMW), (2) polar (NQH), and (3) charged (DEKR)—what percent of the amino acids in each of these classes is buried and what percent is on the surface?

3. (a) Make a table giving the number of side chain degrees of freedom of each of the side chains.

(b) The following table gives the observed statistical propensity for amino acids to appear in α-helices. Is there a correlation with the number of degrees of freedom?

(c) Is any single amino acid a spectacular exception? If so, can you suggest why?

In the following table of helix propensities of amino acids, the *lower* the value the *higher* the helix-forming propensity. (From Pace, C.N. and Scholtz, J.M. (1998). A helix propensity scale based on experimental studies of peptides and proteins. *Biophys. J.*, **75**, 422–7.)

Helix propensity			
Ala	0.00	Phe	0.54
Leu	0.21	Val	0.61
Met	0.24	Thr	0.66
Arg	0.21	His^0	0.56
Lys	0.26	His^+	0.66
Gln	0.39	Cys	0.68
Glu	0.40	Asn	0.65
Ile	0.41	Asp	0.69
Ser	0.50	Gly	1.00
Trp	0.49	Pro	3.10
Tyr	0.53		

4. The accessible surface area (ASA) of small monomeric proteins varies with r.m.m. M according to the relationship: ASA = 11.1 $M^{2/3}$. If the amino acid sequence of these proteins is placed on a polypeptide in the extended chain conformation, the ASA is, of course, higher, and is given by: $ASA_{\text{extended chain}} = 1.45\ M$.

(a) Explain why the accessible surface area of a native protein varies as the $\frac{2}{3}$ power of M, and that of the extended chain varies linearly with M.

(b) What is the formula for the *buried* surface area—relative to the extended chain—of proteins as a function of M?

(c) What is the expected buried surface area per residue for a monomeric protein of 100 residues? (Assume that the average M of a residue is 110.)

(d) For this example, what is the approximate contribution to the free energy of stabilization of the native state from the hydrophobic effect?

5. Suppose that the rates of a reaction, in the absence and in the presence of an enzyme, are given by the expressions:

$$\text{without enzyme: rate} = ke^{-\Delta G^{\ddagger}/(RT)}$$
$$\text{with enzyme: rate} = ke^{-\Delta G_E^{\ddagger}/(RT)};$$

Here $\Delta G^{\ddagger}$ and $\Delta G_E^{\ddagger}$ are the free energies of activation of the uncatalysed reaction and the enzymatically catalysed reaction, respectively. Assume that the constant of proportionality k is the same in both cases. If $\Delta G_E^{\ddagger}$ is 2 kJ mol lower than $\Delta G^{\ddagger}$, by what factor will the rate be increased at 310 K (=37 °C)? ($R = 8.314\ \text{J K}^{-1}\ \text{mol}^{-1}$.)

6. Briggs and Haldane considered the following rate equation for enzymatic catalysis:

$$E + S \underset{k_{-1}}{\overset{k_1}{\leftrightarrow}} [ES] \overset{k_2}{\rightarrow} E + P$$

This scheme is more general than the Michaelis–Menten model because it does not assume that $k_1 \gg k_2$. Show that this scheme is also consistent with the observed substrate dependence of the rate, and reinterpret the parameters derived from fitting the reaction rates, in terms of k_1, k_{-1}, and k_2. Assume that the initial substrate concentration is much greater than the enzyme concentration, and that the concentration of enzyme–substrate complex, [ES], will be constant during the reaction, until substrate is depleted. Find the initial steady-state rate.

7. G. Bowes and W.L. Ogren determined the initial steady-state rate of CO_2 uptake as a function of CO_2 concentration for a preparation of ribulose-1,5-bisphosphosphate carboxylase from soybean:*

$[CO_2]$/mmolar	0.046	0.093	0.139	0.232	0.463	0.926	2.3125
Rate of CO_2 uptake/nmol s^{-1}	6.90	10.81	13.66	17.54	20.19	23.39	25.11

Determine K_m and V_{max}.

* Bowes, G. and Ogren, W.L. (1972). Oxygen inhibition and other properties of soybean ribulose 1, 5-diphosphate carboxylase. *J. Biol. Chem.*, **247**, 2172–6.

Readers who do not have local access to an appropriate curve-fitting package should search the Web for the following combination of keywords: 'Michaelis Menten gui'.

8. For a protein binding a single ligand, PL = P + L, the dissociation constant $K_d = \frac{[P]\,[L]}{[PL]}$, the total concentration of ligand is $[L]_{tot} = [L] + [PL]$, and the total concentration of protein is $[P]_{tot} = [P] + [PL]$. Suppose $K_d = 10^{-6}$ and $[P]_{tot} = 1\ \text{mmol l}^{-1}$.

(a) For what value of $[L]_{tot}$ is the protein half saturated; that is, $[PL] = 0.5 \times ([P] + [PL])$ or $[P] = [PL]$?

(b) For what value of $[L]_{tot}$ is the protein three-quarters saturated?

9. Plot the following data giving rate enhancements and estimated enzyme-transition state affinities. Do the data confirm the idea that enzymes achieve rate enhancement by stabilizing transition states?

Enzyme	Rate enhancement	Transition-state dissociation constant
β-Amylase	1.0×10^{22}	1.0×10^{-22}
Acetylcholinesterase	2.0×10^{17}	5.0×10^{-18}
AMP nucleosidase	5.0×10^{16}	2.0×10^{-17}
Phosphotriesterase	5.3×10^{15}	1.9×10^{-16}
Triosephosphate isomerase	5.6×10^{13}	1.8×10^{-14}
Chorismate mutase	4.2×10^{10}	2.4×10^{-11}
Carbonic anhydrase	9.2×10^{8}	1.1×10^{-9}

10. 2,3-Diphosphoglycerate (DPG) is an allosteric effector of haemoglobin. The binding site for DPG is at the interface between the two β-subunits. The ligand interacts with the positively charged N-terminus of the β-chain, and the side chains of His 2, Lys 82, and His 143, in *both* β-chains.

A mammalian fetus depends for oxygen on transfer from the mother. Several mechanisms ensure adequate oxygen supply to the fetus. Nevertheless, it is surprising that the oxygen affinity of isolated human fetal haemoglobin (containing 2 α-chains and 2 γ-chains) is *lower* than that of adult haemoglobin ($\alpha_2\beta_2$). However, fetal haemoglobin has a lower affinity for DPG, and therefore has a higher oxygen affinity than the maternal haemoglobin.

Consider the sequence alignment of the adult β-chain, and the equivalent in fetal haemoglobin, the γ-chain:

```
                  10         20         30         40         50
                   |          |          |          |          |
adult β   VHLTPEEKSA VTALWGKVNV DEVGGEALGR LLVVYPWTQR FFESFGDLST
fetal γ   GHFTEEDKAT ITSLWGKVNV EDAGGETLGR LLVVYPWTQR FFDSFGNLSS

                  60         70         80         90        100
                   |          |          |          |          |
adult β   PDAVMGNPKV KAHGKKVLGA FSDGLAHLDN LKGTFATLSE LHCDKLHVDP
fetal γ   ASAIMGNPKV KAHGKKVLTS LGDAIKHLDD LKGTFAQLSE LHCDKLHVDP

                 110        120        130        140
                   |          |          |          |
adult β   ENFRLLGNVL VCVLAHHFGK EFTPPVQAAY QKVVAGVANA LAHKYH
fetal γ   ENFKLLGNVL VTVLAIHFGK EFTPEVQASW QKMVTAVASA LSSRYH
```

(a) On a photocopy of the sequence alignment, mark the positions of side chains that interact with DPG. (b) What sequence difference or differences might most obviously affect the binding of DPG?

Weblems

1. Find human proteins containing the peptapeptide Glu–Leu–Val–Ile–Ser.
2. Find five PDB entries containing protein kinase domains determined at 2.1 Å or better. (Remember that a lower number specifies a better (higher) resolution.)
3. (a) What are the conformational angles that specify the common rotamers of the histidine side chain, independent of secondary structure? (b) Which is the most commonly observed rotamer? (c) What is the most common rotamer observed for a histidine residue in an α-helix? Is this the same as the answer to (b), the most commonly observed histidine rotamer independent of secondary structure?
4. What is the total accessible surface area of pepsin?
5. For how many colours of the rainbow (red, orange, yellow, [green], blue, indigo, violet) can you identify variants of green fluorescent protein that fluoresce with that colour?
6. Protein kinase A upregulates the activity of tyrosine hydroxylase by phosphorylating Ser40. Protein kinase C also phosphorylates tyrosine hydroxylase, but does not alter its activity. What other sites on tyrosine hydroxylase are known to be phosphorylated, what enzymes catalyse these modifications, and what is the effect of each on enzymatic activity?

Chapter 4

EVOLUTION OF PROTEIN STRUCTURE AND FUNCTION

LEARNING GOALS

1. **To understand the basis for the classification of protein folding patterns,** distinguishing between structural similarities based on true homology—descent from a common ancestor—and general similarities arising from structural themes common to many unrelated proteins.
2. **To appreciate the distinction between two types of homologues: orthologues and paralogues:** Orthologues are homologous proteins in different species, descended from a single ancestral protein. Paralogues are homologues in the same species arising from gene duplication, and their descendants.
3. **To understand the different features of pairwise sequence alignment, multiple sequence alignment, and structural alignment.**
4. **To be able to define supersecondary structures, domains and modular proteins,** and to relate these terms to the classical hierarchy of primary, secondary, tertiary, and quaternary structure.
5. **To recognize domain swapping as a general mechanism of the formation of oligomeric proteins.**
6. **To gain familiarity with Web sites offering classifications of proteins based on their folding patterns.**
7. **To understand the relationship between divergence of sequence and divergence of structure in homologous proteins.**
8. **To be able to analyse patterns of conservation in multiple sequence alignment tables.**
9. **To become familiar with evolutionary variations in protein families,** illustrated by globins, NAD-binding domains, and opsins.

Introduction

Evolutionary relationships among protein structures illuminate sequence–structure–function relationships. How are changes in sequence reflected by changes in structure? How do different sequences create similar structures? What is the topography of the evolutionary space of sequences and structures—that is, starting with one protein, what others are readily accessible to evolution?

Folding patterns of protein domains are built by assembling local structures such as α-helices. Some features of protein structures are what they are because the laws of physics and chemistry would not allow them to be otherwise. Some are necessitated by the mechanism of evolution. In addition, historical accident has played a large role in creating the roster of folding patterns observed in Nature. It is by no means easy to sort out these effects. A creative tension among them pervades and animates the field of protein science.

Protein structure classification

Homologues can appear within the same species or in different species (see Box).

A *family* of proteins is a set for which similarities of sequence and/or structure provide evidence that they are related by evolution. Recall (see p. 45) that homology means descent from a common ancestor. Depending on historical events, homology cannot, in general, be directly observed. Similarity, of sequence and structure, can be observed and measured. Most statements about homology are inferences from observed similarities.

Proteins from the same family have similar structures. However, proteins from different families often contain recurrent structural themes. We can compare and classify the conformations of apparently unrelated proteins, on the basis of secondary structures and their folding patterns. In this way we can achieve a classification encompassing all known protein structures, a useful thing to have. However, within the hierarchy of such a classification, only the relationships among classes of proteins *within* the same family reflect evolutionary divergence. At higher levels of the hierarchy, the classification is based purely on architectural similarity, independent of provable evolutionary history and relationship.

In practice, the finest levels of the classification are the easiest. Given a set of closely related homologous proteins, computer programs can detect, and measure quantitatively, degrees of similarity. Similarities among sequences of orthologues correlate well with similarities of the corresponding structures, and with the classifications of species of origin established by classical methods of comparative anatomy, embryology, and palaeontology.

Extending the classification scheme to more distant relatives is more difficult. In many cases, alignment methods based solely on sequences are uninformative, because the sequences have diverged too far. But indications of relationship persist longer in structures than in sequences. Therefore, for distantly related proteins it is sometimes possible to align the sequences via the structures (see Box, p. 127–8). This is called '*structural alignment*'.

Evolutionary relationships among proteins: homologues, orthologues, and paralogues

- Proteins are homologous *if, and only if,* they are descended from a common ancestor.
- Homologues in different species, descended from single ancestral protein, are **orthologues**.
- Homologues in the same species, arising from gene duplication, are **paralogues**. Their descendants are also paralogues. After gene duplication, one of the resulting pairs of proteins can continue to provide its customary function, releasing the other to diverge, to develop new functions. Therefore inferences of function from homology are more secure for orthologues than for paralogues.

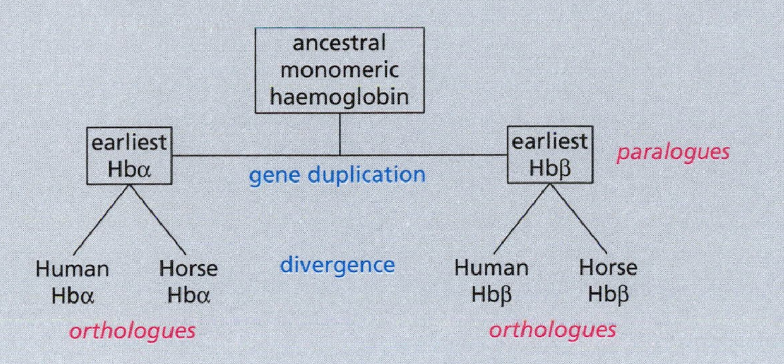

The globin family contains both orthologues and paralogues. An ancestral monomeric haemoglobin gene duplicated about 450–500 million years ago to form paralogous α- and β-chains. Subsequent divergence produced orthologous α-chains and orthologous β- chains in horse and human (and many other species).

Differential gene loss can make it difficult to distinguish orthology from paralogy. What if humans had lost their α-chains and horses their β-chains? The remaining human β- and horse α-chains would appear to be orthologues. Full-genome information can often resolve such ambiguities.

To measure the similarities of proteins, align their sequences

- **Alignment is the assignment of residue–residue correspondences.**
- **Sequence alignment**: for closely related homologues, alignment can be based on sequences alone. The correct alignment of the following character strings is clear:

```
a b c d e f g
a b z d — f g
```

There is one substitution (c↔z) and one deletion.

Computer programs for sequence alignment also provide measures of the similarity of the sequences. A popular index of sequence similarity is the *percent identical residues in an optimal alignment*.

continues...

- **A multiple sequence alignment** is a mutual alignment of three or more sequences. Multiple sequence alignment is more informative than pairwise sequence alignment: it reveals *patterns of conservation*. It identifies regions commonly subject to insertions and deletions, which usually correspond to peripheral loops in the structure.
- **Structural alignment**: by superposing two protein structures, one can identify residues that occupy the equivalent positions in molecular space, and base the alignment on this. Although structural alignment depends on structural similarity, the result is still an alignment of the *sequences*. Because structure diverges more slowly than sequence in protein evolution, structural alignment succeeds for pairs of homologous proteins too distantly related for methods based purely on the sequences to give accurate results.

  The results of a structural alignment are:

  (a) *The alignment itself*—that is, the set of residue–residue correspondences. In some cases, parts of the structure may have changed so much that no residue–residue correspondence can be made (Fig. 4.1). In these cases, we can distinguish between *alignable* regions of the sequence and *non-alignable* regions. This distinction cannot be made by purely sequence-based approaches.

  (b) *A measure of structural similarity*—the average (root-mean-square) deviation of the alignable atoms.

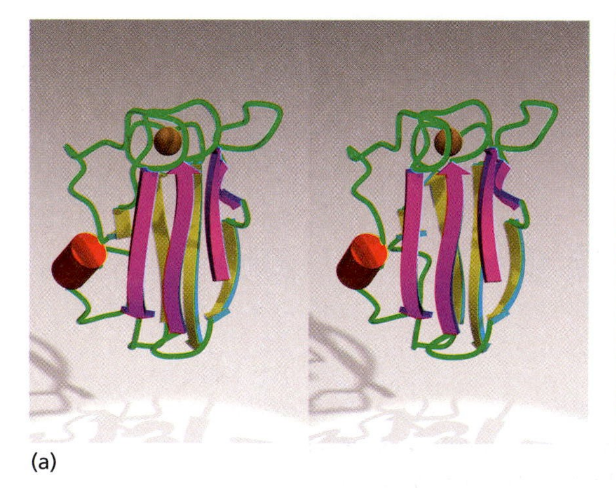

(a)

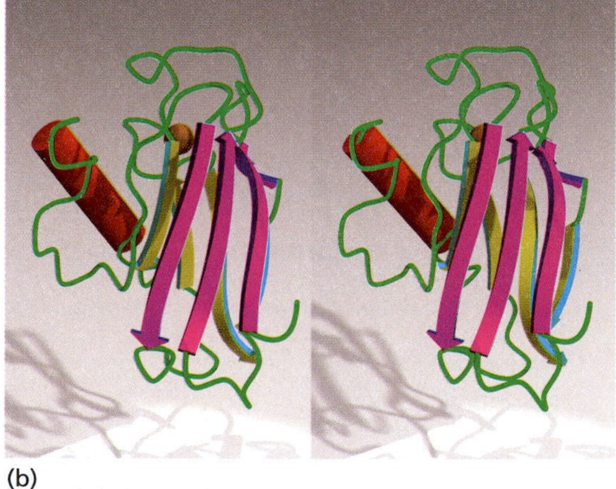

(b)

Fig. 4.1 Two copper-binding electron-transport proteins: (a) Poplar-leaf plastocyanin [1PCY], (b) azurin [2AZA]. These distant homologues have similar folding patterns. Both contain a double-β-sheet structure, with the sheets packed face-to-face, with the strands in the two sheets running in approximately the same directions. However, the two long loops at the left in each molecule have completely changed conformation, and the corresponding regions of the sequences are *not alignable*. The long loop at the left contains a helix in both molecules, but there is no reason to believe that these helices are homologous.

CASE STUDY Superpositions and alignments of pairs of proteins with increasingly more distant relationships

Sulphydryl proteinases are a family of enzymes found in prokaryotes, animals, and plants. In humans, cathepsins are lysosomal proteinases, caspases are involved in apoptosis (programmed cell death), and calpains cleave proteins involved in cell motility and adhesion. Plants also contain sulphydryl proteinases, which explains the effectiveness of papaya or pineapple juice in cooking as a meat tenderizer.

Figure 4.2(a–d) shows sequence alignments and superpositions of papaya papain, and four homologues—the close relative, kiwi fruit actinidin, and increasingly more distant relatives: human procathepsin L, human cathepsin B, and *Staphylococcus aureus* staphopain. The more distant the relationship, the lower the similarity in both sequence and structure. This series of alignments and superpositions shows the progressive divergence of the sequences and structures.

Figure 4.2(e), showing the structure of papaya papain, provides a key to navigate between the sequence alignments and structure superpositions.

```
9pap IPEYVDWRQKGAVTPVKNQGSCGSCWAFSAVVTIEGIIKIRTGNLNQYSEQEL
      | |||||  |||   | || || ||||||  | ||| ||  | |   |||||
2act LPSYVDWRSAGAVVDIKSQGECGGCWAFSAIATVEGINKITSGSLISLSEQEL

9pap LDCDRR--SYGCNGGYPWSALQLVAQY-GIHYRNTYPYEGVQRYCRSR-EKGP
      || |     || |||     |      ||     |||      |
2act IDCGRTQNTRGCDGGYITDGFQFIINDGGINTEENYPYTAQDGDCD--VAL--

9pap ---YAAKTDGVRQVQPYNQGALLYSIANQPVSVVLQAAGKDFQLYRGGIFVGP
             |    |   |  ||      ||||| | |||  |  |  ||| ||
2act QDQKYVTIDTYENVPYNNEWALQTAVTYQPVSVALDAAGDAFKQYASGIFTGP

9pap CGNKVDHAVAAVGYGP-----NYILIKNSWGTGWGENGYIRIKRGTGNSYG
     ||  ||||   ||||       |   |||| | ||| || || |      |
2act CGTAVDHAIVIVGYG-TEGGVDYWIVKNSWDTTWGEEGYMRILRNV-GGAG
```

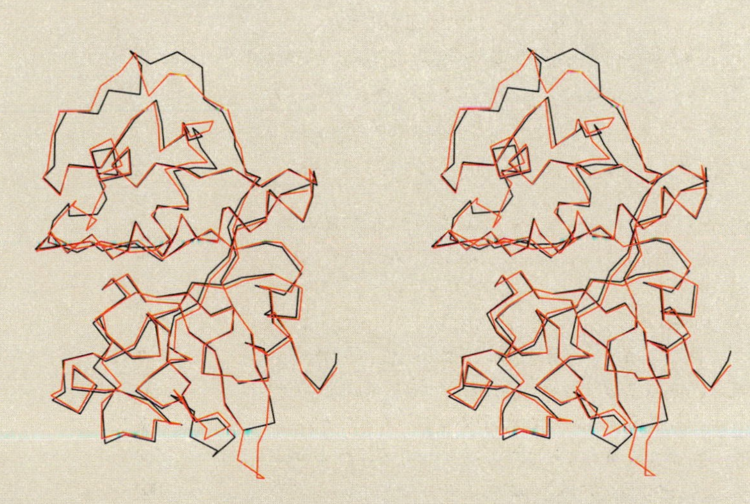

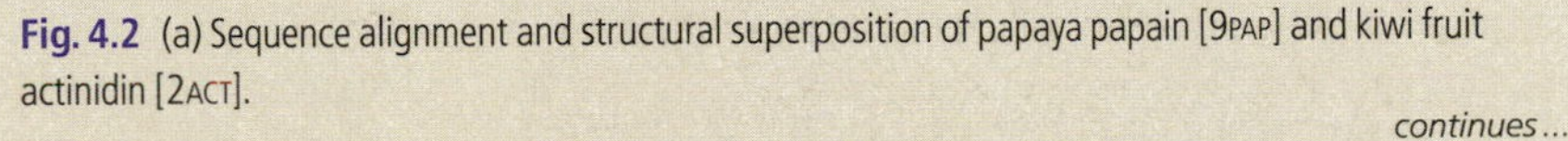

Fig. 4.2 (a) Sequence alignment and structural superposition of papaya papain [9PAP] and kiwi fruit actinidin [2ACT].

continues...

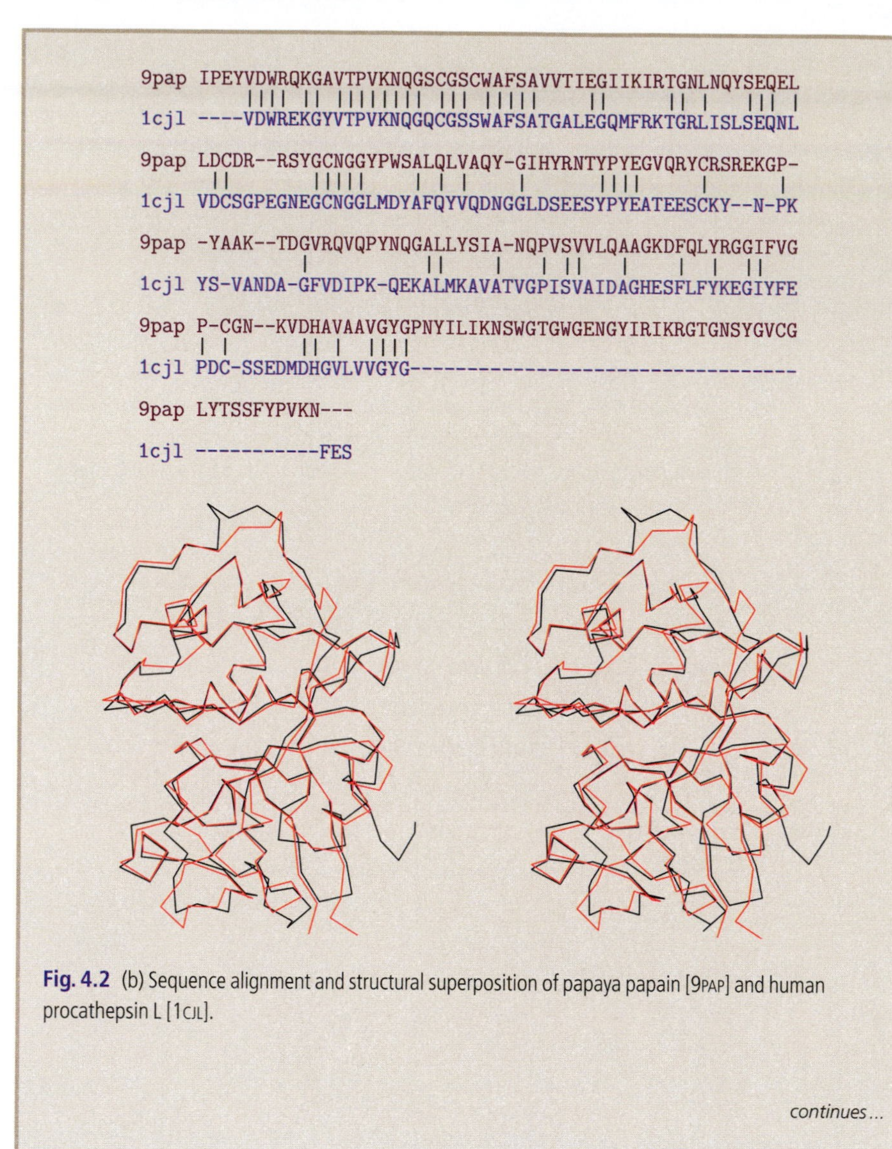

Fig. 4.2 (b) Sequence alignment and structural superposition of papaya papain [9PAP] and human procathepsin L [1CJL].

continues...

```
9pap IPEYVDWRQ-KG-A--VT-PVKNQGSCGSCWAFSAVVTIEGIIKIRTG-NLNQY
      |   | |               |||||||||| ||  |   | | |
1huc LPASFDAREQWPQCPTI-KEIRDQGSCGSCWAFGAVEAISDRICIHT-NVSVEV

9pap SEQELLDCDR-R-SYGCNGGYPWSALQLVAQYGIHYR-------NTYPYEGV--
     |   || |       |||||||  |       |              ||
1huc SAEDLLTCCGSMCGDGCNGGYPAEAWNFWTRKGLVSGGLYESHVGCRPYSI-PP

9pap -----------------Q-RYCRSRE--------KGP-YAAKTDGVRQVQPYNQ
                          |                          |
1huc CEHHVNGSRPPCTGEGDTPK-CSK-ICEPGYSPTYKQDK-HYGYNSYSVSN-SE

9pap GALLYSIAN-QPVSVVLQAAGKDFQLYRGGIFVGPCGNKV------DHAVAAVG
           |    ||         || ||  |                 ||    |
1huc KDIMAEIYKNGPVEGAFSV-YSDFLLYKSGVYQHV-----TGEMMGGHAIRILG

9pap YGP----NYILIKNSWGTGWGENGYIRIKRGTGNSYGVCGLYTSSFYPVKN---
      |      | |  ||| | || ||   | ||       ||
1huc WGVENGTPYWLVANSWNTDWGDNGFFKILRG--Q--DHCGIESEVVAGIP-RTD
```

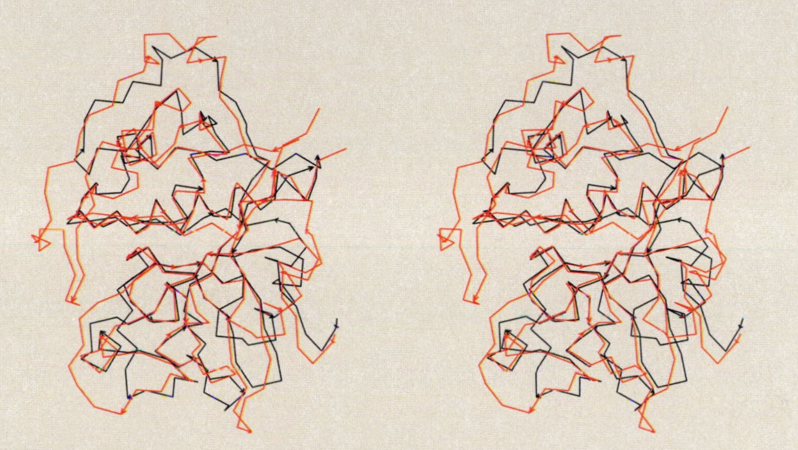

Fig. 4.2 (c) Sequence alignment and structural superposition of papaya papain [9PAP] and human liver cathepsin B [1HUC]. Note, in *both* the sequence alignment and the superposition, the higher similarity at the beginning and end of the sequences than in the middle region.

continues...

```
9pap IPE----YVDWRQKGAVTPVKNQGSCGSCWAFSAVVTIEGIIKIRTGNLNQ-YS
                           ||  | |                       |
1cv8 ---NEQYVNKL--E-NFKIRETQGNNGWCAGYTMSALLNATYN-----T-NKYH

9pap EQELLDCDRRSYG---------CNGGYPW-S---ALQLVAQYGIHYRNTYPYEG
                             |           |
1cv8 AEAVMRFLH----PNLQGQQFQFTGLT-PREMIYFGQ--T--------------

9pap VQRYCRSREKGPYAAKTDGVRQVQP-Y-NQ--GALL-YSIA-NQ-PVSVVLQAA
                          |                    |
1cv8 ---------------QG-RSPQL-LNRMTTYNE--VDNL-TKNNKGIAILG---

9pap GKDFQLYRGGIFVGPCGNKV-----------DHAVAAVGYGP-----NYILIKN
                                     || | ||          | | |
1cv8 --------------------SRVESRNGMHAGHAMAVVGNAKLNNGQEVIIIWN

9pap SWGTGWGENGYIRIKRGTGNSYG-VCGLY------------TSSFYPVKN
      |                                         |
1cv8 PWDN----G-FMTQDA-K-----NN----VIPVSNGDHYQWYSSIYGY---
```

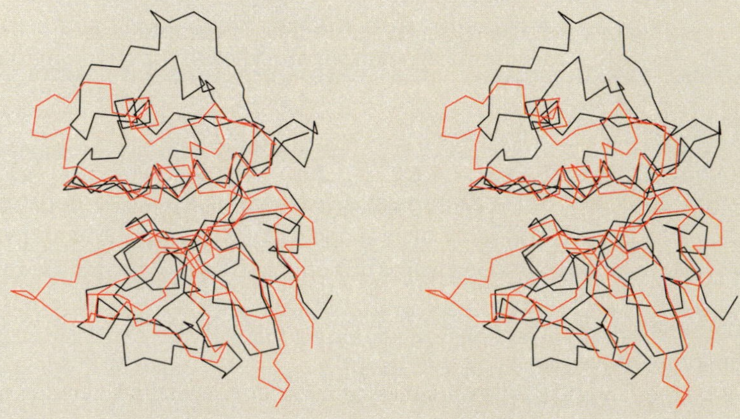

Fig. 4.2 (d) Sequence alignment and structural superposition of papaya papain [9PAP] and *S. aureus* staphopain [1CV8].

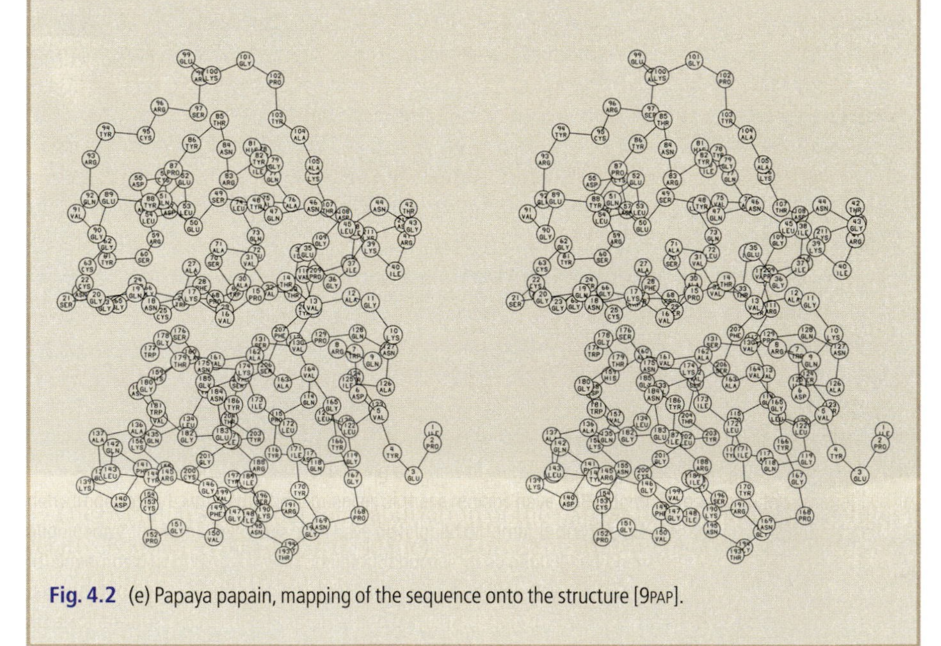

Fig. 4.2 (e) Papaya papain, mapping of the sequence onto the structure [9PAP].

Secondary, tertiary, and quaternary structure

Recall that:

- The set of chemical bonds in a protein is the **primary structure**.
- The distribution of helices and sheets—the thematic hydrogen-bonding patterns of the main chain—is the **secondary structure**.
- The spatial assembly and interactions of the helices and sheets is the **tertiary structure**.
- For proteins composed of more than one subunit, the assembly of the monomers is the **quaternary structure**.

It has proved useful to add additional levels to this hierarchy:

- **Supersecondary structures.** Proteins show recurrent patterns of interaction between helices and sheets close together in the sequence. These *supersecondary structures* include the α-helix hairpin, the β-hairpin, and the β–α–β-unit (Fig. 4.3). Examples appear in structures previously illustrated.

  Supersecondary structures are local structures, formed by residues in a contiguous segment of the sequence.

  Short β-hairpins, up to about six residues long, fall into well-defined families of structures. To make a tight turn, at least one residue must escape from the standard conformations—α_R or β-region of the Sasisekharan–Ramakrishnan–Ramachandran plot, and

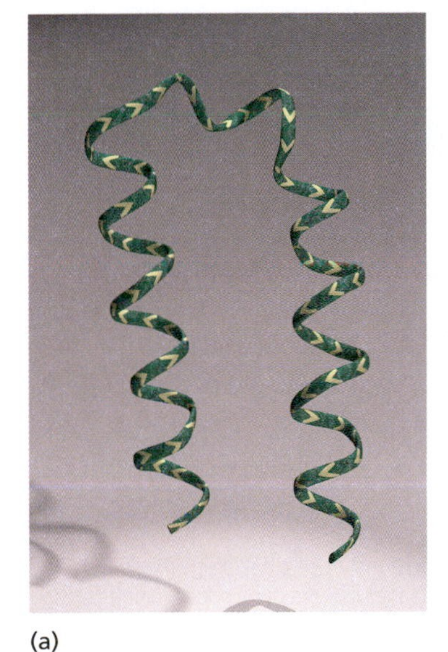

(a)

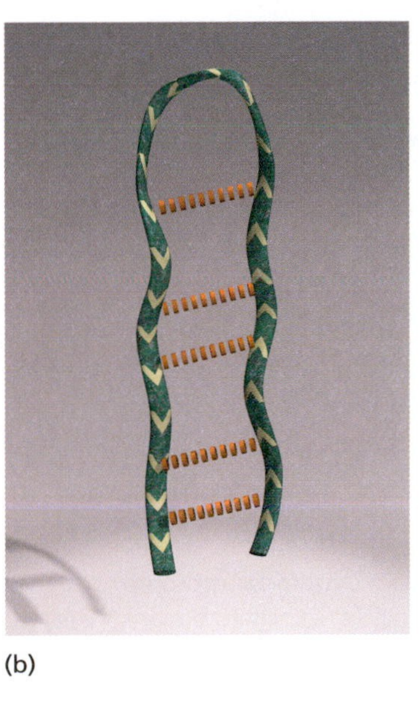

(b)

(c)

Fig. 4.3 Supersecondary structures: (a) helix hairpin; (b) β-hairpin; (c) β–α–β-unit.

trans peptide bond. This can be a glycine in the α_L conformation, or a proline preceded by a *cis* peptide bond. The conformations of short β-hairpins can be predicted from the position within the loop of a glycine or proline residue.

In the hierarchy, supersecondary structures fall between secondary structure and domain.

- **Domains**. Many proteins contain compact units within the folding pattern of a single chain, that look as if they should have independent stability, although this is rarely demonstrated experimentally. Phosphoglycerate kinase is an enzyme with its active site in a cleft between two domains (Fig. 4.4(a)). Aspartate carbamoyltransferase is composed of subunits containing a catalytic and a regulatory chain, each composed of two domains (Fig. 4.4(b)). In the hierarchy, domains fall between supersecondary structures and the tertiary structure of a complete monomer.
- **Modular proteins**. Modular proteins often contain many copies of closely related domains (recall Fig. 2.24). The domains can appear in different structural contexts; that is, proteins can 'mix and match' sets of domains. For example, fibronectin, a large extracellular protein involved in cell adhesion and migration, contains 29 domains including multiple tandem repeats of three types of domains called F1, F2, and F3. It is a linear array of the form: $(F1)_6(F2)_2(F1)_3(F3)_{15}(F1)_3$. Fibronectin domains also appear in other modular proteins. Recombination of modules is an important mode of protein evolution.

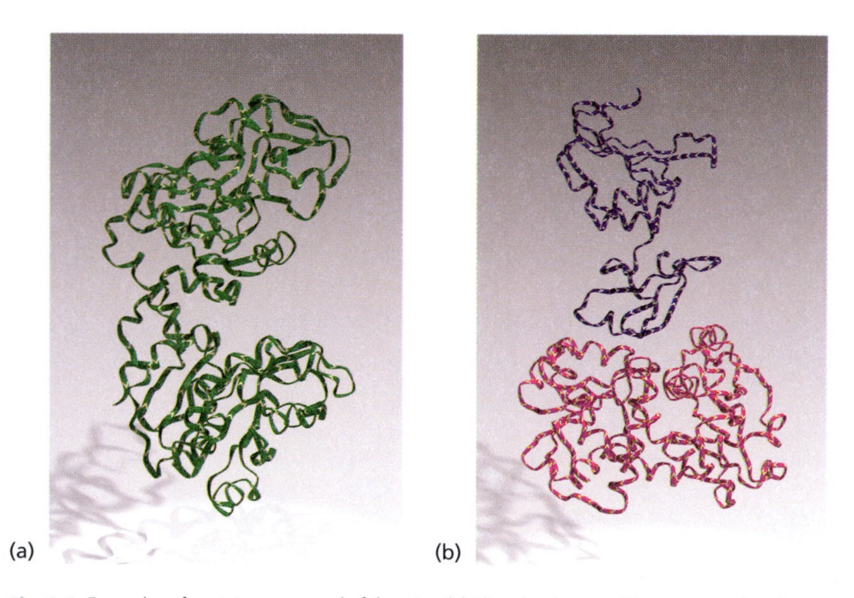

Fig. 4.4 Examples of proteins composed of domains. (a) Phosphoglycerate kinase, a two-domain enzyme with the binding site in the cleft between domains. This molecule changes conformation upon binding substrate [3PGK]. (b) Two subunits of *E. coli* aspartate carbamoyltransferase, each comprising two domains. This figure contains one regulatory subunit (blue) and one catalytic subunit (red). The full molecule contains six catalytic and six regulatory subunits.

What is the difference between a module and a domain?

The terms are nearly synonomous, and usage is not consistent.

- A **domain** is defined in the context of the structure of an individual polypeptide chain in a protein, as a compact subunit.
- A **module** is a protein subunit that stays together as a unit during evolution, appearing in different structural contexts.

Domain swapping

Domain swapping is a general mechanism for forming an oligomer from a multidomain protein. Suppose a monomer contains two domains, A and B, connected by a flexible linker, with a well-developed interface between A and B (Fig. 4.5). A dimer, containing four domains, can be stabilized by two copies of the *same* A–B interface that appears in the monomer.

Pig odorant-binding protein is a monomer (Fig. 4.6(a)). The C-terminal segment contains a helix and strand of sheet (red in Fig. 4.6(a)). In contrast, cow odorant-binding protein is a dimer (Fig. 4.6(b)). In this dimer, the C-terminal helix and strand of each monomer flip over to interact with the partner. The other monomer provides these residues with an equivalent environment and interactions (Fig. 4.6(c)).

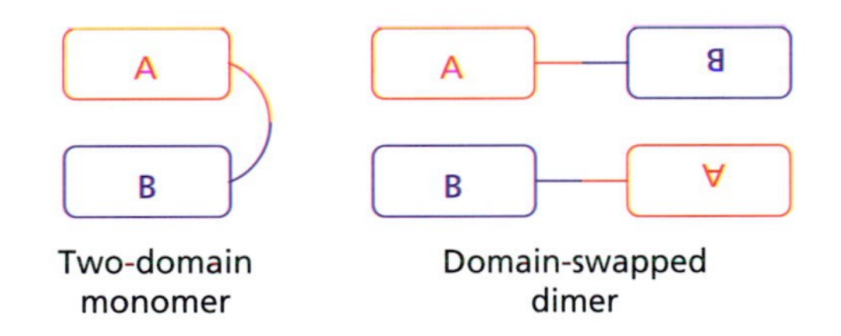

Fig. 4.5 Domain swapping. The reader should verify that the two A–B interfaces in the dimer are formed from domains in the same relative position and orientation as the single interface in the monomer.

Classifications of protein folding patterns

Catalogues of protein structures

There are now so many known protein structures that we need directories and catalogues to keep track of them. 'White pages', listing the structures by name, are useful if we know what we are looking for. More generally helpful are 'Yellow pages', which organize the entries into some reasonable classification.

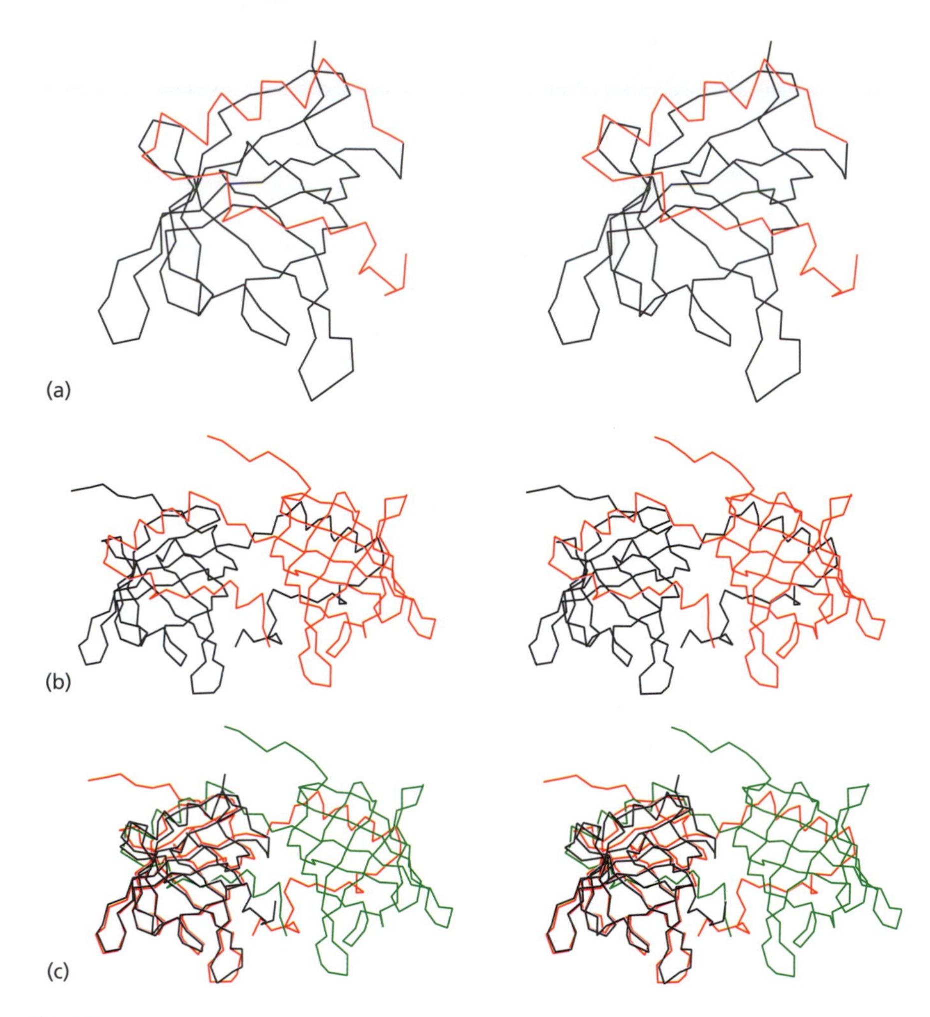

Fig. 4.6 (a) Pig odorant-binding protein, a monomeric protein [1A3Y]. The C-terminal region, containing a helix and a strand, are shown in red. (b) Cow odorant-binding protein, a dimeric protein [1OBP]. The C-terminal helix and strand are exchanged between different monomers. (c) Superposition of pig (black) and cow (green and red) odorant-binding proteins shows that the structure of each monomer is effectively intact. Significant main chain conformational changes involve only a few residues in a hinge that allows the C-terminal region of each monomer to flip over to interact with the other.

Several databases present classfications of proteins based on the similarities of their folding patterns. Accessible over the Web, their many useful features include: search by keyword or sequence; presentation of structure pictures; and links to related sites, including bibliographical databases.

Structural Classification of Proteins (SCOP)

SCOP organizes protein structures in a hierarchy according to evolutionary relationship and structural similarity. At the lowest level of the SCOP hierarchy are the individual *domains*, extracted from the Protein Data Bank entries. Sets of domains are grouped into

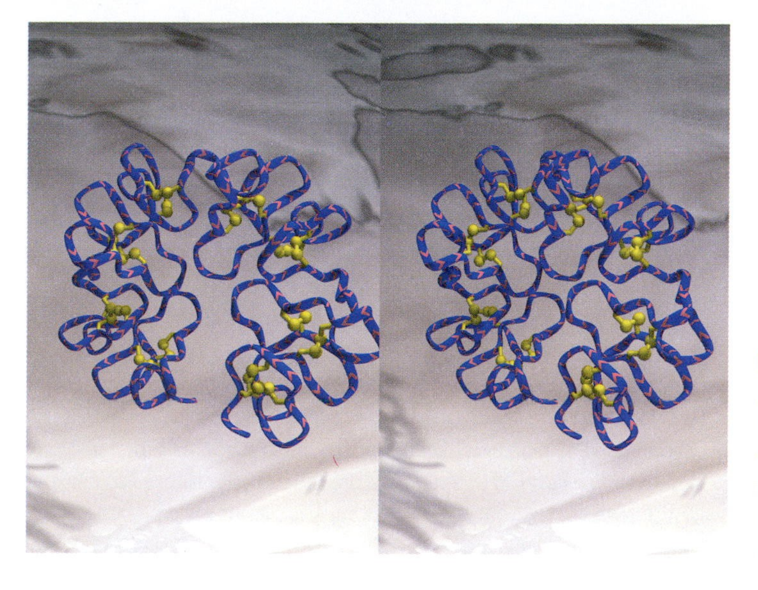

Fig. 4.7 Wheat-germ agglutinin, a protein with little secondary structure. Its native conformation is held together by many disulphide bridges.

The SCOP hierarchy
(on 15 Dec 2003)

Level	Number of cases
Major Classes	7
Fold	800
Superfamilies	1294
Families	2327
Domains	54 745

SCOP: Structural Classification of Proteins
15 Dec 2003 release: 20 619 PDB entries, 54 745 domains

Class	Number of folds	Number of superfamilies	Number of families
All α proteins	179	299	480
All β proteins	126	248	462
α and β proteins (α/β)	121	199	542
α and β proteins ($\alpha + \beta$)	234	349	567
Multidomain proteins	38	38	53
Membrane and cell-surface proteins	36	66	73
Small proteins	66	95	150
Total	800	1294	2327

CASE STUDY The SCOP classification of thioredoxin from *E. coli*

Class	α and β proteins (α/β)
Fold	Thioredoxin fold
Superfamily	Thioredoxin-like
Family	Thioltransferase
Protein	Thioredoxin
Species	*E. coli*

To give some idea of the nature of the similarities expressed by the different levels of the hierarchy, Figs 4.8 through 4.11 show pairs of protein domains that are classified, in SCOP, into the same family, superfamily, fold, and class, as *E. coli* thioredoxin.

Thioredoxin from *E. coli* [2TRX] and a human homologue [1GH2] are in the same family. The calcium-binding protein calsequestrin [1A8Y] is in the same superfamily as thioredoxin, but in a different family. An RNA 3′-terminal phosphate cyclase [1QMH] contains a domain with a fold similar to thioredoxin but in a different superfamily. All the proteins illustrated, and many more with entirely unrelated topologies, are in the general SCOP class of α/β proteins.

families of homologues. These comprise domains for which the similarities in structure, function, and sequence imply a common evolutionary origin. Families that share common structure and function, but for which the evidence for evolutionary relationship is suggestive but not compelling, are grouped into *superfamilies*. Superfamilies that share a common folding topology, for at least a substantial portion of the structure, are grouped as *folds*. Finally, each fold group falls into one of the general *classes*. The major classes in SCOP are α, β, $\alpha + \beta$, α/β, and 'small proteins', which often have little secondary structure and are held together by disulphide bridges or ligands. Wheat-germ agglutinin is an example (Fig. 4.7).

FSSP and the DALI domain dictionary

The program DALI provides a general and sensitive method for comparing protein structures. Given two sets of coordinates, it determines the maximal common substructure and provides an alignment of the common residues. DALI is the best of many programs that address this problem, because of its ability to recognize distant relationships, and its speed of execution. DALI is fast enough to scan the entire PDB for proteins similar to a probe structure. Crystallographers and NMR spectroscopists who solve a new structure routinely run the coordinates through DALI, to detect similarities to known structures.

Application of DALI to the entire Protein Data Bank produces two classifications:

- **FSSP** (Fold classification based on Structure–Structure alignment of Proteins) presents the results of applying DALI to all *chains* from proteins in the PDB. The FSSP entry for each chain includes its alignment with proteins of similar structure, and

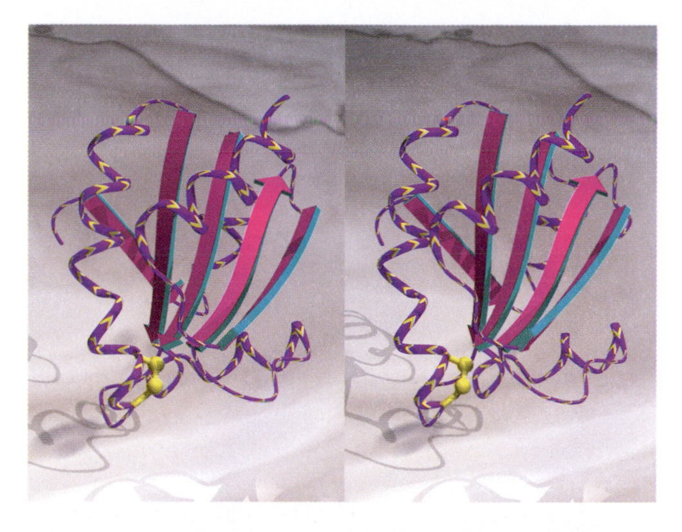

Fig. 4.8 *E. coli* thioredoxin [2TRX].

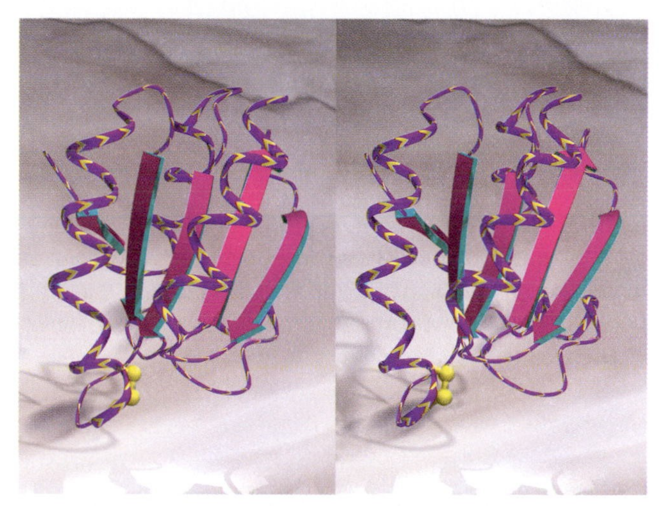

Fig. 4.9 Human thioredoxin homologue [1GH2], a member of the same family as *E. coli* thioredoxin.

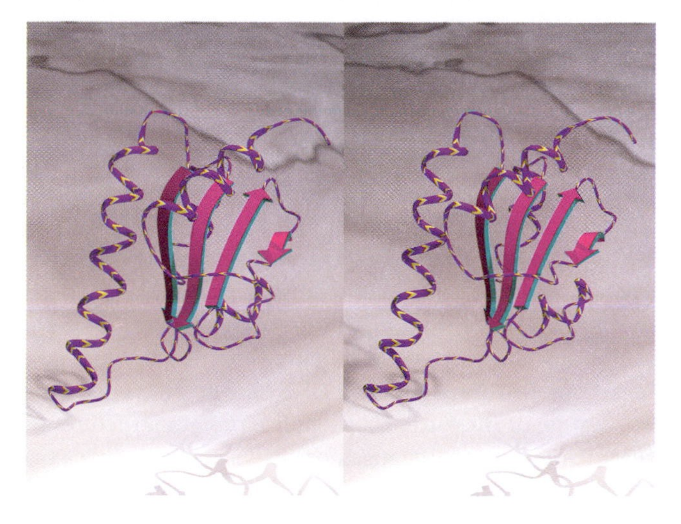

Fig. 4.10 Rabbit calsequestrin [1A87], in the same superfamily as *E. coli* thioredoxin, but in a different family.

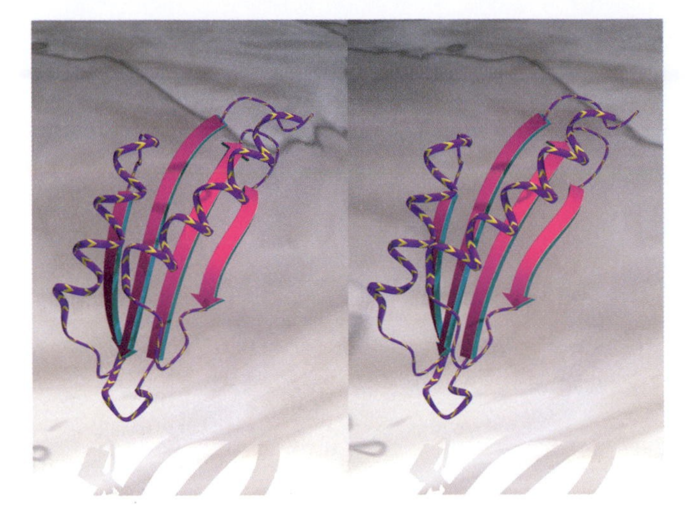

Fig. 4.11 A domain from RNA 3′-terminal phosphate cyclase [1QMH], with a fold similar to thioredoxin but in a different superfamily.

reports the structurally equivalent residues. From the Web site the user can display multiple sequence alignments and superimposed structures.

- The **DALI domain dictionary** is a corresponding classification of recurrent **domains** automatically extracted from known proteins.

Numerous other Web sites offer classifications of protein structures. Do the different schemes agree? They do agree, for the most part, on what is similar. However, a measure of similarity does not specify how to define different levels of a hierarchy—for instance, family or superfamily—and different classification schemes differ to some extent in the division of their hierarchies into levels.

Structural relationships among homologous proteins

Included in the 15 000 protein structures now known are several families in which the molecules maintain the same basic folding pattern over ranges of sequence similarity, from near-identity down to well below 20% conservation, even below the so-called 'twilight zone'. Plastocyanin and azurin (Fig. 4.1) and papain and its homologues (Fig. 4.2) are typical examples.

Comparisons of homologous proteins reveal how structures accommodate changes in amino acid sequence. Surface residues not involved in function are usually free to mutate. Loops on the surface can often accommodate changes by local refolding. Mutations that change the volumes of buried residues generally do not change the local conformations of individual helices or sheets, but distort their spatial assembly. The nature of the forces that stabilize protein structures sets general limitations on these conformational changes; particular constraints derived from function vary from case to case (see Box).

Stasis and change in the evolution of protein structures

1. Related structures retain most elements of secondary structure: the helices and sheets.
2. A *core* of the structure—the assembly of the central helices and/or sheets—retains its topology or folding pattern. For closely related proteins the core comprises almost the entire structure. For distantly related proteins, the core may contain only half the residues, or even fewer.
3. Peripheral regions, outside the core, may change their folding pattern entirely.
4. The relative geometry of the secondary structures, even in the core, is variable. As a result of mutations, helices and sheets can shift and rotate with respect to one another.
5. For evolution with retention of function, the structural changes are subject to constraints that conserve function, for example to maintain the integrity of the active site. For evolution with change in function, these constraints are replaced by other constraints required by the altered function, producing greater structural change.

The general response to mutation is structural change. Families of related proteins tend to retain common folding patterns. However, although the general folding pattern is preserved, there are distortions which increase as the amino acid sequences progressively diverge. These distortions are not uniformly distributed throughout the structure. A large central core of the structure usually retains the same qualitative fold, and other parts of the structure change conformation more radically. (The letters B and R, considered as structures, have a common core which corresponds to the letter P. Outside the common core they differ: at the bottom right, B has a loop and R has a diagonal stroke.)

There is a quantitative relationship between the divergence of the amino acid sequences of the core of a family of proteins, and the divergence of the structures. As the sequence diverges, there are progressively increasing distortions in the main chain conformation, and the fraction of the residues in the core usually decreases. Until the fraction of identical residues in the sequence drops below about 40–50%, these effects are relatively modest. Almost all the structure remains in the core, and the deformation of the main chain atoms is, on average, no more than 1.0 Å. With increasing sequence divergence, some regions refold entirely, reducing the size of the core, and the distortions of the remaining core residues increase in magnitude.

A correlation between the divergence of sequence and structure applies to all families of proteins. Figure 4.12(a) shows the changes in structure of the core, expressed as the root-mean-square deviation of the backbone atoms after optimal superposition, plotted against the sequence divergence: the percent of conserved amino acids of the core after optimal alignment. The points correspond to pairs of homologous proteins from many related families. (Those at 100% residue identity are proteins for which the structure was determined in two or more crystal environments. The deviations show that crystal packing forces—and, to a lesser extent, solvent and temperature—can slightly modify the conformation of the proteins.) Figure 4.12(b) shows the changes in the fraction of

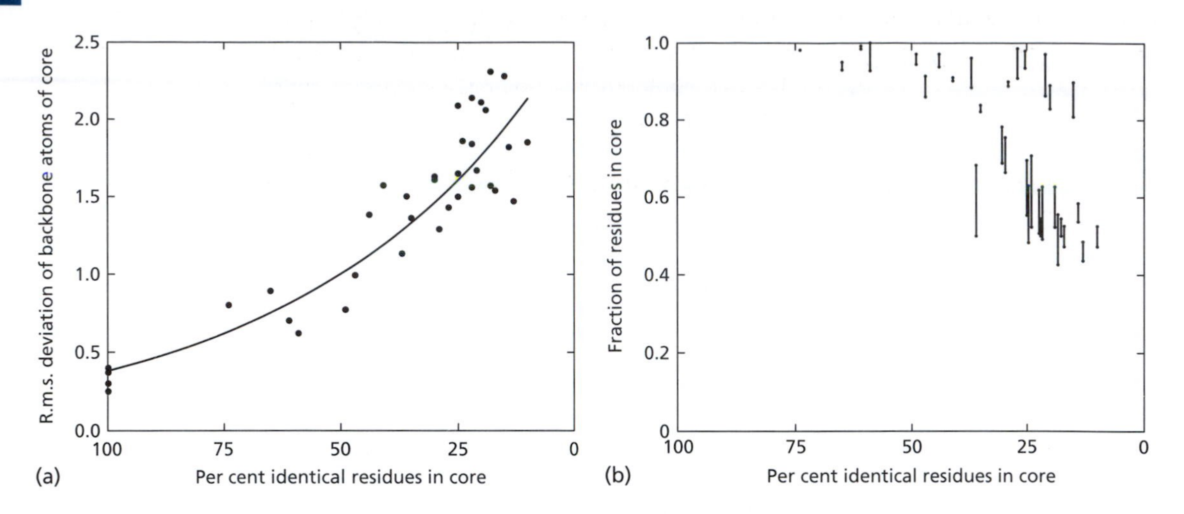

Fig. 4.12 Relationships between divergence of amino acid sequence and three-dimensional structure of the core, in evolving proteins. (a) Variation of r.m.s. deviation of the core (Å) with the percent identical residues in the core. (b) Variation of size of the core with the percent identical residues in the core. This figure shows results calculated for 32 pairs of homologous proteins of a variety of structural types. (Adapted from Chothia, C. and Lesk, A.M. (1986). Relationship between the divergence of sequence and structure in proteins. *EMBO J.*, **5**, 823–6.)

For larger structural changes, see p. 248 and 263.

residues in the core as a function of sequence divergence. The fraction of residues in the cores of distantly related proteins can vary widely: in some cases the fraction of residues in the core remains high, in others it can drop to below 50% of the structure.

Changes in proteins during evolution give clues to the roles of residues at different positions

Residues at different positions in proteins play different roles in structure and function. Some residues participate directly in function. Some are essential for creating the structure. Some are not so imporant to the structure of the domain in which they appear, but mediate intramolecular or intermolecular interactions. Others are not subject to any obvious constraints. If we were smart enough, we could look at the sequence and structure of a single protein, and be able to assign the roles of the different residues.

But that's very difficult.

It is much easier to let Nature do the work for us. Align the amino acid sequences, and superpose the structures, of proteins from a family. Patterns of conservation will then declare themselves, in both the sequences and the structures. These patterns of conservation provide important clues to the roles of the residues.

Many residues involved directly in function are tightly conserved. Many residues buried in protein interiors are largely restricted to hydrophobic amino acids. Many positions on the surface are relatively free to vary. Many surface loops readily tolerate insertions and deletions. However, looking at individual positions provides much valuable information but does not tell the whole story. Subtle linkages among inter-residue interactions restrict the *combinations* of amino acids that can form a viable sequence of a protein in a family. Moreover, our analysis is limited by the range of proteins in the family that are known. As we shall see, inferences from a set containing only closely related proteins are often misleading.

Evolution of the globins

Globins are an ancient family of proteins, appearing in prokaryotes; animals, including vertebrates and invertebrates; and in plants.

Haemoglobin in our blood transports oxygen and carbon dioxide, delivering oxygen to myoglobin in other tissues. To promote this process, tetrameric haemoglobin arose by gene duplication and divergence from monomeric globins. We discussed the allosteric change in haemoglobin in Chapter 3.

The sequencing of the human genome has turned up additional globins. For example, neuroglobin is expressed in the brain, although its precise function remains obscure.

Sperm-whale myoglobin, the first protein structure to be solved by X-ray crystallography, shows a characteristic globin folding pattern (Fig. 4.13) containing eight helices enfolding a haem group (Fig. 4.14).

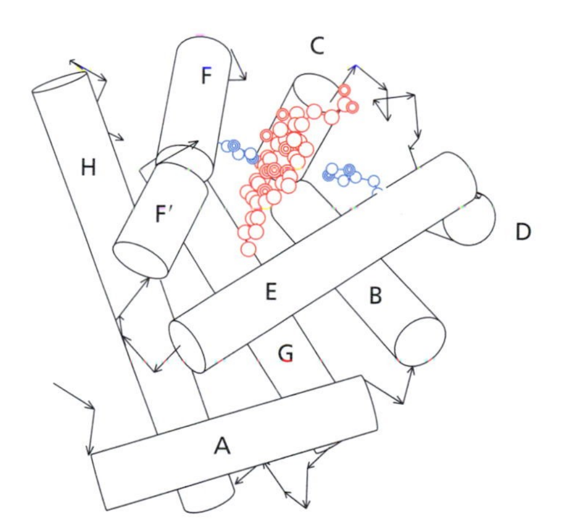

Fig. 4.13 The 'globin fold' consists of a characteristic arrangement of helices, surrounding a haem group. The helices are denoted A, B, C (a 3_{10} helix), D, E, F (broken in mammalian globins into two consecutive helices, F′ and F), G, H. Many globins lack the D helix. In most globins, the closest neighbours of the iron are the four pyrrole rings of the haem group and two histidine side chains, shown here in blue, on the E and F helices. Until recently, it was thought that this overall architecture characterized the entire globin family. The discovery of *truncated globins*—about 120 residues long instead of the ~150 residues characteristic of full-length globins—overturned this idea.

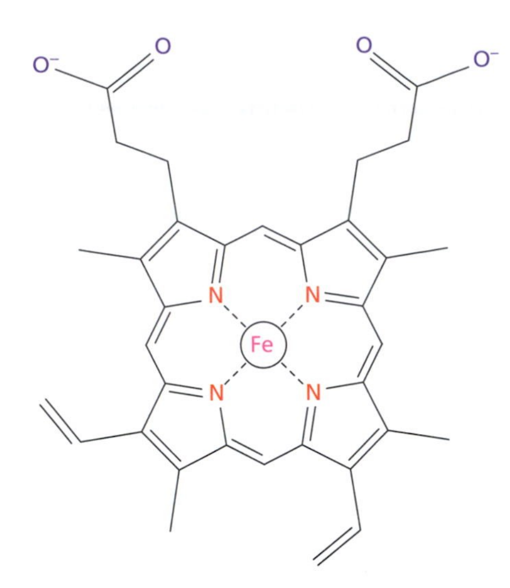

Fig. 4.14 The haem group contains an iron at the centre of a porphyrin ring.

Historically, analysis of the globin structures began with the mammalian structures, the first to be solved. One then went on to treat a wide range of globins from eukaryotes, which had diverged in sequence but retained a generally similar folding pattern. Many years later, structures of a new class of *truncated globins* emerged—substantially shorter than those previously considered. This reopened questions about globin evolution.

In this section we shall follow the history of investigations of sequence–structure relationships and evolution of the globin structures. This is an example of a question that was widely assumed to be 'done and dusted' many years ago, but has been reopened by new discoveries.

Mammalian globins

Not quite true—see Box, p. 146

The structures of sperm-whale myoglobin and human haemoglobin, first solved almost 50 years ago, gave the first direct glimpse of the evolutionary divergence of protein structures.

We now recognize that most mammalian globins form a very closely related subfamily, its members all highly similar in both sequence (Fig. 4.15(a)) and structure (Fig. 4.16). (For comparison, Fig. 4.15(b) extends this sequence alignment to non-mammalian globins.) At 25 positions in the alignment of mammalian globins shown in Fig. 4.15(a), a unique amino acid is conserved in all five sequences. Other positions show changes only among residues with very similar physicochemical properties: for example, position 4 contains only serine or threonine, both polar side chains. Position 119 contains only valine, isoleucine, or leucine, hydrophobic side chains of moderate size. But other positions contain residues showing very wide variations in side chain size and polarity; for example, position 32 contains glutamic acid (negatively charged), glycine (uncharged, small), or isoleucine (uncharged, medium-sized).

Even within mammalian globins, the patterns of residue conservation and change suggest a hierarchical classification. Position 44, for example, contains tyrosine in the

Mammalian Globin Sequences

```
                                           10        20        30        40        50        60
Human Haemoglobin α chain           VLSPADKTNVKAAWGKVGA-HAGEYGAEALERMFLSFPTTKTYFPHF-DLS-----HGS
Horse Haemoglobin α chain           VLSAADKTNVKAAWSKVGG-HAGEYGAEALERMFLGFPTTKTYFPHF-DLS-----HGS
Human Haemoglobin β chain          VHLTPEEKSAVTALWGKV---NVDEVGGEALGRLLVVYPWTQRFFESFGDLSTPDAVMGN
Horse Haemoglobin β chain          VQLSGEEKAAVLALWDKV---NEEEVGGEALGRLLVVYPWTQRFFDSFGDLSNPGAVMGN
Sperm whale myoglobin               VLSEGEWQLVLHVWAKVEA-DVAGHGQDILIRLFKSHPETLEKFDRFKHLKTEAEMKAS
                                     L    k  V a W KV      e G eaL R     P T   F  F dLs      g

                                           70        80        90       100       110       120
Human Haemoglobin α chain          AQVKGHGKKVADALTNAVAHV----D-DMPNALSALSDLHAHKLRVDPVNFKLLSHCLLV
Horse Haemoglobin α chain          AQVKAHGKKVGDALTLAVGHL----D-DLPGALSNLSDLHAHKLRVDPVNFKLLSHCLLS
Human Haemoglobin β chain          PKVKAHGKKVLGAFSDGLAHL----D-NLKGTFATLSELHCDKLHVDPENFRLLGNVLVC
Horse Haemoglobin β chain          PKVKAHGKKVLHSFGEGVHHL----D-NLKGTFAALSELHCDKLHVDPENFRLLGNVLVV
Sperm whale myoglobin              EDLKKHGVTVLTALGAILKK-----KGHHEAELKPLAQSHATKHKIPIKYLEFISEAIIH
                                     vK HGkkV         h     d         Ls lH  Kl vdp nf ll   l

                                          130       140       150       160
Human Haemoglobin α chain          TLAAHLP-A-EFTPAVHASLDKFLASVSTVLTSKYR
Horse Haemoglobin α chain          TLAVHLP-N-DFTPAVHASLDKFLSSVSTVLTSKYR
Human Haemoglobin β chain          VLAHHFG-K-EFTPPVQAAYQKVVAGVANALAHKYH
Horse Haemoglobin β chain          VLARHFG-K-DFTPELQASYQKVVAGVANALAHKYH
Sperm whale myoglobin              VLHSRHP-G-DFGADAQGAMNKALELFRKDIAAKYKELGYQG
                                   La  h      Ftp   a   K    v   l  KY
```

(a)

Eukaryote and Prokaryote Full-length Globin Sequences

```
                                           10        20        30        40        50        60
Human Haemoglobin α chain          -VLSPADKTNVKAAWGKVGA-HAGEYGAEALERMFLSFPTTKTYFPHF-DLS-----HGS
Horse Haemoglobin α chain          -VLSAADKTNVKAAWSKVGG-HAGEYGAEALERMFLGFPTTKTYFPHF-DLS-----HGS
Human Haemoglobin β chain          VHLTPEEKSAVTALWGKV---NVDEVGGEALGRLLVVYPWTQRFFESFGDLSTPDAVMGN
Horse Haemoglobin β chain          VQLSGEEKAAVLALWDKV---NEEEVGGEALGRLLVVYPWTQRFFDSFGDLSNPGAVMGN
Sperm whale myoglobin              -VLSEGEWQLVLHVWAKVEA-DVAGHGQDILIRLFKSHPETLEKFDRFKHLKTEAEMKAS
Chironomus erythrocruorin          --LSADQISTVQASFDKVKG-----DPVGILYAVFKADPSIMAKFTQFAG-KDLESIKGT
Lupin leghaemoglobin               GALTESQAALVKSSWEEFNA-NIPKHTHRFFILVLEIAPAAKDLFS-FLK-GTSEVPQNN
Bacterial globin (Vitroscilla sp.) --MLDQQTINIIKATVPVLKEHGVTITTTFYKNLFAKHPEVRPLFD-M----------GR
                                     l       v                           P     F  f

                                           70        80        90       100       110       120
Human Haemoglobin α chain          AQVKGHGKKVADALTNAVAHV----D-DMPNALSALSDLHAHKLRVDPVNFKLLSHCLLV
Horse Haemoglobin α chain          AQVKAHGKKVGDALTLAVGHL----D-DLPGALSNLSDLHAHKLRVDPVNFKLLSHCLLS
Human Haemoglobin β chain          PKVKAHGKKVLGAFSDGLAHL----D-NLKGTFATLSELHCDKLHVDPENFRLLGNVLVC
Horse Haemoglobin β chain          PKVKAHGKKVLHSFGEGVHHL----D-NLKGTFAALSELHCDKLHVDPENFRLLGNVLVV
Sperm whale myoglobin              EDLKKHGVTVLTALGAILKK-----KGHHEAELKPLAQSHATKHKIPIKYLEFISEAIIH
Chironomus erythrocruorin          APFETHANRIVGFFSKIIGEL---P--NIEADVNTFVASHKPRG-VTHDQLNNFRAGFVS
Lupin leghaemoglobin               PELQAHAGKVFKLVYEAAIQLEVTGVVVSDATLKNLGSVHVSKG-VADAHFPVVKEAILK
Bacterial globin (Vitroscilla sp.) QESLEQPKALAMTVLAAAQNI--ENLPAILPAVKKIAVKHCQAG-VAAAHYPIVGQELLG
                                        h                                 H

                                          130       140       150       160
Human Haemoglobin α chain          TLAAHLP-A-EFTPAVHASLDKFLASVSTVLTSKYR
Horse Haemoglobin α chain          TLAVHLP-N-DFTPAVHASLDKFLSSVSTVLTSKYR
Human Haemoglobin β chain          VLAHHFG-K-EFTPPVQAAYQKVVAGVANALAHKYH
Horse Haemoglobin β chain          VLARHFG-K-DFTPELQASYQKVVAGVANALAHKYH
Sperm whale myoglobin              VLHSRHP-G-DFGADAQGAMNKALELFRKDIAAKYKELGYQG
Chironomus erythrocruorin          YMKAHT-----DFAGAEAAWGATLDTFFGMIFSKM
Lupin leghaemoglobin               TIKEVVG-A-KWSEELNSAWTIAYDELAIVIKKEMDDAA
Bacterial globin (Vitroscilla sp.) AIKEVLGDAA--TDDILDAWGKAYGVIADVFIQVEADLYAQAVE
```

(b)

Fig. 4.15 (a) Multiple sequence alignment of five mammalian globins: sperm-whale myoglobin, and the α- and β-chains of human and horse haemoglobin. Each sequence contains approximately 150 residues. In the line below the tabulation, upper-case letters indicate residues that are conserved in all five sequences, and lower-case letters indicate residues that are conserved in all but sperm-whale myoglobin. (b) Multiple sequence alignment of full-length globins from eukaryotes and prokaryotes. Many fewer positions are conserved than in the mammals-only alignment. In the line below this tabulation, upper-case letters indicate residues that are conserved in all eight sequences, and lower-case letters indicate residues that are conserved in all but the bacterial globin.

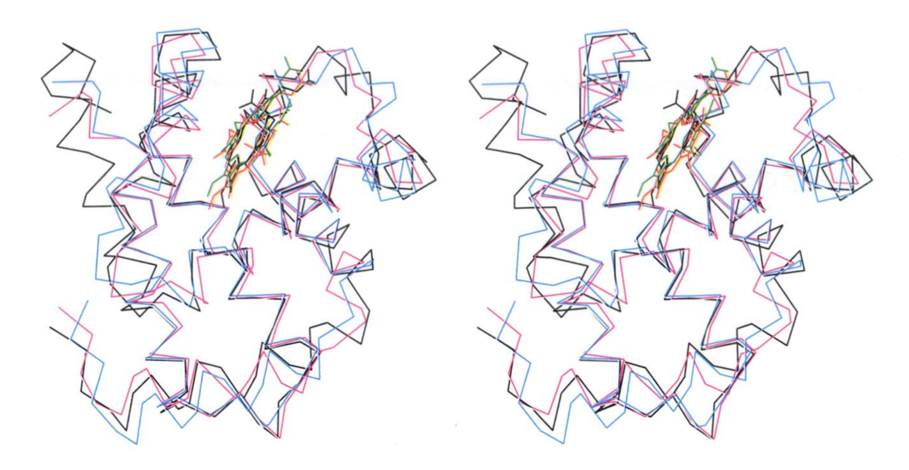

Fig. 4.16 Superposition of three closely related mammalian globins: sperm-whale myoglobin (black), human haemoglobin, α-chain (cyan) and β-chain (magenta).

Evolutionary relationships among proteins: haemoglobin crystallography a century ago

In 1909, E.T. Reichert and A.P. Brown published a study of crystals of haemoglobin isolated from different species of fishes. Haemoglobin crystallography, three years before the discovery of X-ray diffraction, was limited to measuring the angles between crystal faces. Stenö's law (1669) states that the interfacial angles of all crystals of a substance are the same, independent of the size and macroscopic shape of the crystal. Therefore these angles characterize the substance. Reichert and Brown found that the patterns of divergence of these angles correlated with the evolutionary tree of the species of fishes. They even found differences between crystals of deoxy- and oxyhaemoglobin.

We can now interpret and appreciate these observations. The formation of crystals implies that the molecules can take up a definite structure, able to pack into regular arrays. The differences in interfacial angles imply that crystals of haemoglobins from different fishes have different structures. The correlation of the divergence patterns of the crystals and the species implies that evolution is shaping molecules as well as bodies, in parallel processes. The differences between crystals of deoxy- and oxyhaemoglobin imply that the protein undergoes a conformational change upon binding oxygen.

Some 50 years later, Perutz announced the solution of the X-ray crystal structure of haemoglobin.

human and horse α-chains, phenylalanine in the human and horse β-chains, and lysine in sperm-whale myoglobin. The reader can easily identify other such positions. The α-chains are more similar to each other than to the β-chains or to myoglobin, the β-chains are more similar to each other than to the α-chains or to myoglobin, but the α- and β-chains are more similar to each other than either is to myoglobin. Given another mammalian globin sequence, it would be easy to identify it as a haemoglobin α-chain, a haemoglobin β-chain, or a myoglobin. This is consistent with the evolutionary tree shown in Fig. 4.17. Because myoglobin and the α- and β-chains of haemoglobin

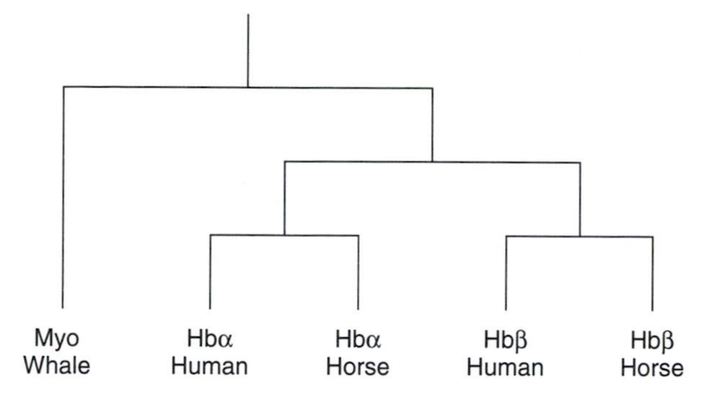

Fig. 4.17 Evolutionary tree of five mammalian globins. This tree shows only the topology of the relationships. By varying the lengths of the links it would be possible to suggest the relative evolutionary distances also.

The split between myoglobin and haemoglobin probably occurred about 800 million years ago. The split between horse and human probably occurred about 70 million years ago.

are **paralogues**, this tree represents the evolutionary divergence of the molecules, not the species.

Orthologue–paralogue distinction: see page 127.

The variation in structure confirms these implications of the patterns of sequence variation. The structures also permit interpretation of the roles of conserved residues. For example, the iron-linked proximal histidine, in the F-helix, is absolutely conserved, for functional reasons. Some residues, including the Phe at position 46 in the tables, are required for haem binding. A mutant human haemoglobin, Hb Hammersmith, having the mutation Phe→Ser at this residue in the β-chain, is unstable and easily loses its haem group. A feature of the pattern of sequence conservation is that several sets of conserved residues are separated by 3, 4, or 7 positions in the sequence. Because the α-helix has a periodicity of 3.6 residues per turn, these residues appear on the same face of a helix. This suggests that this face is involved in an important and conserved structural interaction.

Are the conservation patterns of mammalian globins intrinsic to all globins or are many of them common to only this narrow set of close relatives? Can we say what detailed intramolecular interactions create the globin fold, and how they are compatible with so many different amino acid sequences?

The sequences in Fig. 4.15(b) include the same five mammalian globins, plus additional homologues from an insect, a plant, and a bacterium. Figure 4.18 shows the superposition of three of the structures. The sequences shown in Fig. 4.15(b) are much more diverse—the bacterial globin is the most distant from the others—and indeed only four positions are conserved in all eight. Could these be the irreducible requirements for the globin structure?

The structures of globins

Protein architectures are stabilized by packing of residues in the protein interior. In the case of globins, most buried residues appear at interfaces between helices (see Box, p. 148). Five helix packings occur, with extensive interfacial contact, in full-length globins: A/H, B/E, B/G, F/H, and G/H. Other helix packings occur in some, but not all. In full-length

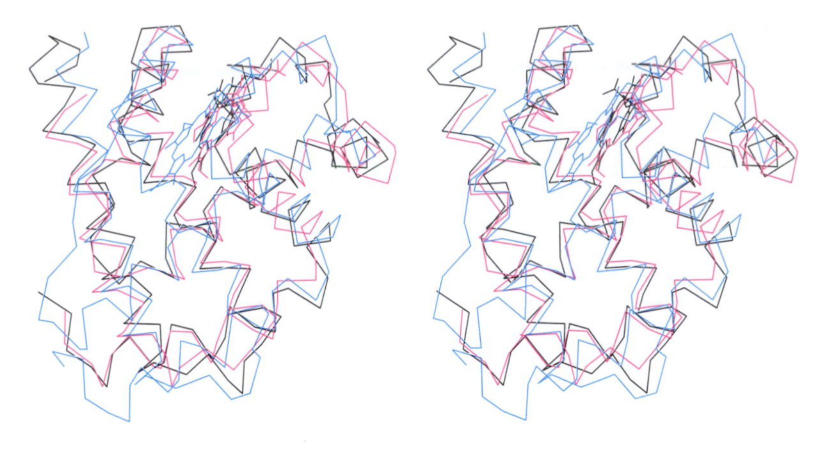

Fig. 4.18
Superposition of three more distantly related globins from: sperm-whale (black), insect (*Chironomus*) (magenta), and plant (yellow lupin) (cyan).

Structure of helix–helix interfaces

Helix–helix packings in globular proteins generally follow a *ridges-into-grooves model*:

- Side chains on the surfaces of helices form ridges protruding from the helix surface.
- These ridges may form from residues separated by 4 positions in the sequence (most common), 3 positions, or 1 position.
- Ridges on the surface of one helix pack into grooves of the other.
- Different combinations of ridge–groove structures produce different interaxial angles in packed helices. The most common packing—$i \pm 4$ ridges from both helices—corresponds to an interaxial angle of about $-40°$ (see Fig. 4.19).

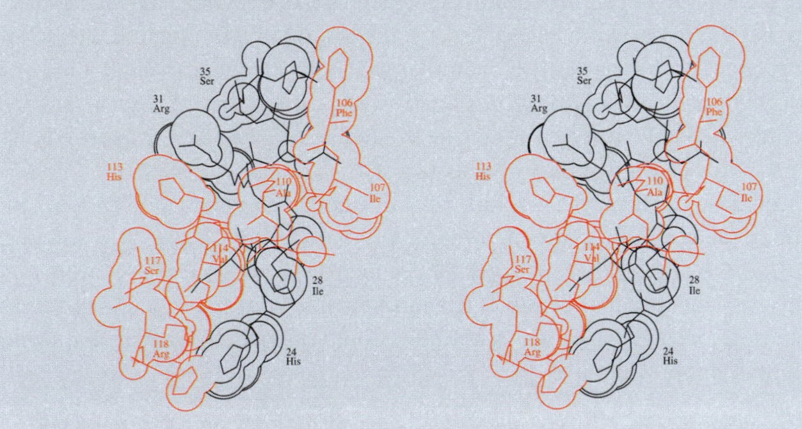

Fig. 4.19 The helix-B–helix-G contact in sperm-whale myoglobin. Residues from helix-B in black; residues from helix-G in red. The axis of the B-helix is vertical. The axis of the G-helix runs from lower left to upper right. Note the diagonal ridge formed by the side chains of helix-G residues 106–110–114–118, packed in a groove between ridges formed by helix-B residues 24–28 and 31–35. This packing of ridges formed by residues separated by four positions in the sequence from both helices is the most common packing pattern observed at interfaces between α-helices.

globins the structures of the helix interfaces—the ridge and groove patterns (see Box)—are conserved, conferring sequence and structural specificity.

Comparative analyses of structures from many different organisms suggested the following picture of sequence–structure relationships and evolution in full-length globins:

1. The principal determinants of the three-dimensional structure of the full-length globins lie in approximately 60 residues that are involved in the packing of helices and in the interactions between the helices and the haem group.
2. Although mutations of the buried residues keep the side chains non-polar, the side chains vary in *size*.
3. In response to mutations at interfaces, the assembly of helices makes structural adjustments. Shifts in the relative position and orientation of homologous pairs of packed helices may be as much as 7 Å and 30°.
4. Despite the change in volume of residues at helix interfaces, and the relative shifts and rotations of the helices packed, there is substantial conservation of the reticulation of the residues; that is, homologous residues tend to make homologous contacts. This includes the retention of ridge–groove patterns at the helix interfaces.
5. Despite the large changes in the relative positions and orientations of the helices, the structures of the nine haem pockets are very similar. The shifts in the helix packings produced by mutations are coupled to maintain the relative geometry of the residues that form the haem pocket.

These conclusions, first proposed almost 30 years ago, were challenged recently when the structures of a class of shortened globin structures appeared. These may be as small as 116 residues.

Truncated globins

Truncated globins are short proteins, occurring in prokaryotes and eukaryotes, that maintain a recognizable globin fold despite containing only ~120 residues, substantially smaller than the ~150 residues of typical full-length globins. They have been implicated in diverse functions, including detoxification of NO, and photosynthesis.

Which residues, and which structural elements, are sacrificed? Truncated globins retain most, but not all, of the helices of the standard globin fold, with the notable exception of the loss of the F-helix, which contains the iron-linked histidine. They show a shortening of the A-helix and of the CD region. Of the 59 sites involved in conserved helix-to-helix or helix-to-haem contacts in full-length globins, 41 of them appear, with conserved contacts, in truncated globins. The helix–helix interfaces have ridge–groove packing patterns similar to those of the full-length globins, with the exception of the B/E contact, which has an unusual crossed-ridge structure in full-length globins but is normal in truncated globins.

A structural difference clearly affecting function involves the accessibility of the ligand-binding site. In typical full-length globins such as myoglobin the O_2 binding site is blocked, and the protein must partially unfold to permit ligand entry and exit. In contrast, in the truncated globins, channels link the ligand-binding site to the surface of the molecule.

Despite the differences, it is possible to align the truncated and full-length globins (Figs 4.20 and 4.21).

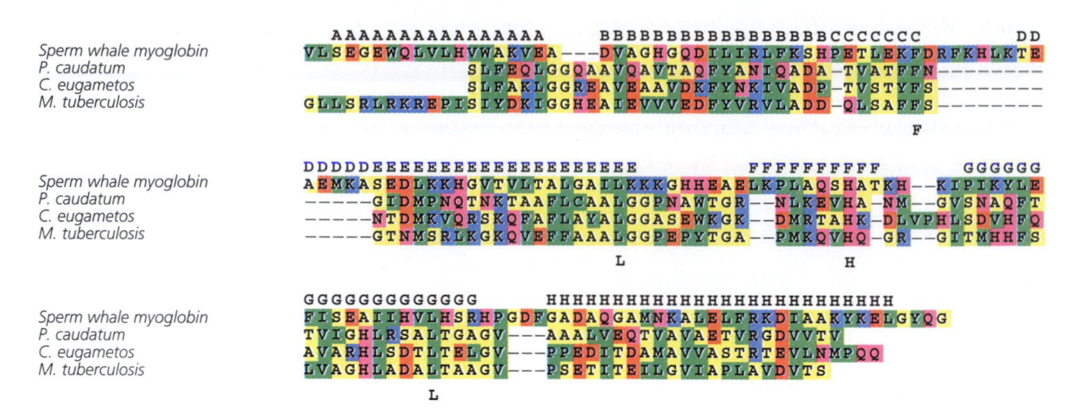

Fig. 4.20 Alignment of the sequences of sperm-whale myoglobin, and truncated globins from *Paramecium caudatum, Chlamydomonas eugametos*, and *Mycobacterium tuberculosis*. Letters on the top line indicate the extents of the helices in the sperm-whale myoglobin structure.

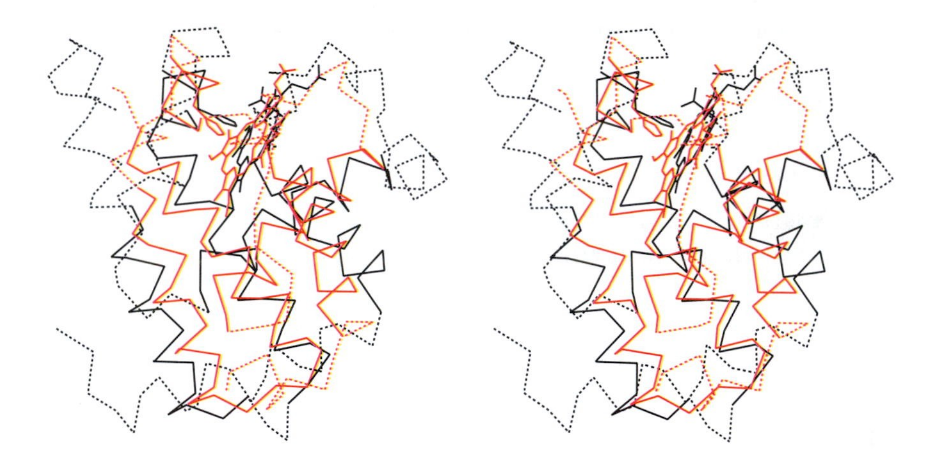

Fig. 4.21 Superposition of the structures of sperm-whale myoglobin [1MBO] and truncated globin from *Paramecium caudatum* [1DLW]. Solid lines indicate the parts of the structure common to full-length and truncated globins. Broken lines indicate the parts of the structure absent from truncated globins or changed in conformation.

Evolution of NAD-binding domains of dehydrogenases

In 1973, knowing the structures of lactate, malate, and alcohol dehydrogenases, C.-I. Brändén, H. Eklund, B. Nordström, T. Boiwe, G. Söderlund, E. Zeppezauer, I. Ohlsson, and Å. Åkeson wrote:

> The coenzyme binding region [of horse liver alcohol dehydrogenase] has a main chain conformation very similar to a corresponding region in lactate and malate dehydrogenase. It is suggested that this substructure is a general one for binding of nucleotides and, in particular, the coenzyme NAD^+.

Since then, many additional crystal structures have confirmed and extended this principle. The paradigm nucleotide-binding domain in horse liver alcohol dehydrogenase contains two sets of β–α–β–α–β-units, together forming a single parallel β-sheet flanked by α-helices (Fig. 4.22). The strands appearing from right to left in Fig. 4.22 appear in the *sequence* in the order 6–5–4–1–2–3 (see Fig. 4.23). There is a long loop, or crossover, between strands 3 and 4. As Brändén described in 1980, this feature of the fold creates a natural cavity for binding of the adenine ring of the NAD and, in other proteins with similar supersecondary structures, for other nucleotide-containing fragments.

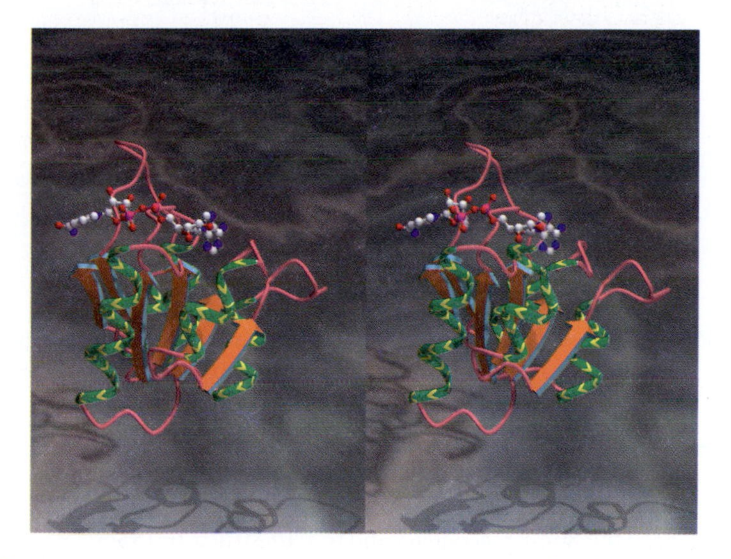

Fig. 4.22 NAD-binding domain of horse liver alcohol dehydrogenase [6ADH]. The folding pattern includes a central parallel β-sheet flanked by helices. Domains with this folding pattern appear in many dehydrogenases and in other proteins that bind related molecules.

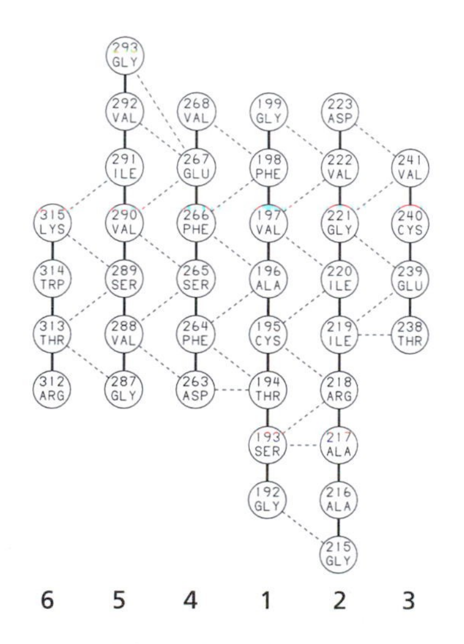

Fig. 4.23 Hydrogen-bonding pattern of the β-sheet in horse liver alcohol dehydrogenase [6ADH].

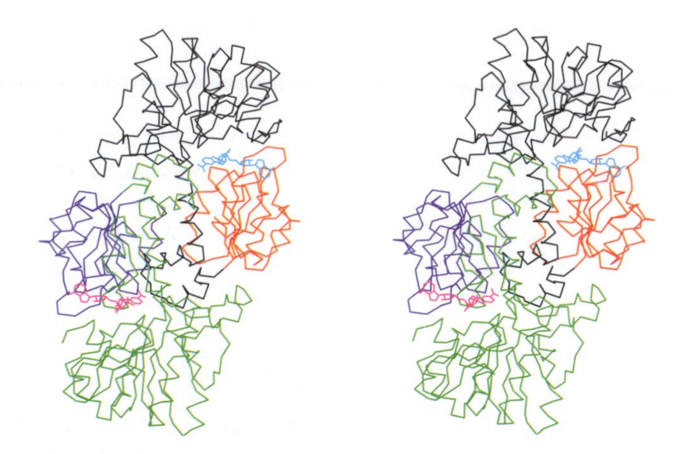

Fig. 4.24 The dimeric structure of formate dehydrogenase, viewed down the axis of symmetry [2NAD]. One subunit is black, with its NAD-binding domain highlighted in red. The other is green, with its NAD-binding domain highlighted in blue.

NAD-binding domains combine with other domains in enzymes with very different substrate specificities. They occur as oligomers of different sizes; many are tetramers. Figure 4.24 shows formate dehydrogenase, a dimer. The NAD-binding domains in each monomer are highlighted.

Proteins other than dehydrogenases bind nucleotide-containing cofactors in a manner similar to the binding of NAD to dehydrogenases, and create the binding site from domains with generally similar secondary and tertiary structure. Examples include the FMN in flavodoxin and the FAD in pyruvate oxidase. Conversely, other structures have extended the repertoire of modes of binding NAD and related ligands. Some have folding patterns very different from the NAD-binding domains of dehydrogenases. These include proteins from other general topological classes: all-β and $\alpha + \beta$ as well as proteins in the α/β class but unrelated to the dehydrogenase NAD-binding fold.

Comparison of NAD-binding domains of dehydrogenases

Figures 4.25–4.30 illustrate six NAD-binding domains, showing the folding of the chain and the hydrogen-bonding net of the sheet. These give some idea of the observed structural variety.

In each of these domains, the sheets contain the canonical six strands, but are extended by additional strands. Dihydropteridine reductase has a seventh strand adjacent and parallel to the sixth strand, and an eighth strand forming a hairpin with the seventh. 6-Phosphogluconate dehydrogenase and formate dehydrogenase each have a seventh strand adjacent but parallel to the sixth. Glyceraldehyde-3-phosphate dehydrogenase has a short stretch of antiparallel sheet between the third and fourth strand, before the 'crossover'. 3α,20β-hydroxysteroid dehydrogenase has a seventh strand adjacent and parallel to the sixth.

The helices also differ considerably in length, and can shift relative to the sheet. In 3α,20β-hydroxysteroid dehydrogenase and dihydropteridine reductase the two

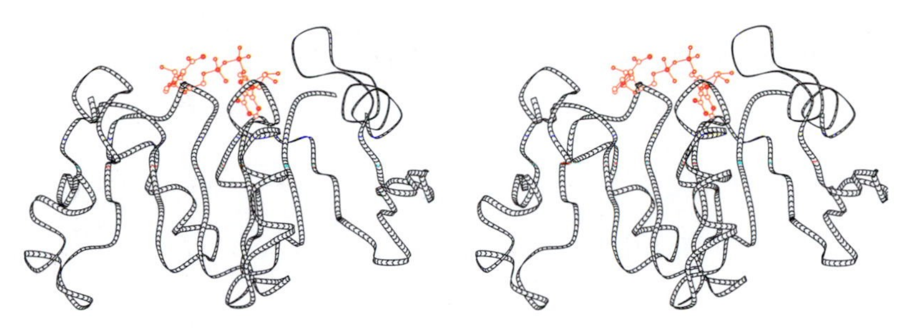

(a)

(b)

6 5 4 1 2 3

Fig. 4.25 (a) NAD-binding domain of glyceraldehyde-3-phosphate dehydrogenase [1GD1]. (b) Hydrogen-bonding pattern of the β-sheet.

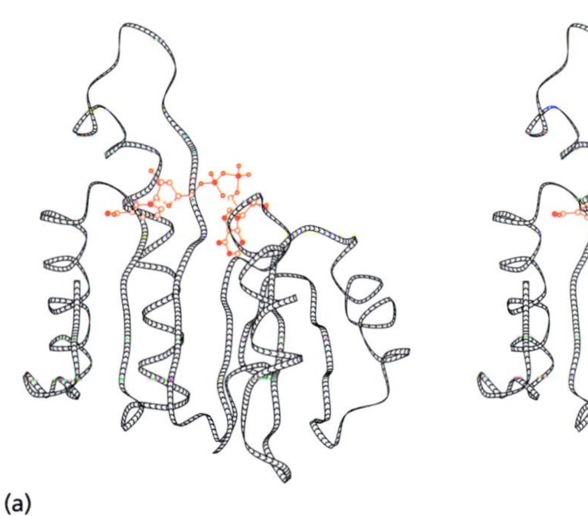

(a)

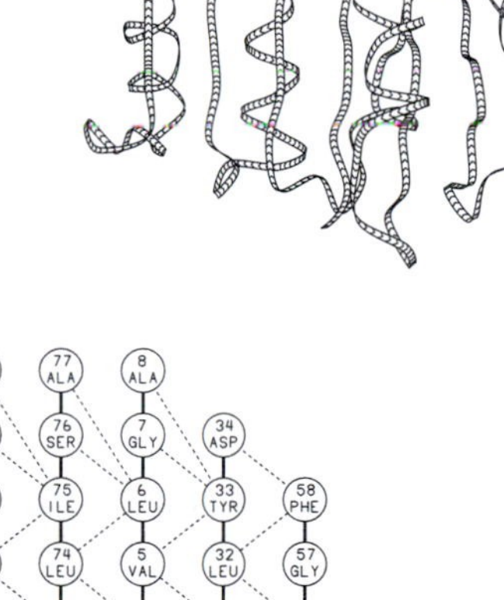

(b)

6 5 4 1 2 3

Fig. 4.26 (a) NAD-binding domain of malate dehydrogenase [1EMD]. (b) Hydrogen-bonding pattern of the β-sheet.

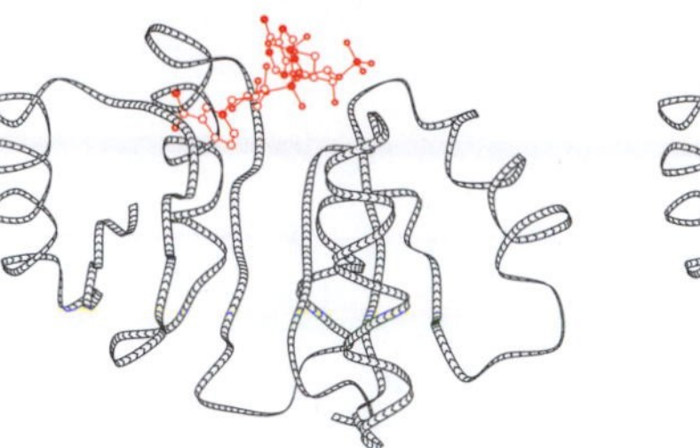

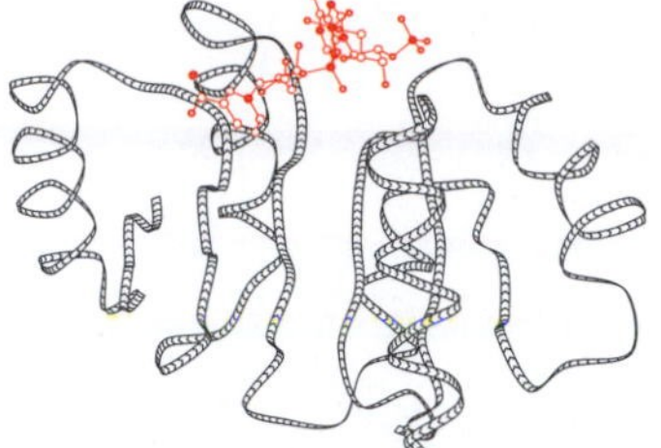

(a)

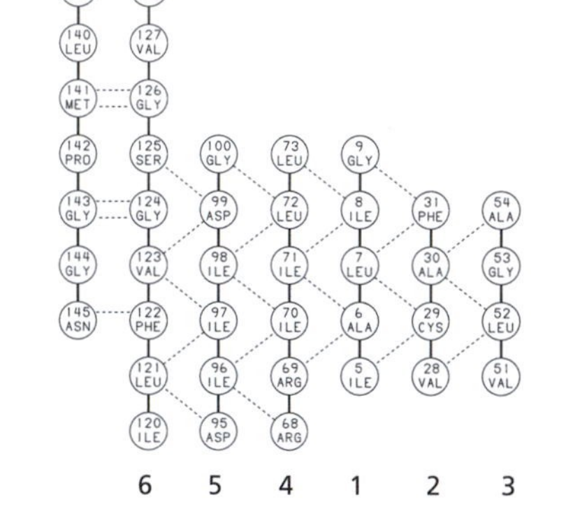

(b) 6 5 4 1 2 3

Fig. 4.27
(a) NAD-binding domain of 6-phosphogluconate dehydrogenase [1PGO]. (b) Hydrogen-bonding pattern of the β-sheet.

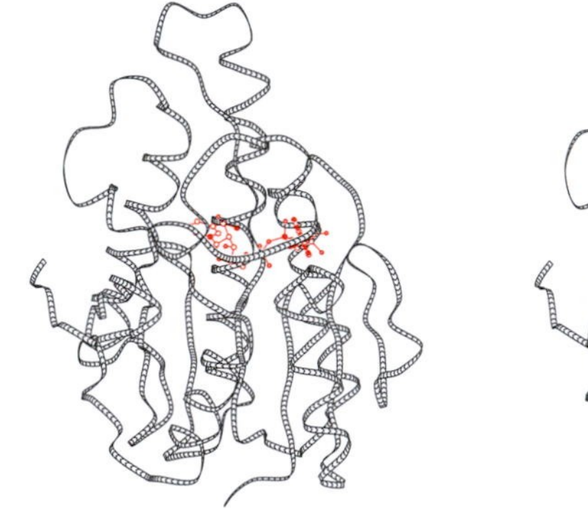

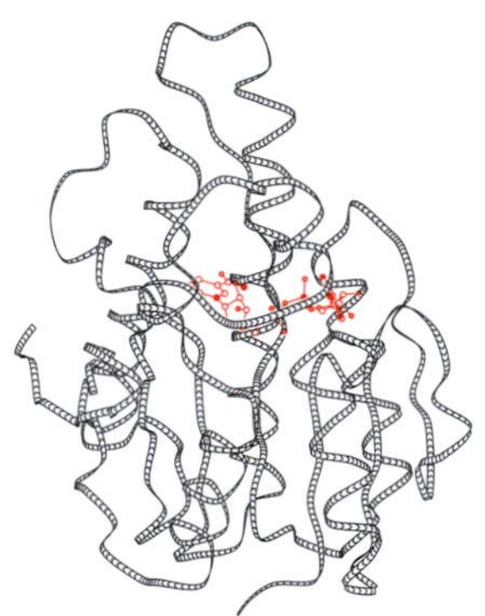

(a)

(b) 6 5 4 1 2 3

Fig. 4.28
(a) NAD-binding domain of dihydropteridine reductase [1DHR]. (b) Hydrogen-bonding pattern of the β-sheet.

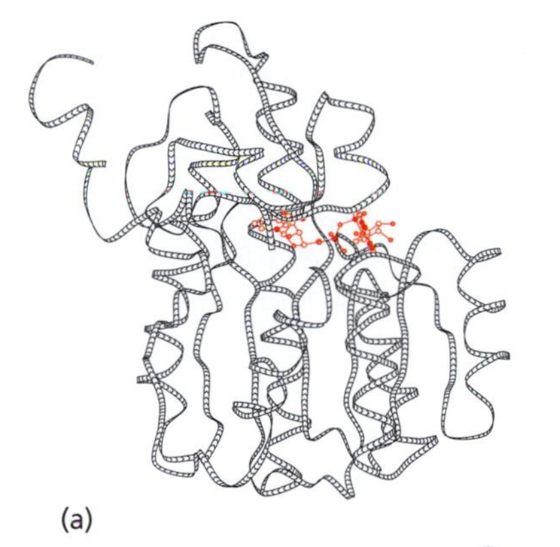

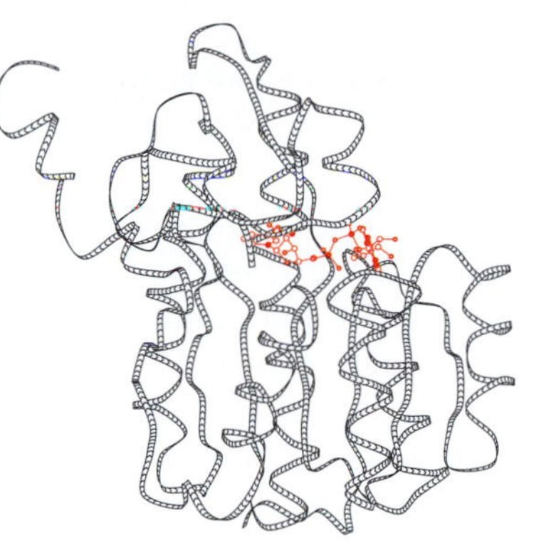

(a)

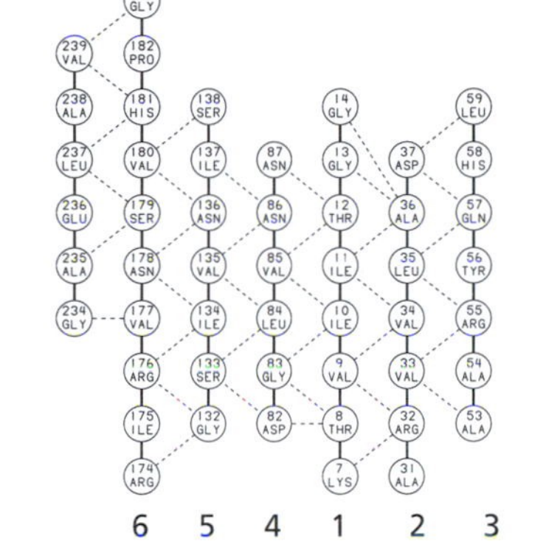

(b)

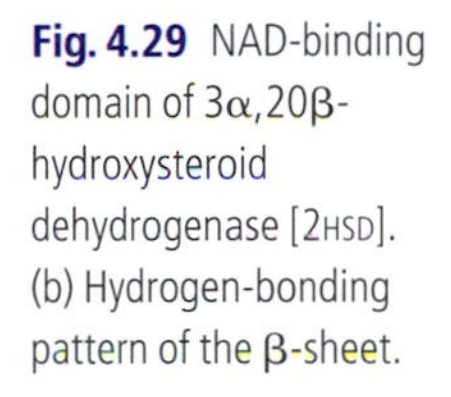

Fig. 4.29 NAD-binding domain of $3\alpha,20\beta$-hydroxysteroid dehydrogenase [2HSD]. (b) Hydrogen-bonding pattern of the β-sheet.

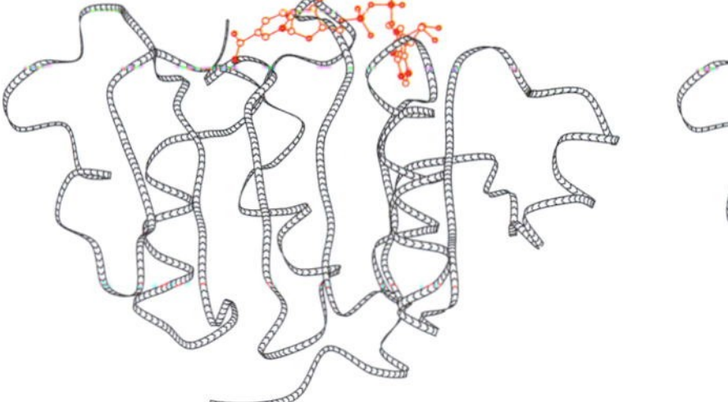

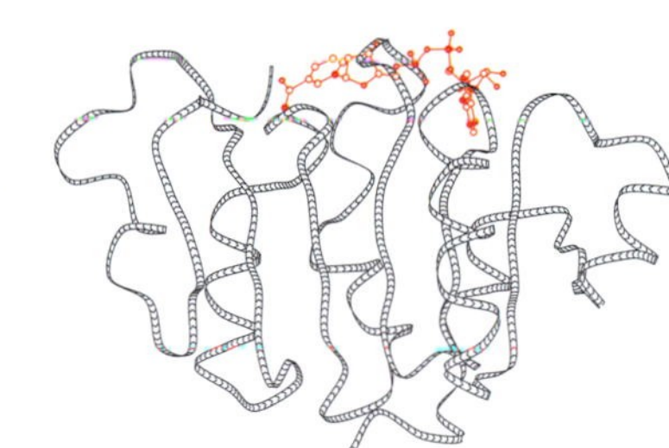

(a)

254 ASN, 198 ALA
281 ASN, 253 LEU, 197 VAL, 220 THR, 238 HIS
328 GLY, 306 ALA, 280 VAL, 252 THR, 196 THR, 219 TYR, 237 TRP
327 ASN, 305 TYR, 279 ILE, 251 VAL, 195 GLY, 218 HIS, 236 THR
326 TYR, 304 GLY, 278 TYR, 250 VAL, 194 VAL, 217 LEU, 235 LEU
303 ALA, 277 ALA, 249 ASP, 193 HIS, 216 HIS
192 MET, 215 VAL

6 5 4 1 2 3

(b)

Fig. 4.30 (a) NAD-binding domain of formate dehydrogenase [2NAD]. (b) Hydrogen-bonding pattern of the β-sheet.

helices in the C-terminal portion of the domain (between strands 4 and 5, and between strands 5 and 6) are elongated. There is also a helix in the crossover region, which appears in many NAD-binding domains. Dihydropteridine reductase has lost the helix between strands 2 and 3.

What is the conserved core of the family? Figure 4.31 shows the superposition of NAD-binding domains of a closely related pair of enzymes (lactate and malate dehydrogenases) and a more distantly related pair (alcohol dehydrogenase and dihydropteridine reductase). The crossover region—the loop between strands 3 and 4—is especially variable in structure among the different enzymes. The maximal common substructure of the distantly related pair corresponds to the core of the members of this family of domains illustrated here.

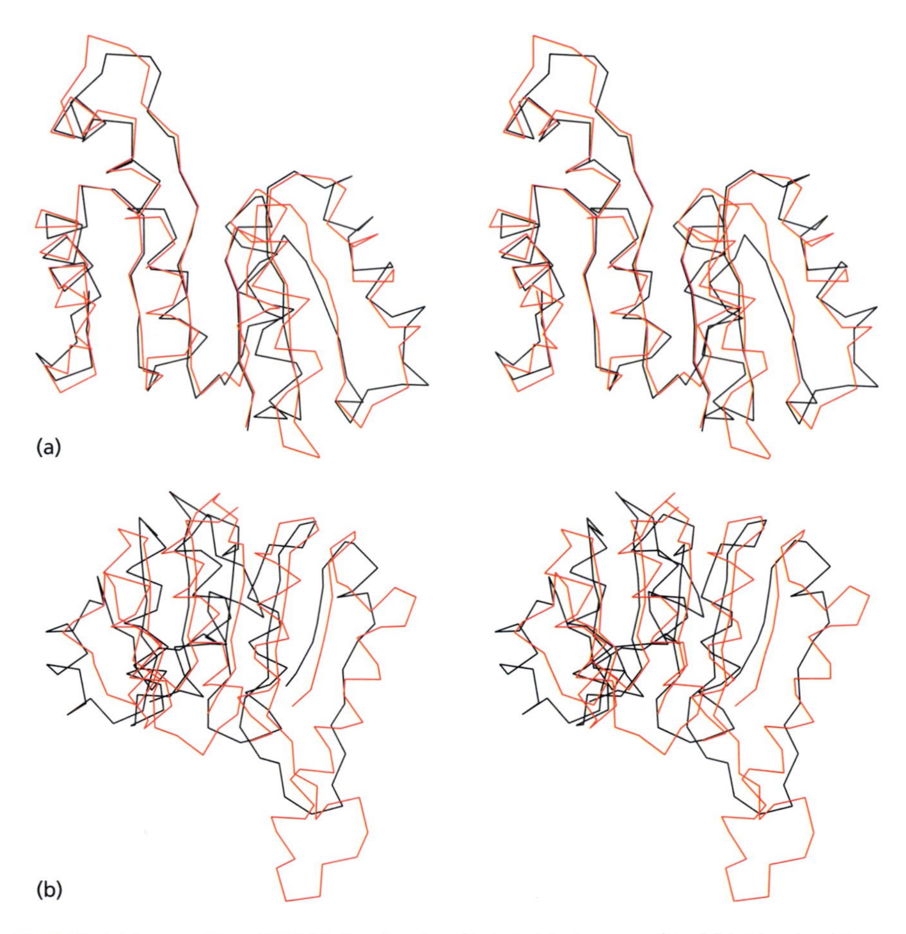

Fig. 4.31 (a) Superposition of NAD-binding domains of lactate dehydrogenase [9LDT] (black) and malate dehydrogenase [1EMD] (red). The sequences of these regions have 23% identical residues upon optimal alignment. Although these molecules have developed different functions, they are still fairly closely related. (b) Superposition of NAD-binding domains of horse liver alcohol dehydrogenase [2OHX] (black) and dihydropteridine reductase [1DHR] (red). These molecules have diverged more radically. There are only 14% identical residues in an optimal alignment.

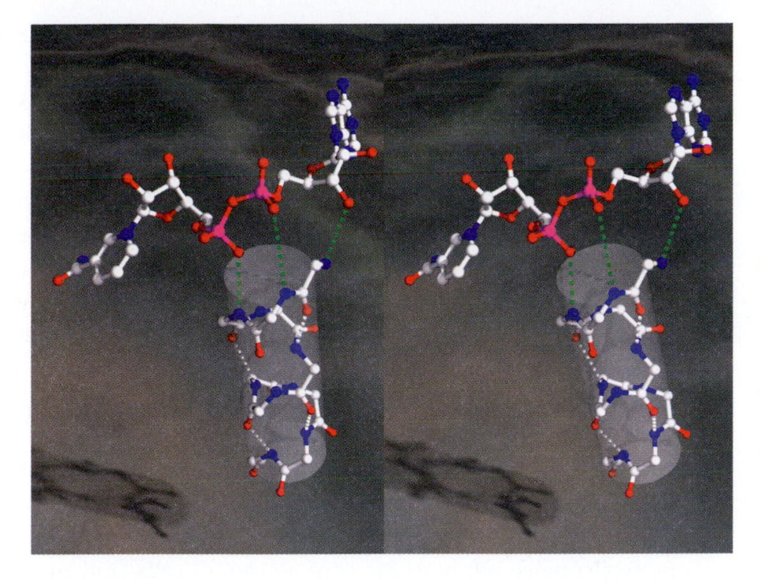

Fig. 4.32 Interactions between protein and ligand in NAD-binding domains commonly include hydrogen bonding between phosphate oxygens and the N–H groups in the last turn of the first helix.

*The sequence motif G*G**G*

The binding of NAD to dehydrogenases involves numerous hydrogen bonds and Van der Waals contacts between the cofactor and the enzyme. In particular, there are usually hydrogen bonds from the residues in the turn between the first strand and the helix that follows it, to one of the phosphate groups of the cofactor (Fig. 4.32). These interactions give rise to a consensus sequence in this region containing the three-glycine pattern G*G**G characteristic of the first β–α–β-unit in the dehydrogenase domain. (This motif signifies a hexapeptide: gly–xxx–gly–xxx–xxx–gly, where xxx represents any residue.)

The first two glycines are involved in nucleotide binding and the third, which is in the helix following the first strand, is involved in the packing of the helix against the sheet. The first glycine is in a ++ conformation integral to the structure of the turn. A Cβ at the position of the second glycine would collide with the cofactor. In horse liver alcohol dehydrogenase, a Cβ in the residue at the position of the third glycine would clash with the carbonyl of the first glycine.

The dehydrogenases contain a well (but not absolutely) conserved aspartate approximately 20 residues C-terminal to the G*G**G motif. This aspartate appears near the C-terminus of the second strand, and forms hydrogen bonds to the ribose of the adenosine moiety of the NAD.

Evolution of visual pigments and related molecules

Visual pigments have evolved to absorb and process ambient light. Their spectroscopic properties are finely tuned. Different species occupy different light environments, and the spectral sensitivities of their visual pigments are adjusted to the quality and intensity of the ambient light. Humans occupy two light environments—day and night—and our visual pigments have diverged to support both cone vision (high illumination, colour-sensitive, high resolution), and rod vision (low illumination, colour-insensitive, lower resolution). Our eyes are sensitive to light throughout the visible region (of course this is a tautology), but our visual pigments would also react to ultraviolet light, were it not for the photoprotective filter of the lens. We can deal with a dynamic range of 11 orders of magnitude, from 10^{10} photons $m^{-2} s^{-1}$—the dimmest stars visible in the night sky—to 10^{21} photons $m^{-2} s^{-1}$—bright sunlight.

The complex of opsin and retinal in vertebrate rods is called '*rhodopsin*', because of its colour.

The photoreceptor proteins in vision are **opsins**, membrane proteins about 360 residues in length with the seven-helix motif common to G-protein-coupled receptors. Opsins combine with **chromophores** to form light-sensitive complexes (Fig. 4.33). In vertebrate visual systems, the chromophore is 11-*cis*-retinal (Fig. 4.34). Opsin complexes with very similar chromophores are common to the visual systems of vertebrates, insects, and cephalopods.

Vision begins with absorption of light by retinal. Within about a picosecond, retinal isomerizes, causing a conformational change in the opsin that initiates a signal cascade resulting, ultimately, in the triggering of a nerve impulse (see Fig. 4.34, and Box).

G-protein-coupled receptors are common to all cells, and their specialization to photoreceptors occurred early in the history of life. Homologous proteins, binding chromophores related to retinal, appear in Bacteria and Archaea as well as in the Eukaryotes. Figure 4.35 shows the relationships among the known homologues in higher eukaryotes. All have the same arrangement of the seven helices, and a common locus and mode of attachment of the chromophore.

Many homologues, even those in simple organisms, have functions related to primitive 'visual' sensory phenomena, for example, phototaxis in algae. In contrast, bacteriorhodopsin from *Halobacterium salinarum* is not a sensory transducer but a light-driven proton pump, converting light energy to an electrochemical gradient across the bacterial plasma membrane. This electrochemical potential energy can be harvested as ATP. Although the *H. salinarum* system is the best known from laboratory

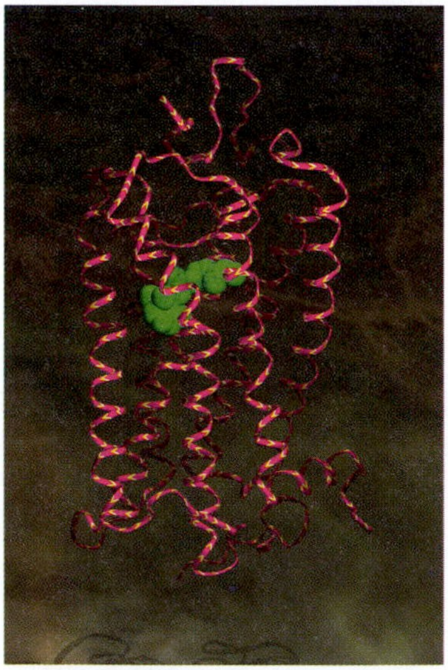

Fig. 4.33
Bovine opsin, a 7-transmembrane helix G-protein-coupled receptor [1F88].

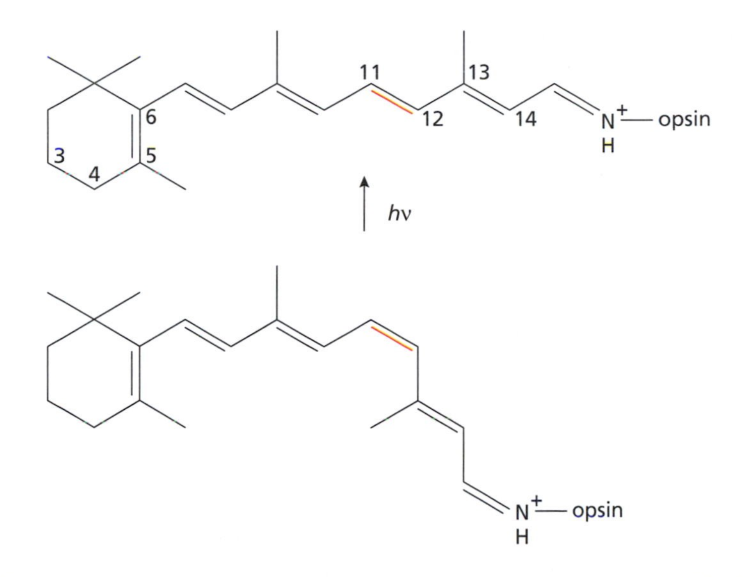

Fig. 4.34 Conformational change in retinal attached to opsin upon absorption of light. One of the double bonds, indicated in red, changes from the 11-*cis* conformation in the resting state (bottom) to the all-*trans* conformation. Some homologues of vertebrate opsins, including bacteriorhodopsin and the photoreceptors of green algae, bind retinal in an alternative conformation, with a different stereochemistry of the attachment of the polyene tail to the ring, the resting state is all-*trans* and the photochemical isomerization takes place at the 13–14 double bond. The visual pigments of many reptiles, amphibia, and fishes contain 3,4-dehydroretinal instead of retinal. *Drosophila* rhodopsin contains 3-hydroxyretinal.

Molecular mechanism of vision taking place as you read this box

- Absorption of light by retinal.
- Isomerization of retinal from 11-*cis* to all-*trans* (Fig. 4.34). The all-*trans* isomer has a different shape, which doesn't fit the binding site well. The protein changes conformation, and releases the retinal.
- The new conformation of the rhodopsin activates the G-protein transducin.
- Transducin activates a phosphodiesterase.
- The phosphodiesterase hydrolyses cyclic GMP.
- cyclic-GMP gated cation channels close.
- There is a build-up of potential difference across the cell membrane.
- The potential difference is transmitted to an adjacent neuron (in vertebrates, via a bipolar cell) triggering a nerve impulse.
- Feedback mechanisms affecting each of these steps restore the system to its resting state.

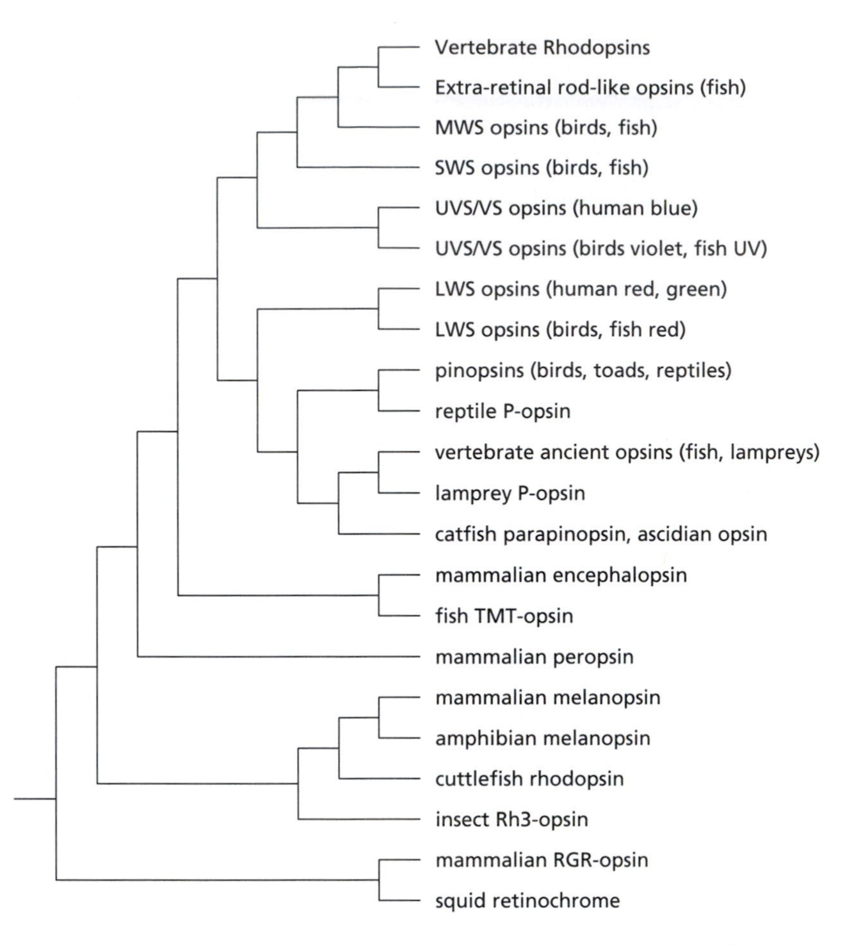

Fig. 4.35 Relationships among visual opsins and other non-image-forming photoreceptors of higher eukaryotes. *Vertebrate rhodopsins*—visual pigments of rods in vertebrates. *Extra-retinal rod-like opsins*—pigments related to rhodopsins but probably not involved directly in image formation. *MWS opsins*—medium-wavelength sensitive opsins in cone vision of birds and fish. *SWS opsins*—short-wavelength sensitive opsins in cone vision of birds and fish. *UVS/VS opsins*—ultraviolet-and violet-sensitive opsins in cone vision of vertebrates. Human blue cone pigment is related to these. *LWS opsins*—long-wavelength sensitive opsins in cone vision of vertebrates. Human red and green cone pigments are related to these. *Pinopsins*—light detectors in the pineal glands of birds, reptiles, and amphibia; function related to circadian rhythms. *Vertebrate ancient options*—opsin expressed in some retinal cells and pineal gland; may function in non-image-forming light detection. *Mammalian encephalopsin*—opsin expressed in mammalian brains; function may be related to circadian rhythms. *Mammalian peropsin*—function unknown (some authorities group mammalian peropsin with mammalian RGR – opsin). *Melanopsin*—probably involved in circadian rhythms. *Cuttlefish rhodopsin*—visual pigment in cephalopods. *Insect Rh3 opsin*—visual pigment in insects. *RGR opsins and retinochrome*—function in recovery of receptive state of visual pigments, by photoreversal of the isomerization shown in Fig. 4.34, possibly signalling also.

investigation and structure determinations, light-driven proton pumps appear in numerous other organisms, including eukaryotes.

Many homologues, even those in higher organisms, have non-visual functions. Pinopsins, found in the pineal glands of birds and some toads, are **circadian**

rhythm photoreceptors. A human homologue, encephalopsin, is likely to share this function.

These observations contain the detailed rebuttal to critics of Darwin who argued that it was unreasonable to think that a complex organ such as the eye could have evolved by accumulation of small variations. Their point was that vision is impossible without the full complexity of the system, and that there is no selective advantage to any *parts* of the system that do not provide visual function. Darwin admitted that the problem of how selection could account for such highly adapted organs had sometimes given even him a '*cold shudder*'.* However, the responses to light of primitive organisms—much simpler than our high-resolution sensory imaging systems—are based on homologues of our opsins, and delineate the evolutionary pathway through which the complexity could arise.

In vertebrate visual systems, multiple opsin genes have diverged to create visual pigments with absorption maxima at different wavelengths, to support colour vision. The ancestral vertebrate visual system was probably established early in vertebrate evolution, about 350–400 million years ago. It was based on four types of cone pigments, with absorption maxima in the red, green, blue, and violet/UV, plus the rod pigment rhodopsin. The cone pigments are called SWS, MWS, LWS, and UVS/VS opsins. The abbreviations stand for short-wavelength sensitive, middle-wavelength sensitive, long-wavelength sensitive, and ultraviolet-sensitive/violet sensitive. This full complement survives in some modern fishes, amphibians, reptiles, and birds.

When placental mammals adopted nocturnal habits, two of the four cone opsins were lost. Marsupials retain three. Some nocturnal mammals have lost their cones entirely, for example hedgehogs and bats. Most mammals have the equivalent of human red–green colour blindness. Snakes, during their evolution, went through a nocturnal stage in which cones were lost; then they became diurnal again and recovered cones.

About 35 million years ago, primates reinvented trichromatic colour vision, by gene duplication and divergence of the red pigment (see Fig. 4.36). In the human genome, the genes encoding the red and green cone pigments appear in tandem on the X chromosome. They have 96% sequence identity. These red and green opsins have 40% sequence identity with the opsin of the violet-sensitive pigment. It has been suggested that primate colour vision and the colour changes associated with the ripening of fruit coevolved.

* Yet in the end Darwin got it exactly right: 'To suppose that the eye, with all its inimitable contrivances for adjusting the focus to different distances, for admitting different amounts of light, and for the correction of spherical and chromatic aberration, could have been formed by natural selection, seems, I freely confess, absurd in the highest possible degree. Yet reason tells me, that if numerous gradations from a perfect and complex eye to one very imperfect and simple, each grade being useful to its possessor, can be shown to exist; if further, the eye does vary ever so slightly, and the variations be inherited, which is certainly the case; and if any variation or modification in the organ be ever useful to an animal under changing conditions of life, then the difficulty of believing that a perfect and complex eye could be formed by natural selection, though insuperable by our imagination, can hardly be considered real.' (*Origin of species*, Chapter 6.)

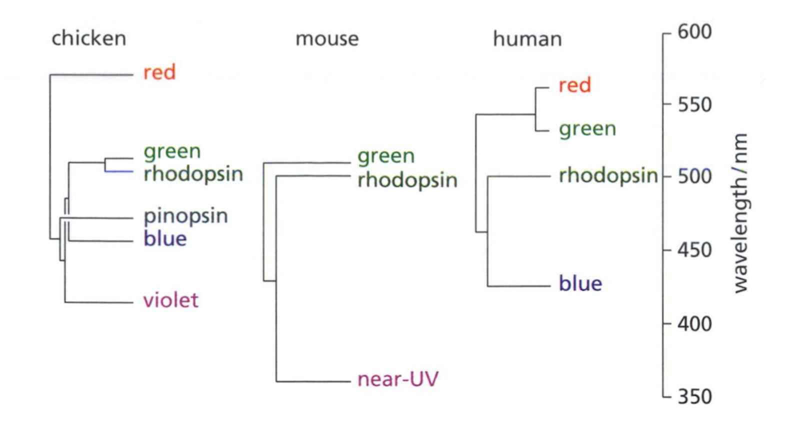

Fig. 4.36 Evolutionary relationships, and distribution of colours, of opsins and related molecules. (After Nathans, J. (1999). The evolution and physiology of human colour vision: insights from molecular genetic studies of visual pigments. *Neuron*, **24**, 299–312.)

Our current colour vision is based on *three* pigments. The visual pigments in human cones have λ_{max} = 420, 530, 558 (552 or 557, depending on the allele). People lacking any of them have deficiencies in colour vision.

It is not unreasonable to think that colour vision, requiring interaction between signals from several visual pigments, might have developed from an ancestral system similar to our colour-insensitive rod vision. However, it seems clear from phylogenetic trees of the sequences that ancestral vertebrates had four groups of cone visual pigments, and that rod opsins evolved from one of these, the middle-wavelength-sensitive opsin, in the Mesozoic.

Selection and vertebrate opsins

Selection has tuned vertebrate opsins so that the absorption maximum matches the light environment. Different qualities of light characterize the habitats of different species of animals. Water filters out red light preferentially. Therefore land animals, fishes living near the surface of the ocean, and deep-sea fishes live in increasingly blue environments. Their rhodopsins are selected for maximum absorbance at the prevailing wavelength of the illumination they experience.

Species	Habitat	λ_{max}/nm
Summer flounder	Shallow water	503
Cod	Intermediate depths	496
Lancet fish	Deep water	480

Correlations of sequences and absorption spectra of vertebrate visual pigments allow:

- identification of the residues responsible for altering the wavelength of the absorption maximum (λ_{max}) of the receptor;

- inference of the mechanism by which mutations affect the energy levels of the chromophore; and
- re-creation of hypothetical sequences of the pigments in ancestral species, even extinct ones.

Phylogenetic analysis suggests that the ancestral vertebrate member of the LWS opsin group contained residues: 180S, 197H, 277Y, 285T, and 308A, and had a λ_{max} of 559 nm. Common mutations in these residues shift the value of λ_{max}:

Mutation	S180A	H197Y	Y277F	T285A	A308S	S180A/H197Y
$\Delta\lambda_{max}$/nm	−7	−28	−8	−15	−27	+11

For instance, a molecule with residues AHYTA would be predicted to have $\lambda_{max} = 559 - 7 = 552$ nm. The different mutations exert their effect independently, except for S180A and H197Y, which give an additional contribution of $\Delta\lambda_{max} = 11$ nm when they occur together. This rule predicts the value of λ_{max} from an amino acid sequence quite well.

Pigment	λ_{max}	**Residue at**					λ_{max}	
	Experiment	**180**	**197**	**277**	**285**	**308**	**Predicted**	**Error**
Human	552	A	H	Y	T	A	553	1
Human	530	A	H	F	A	A	530	0
Marmoset	561	S	H	Y	T	A	559	−2
Marmoset	553	A	H	Y	T	A	553	0
Marmoset	539	A	H	Y	A	A	538	−1
Deer	531	A	H	F	A	A	530	−1
Dolphin	524	A	H	Y	T	S	526	2
Horse	545	A	H	F	T	A	545	0
Cat	553	A	H	Y	T	A	553	0
Guinea pig	516	S	Y	Y	A	A	517	1
Squirrel	532	S	Y	Y	T	A	531	−1
Chicken	561	S	H	Y	T	A	559	−2
Pigeon	559	S	H	Y	T	A	559	0
Zebra finch	560	S	H	Y	T	A	559	−1
Chameleon	560	S	H	Y	T	A	559	−1
Gecko	527	A	H	F	A	A	530	3
Frog	557	S	H	Y	T	A	559	2
Goldfish	559	S	H	Y	T	A	559	0

Substitutions at the five sites critical for adjusting spectra of vertebrate LWS and MWS cone pigments. Comparison of measured λ_{max} with prediction based on amino acid sequence. (After Yokoyama, S. and Radlwimmer, F.B. (2001). The molecular genetics and evolution of red and green color version in vertebrates. *Genetics*, **158**, 1697–710.)

Figure 4.37 shows the side chains of the five residues critical for spectral adjustment, in the structure of bovine opsin.

Identification of the mutations with the most important effects on λ_{max} elucidate the mechanism by which mutations tune the absorption spectrum. The spectrum

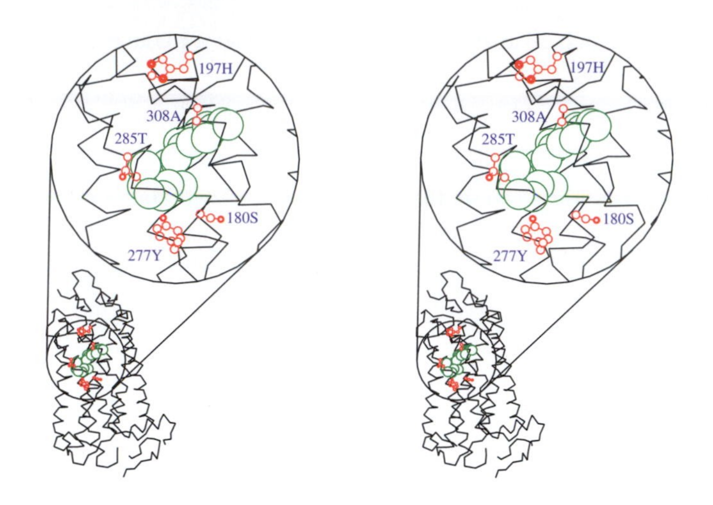

Fig. 4.37 Bovine opsin, and the five side chains that are critical for tuning the spectra of cone pigments. [1F88], modified. Certain residues in the bovine opsin structure have been substituted and renumbered to create this picture.

depends on the polarity of the environment of the chromophore, created by the nature of the side chains lining the binding site. Changing the polarity alters the energy levels of the chromophore. Differences in the charge distribution of the chromophore in ground and excited states means that a change in polarity of the environment will affect the energies of ground and excited states to different extents. This will alter the *difference* in energy between ground and excited states, and hence the wavelength of the transition between them.

The ability to reconstruct ancestral sequences makes it possible to investigate the physiology of extinct species, at the molecular level. For instance, using phylogenetic trees the sequences of visual pigments of dinosaurs can be reconstructed. It *is* possible to synthesize the proteins, and to verify that they function in the absorption of light and in activating transducin. It is *not* possible to verify that they correctly reproduce the sequences of dinosaur proteins.

How do proteins evolve new functions?

Mechanisms of protein evolution that produce altered or novel functions include: (1) divergence, (2) recruitment, and (3) 'mixing and matching' of domains.

1. Divergence

In families of closely related proteins, mutations usually conserve function but modulate specificity. For example, the trypsin family of serine proteinases contains a specificity pocket: a surface cleft complementary in shape and charge distribution to the side chain adjacent to the scissile bond. Mutations tend to leave the backbone conformation of the pocket unchanged but to affect the shape and charge of its lining, altering the specificity.

The change in specificity of the proteases illustrates a common theme: although homologous proteins show a general drifting apart of their sequences as mutations

CASE STUDY Malate and lactate dehydrogenases

Malate and lactate dehydrogenases are related enzymes that catalyse similar reactions (see below). They arose by gene duplication at an early stage of the history of life, and their sequences have diverged. (In an optimal alignment, human malate and lactate dehydrogenases have ~20% identical residues.) Nevertheless, site-directed mutagenesis showed that a single residue change (Gln→Arg) could change the specificity of *Bacillus stearothermophilus* lactate dehydrogenase to malate (Fig. 4.38). (Reports of that work may have been read by a *Trichomonad*, which developed a malate dehydrogenase that, in an evolutionary tree of these enzymes, is much more similar to lactate dehydrogenases than to other malate dehydrogenases.)

Reactions catalysed by lactate dehydrogenase (LDH) and malate dehydrogenase (MDH)

COO^-						COO^-		
I						I		
C=O	+	NADH	+	H^+	=	HCOH	+	NAD
I						I		
CH_3					*LDH*	CH_3		
pyruvate						lactate		
COO^-						COO^-		
I						I		
C=O	+	NADH	+	H^+	=	HCOH	+	NAD
I						I		
CH_2					*MDH*	CH_2		
I						I		
COO^-						COO^-		
oxaloacetate						malate		

Arg171 forms a salt bridge with the carboxyl group common to the substrates of the two enzymes. A proton is transferred from the side chain of His195 to the C=O group common to the substrates, that is the site of reduction. The NADH supplies the other proton.

The mutation that changes the specificity involves 102Arg/Gln at the lower right of Fig. 4.38(a) and (b). The bulkier Arg does not allow room for the larger substrate, oxaloacetate. It is unfortunate that the inhibitor bound in the malate dehydrogenase structure is smaller than the natural substrate, which makes this effect less clear.

Therefore
Exercise 4.18.

accumulate, often a few specific changes account for functional divergence, an idea that initially emerged from study of haemoglobin evolution.

It is arguable that the relationship between malate and lactate dehydrogenases (see box) is really more a change in specificity, analogous to the serine proteinases, than a change in the reaction. A functionally more highly diverged family of enzymes includes enolase, mandelate racemase, muconate lactonizing enzyme I, and D-glucarate dehydratase. From their sequences, these enzymes are fairly close relatives. They have a common structure, closely related to the TIM-barrel fold (Fig. 4.39).

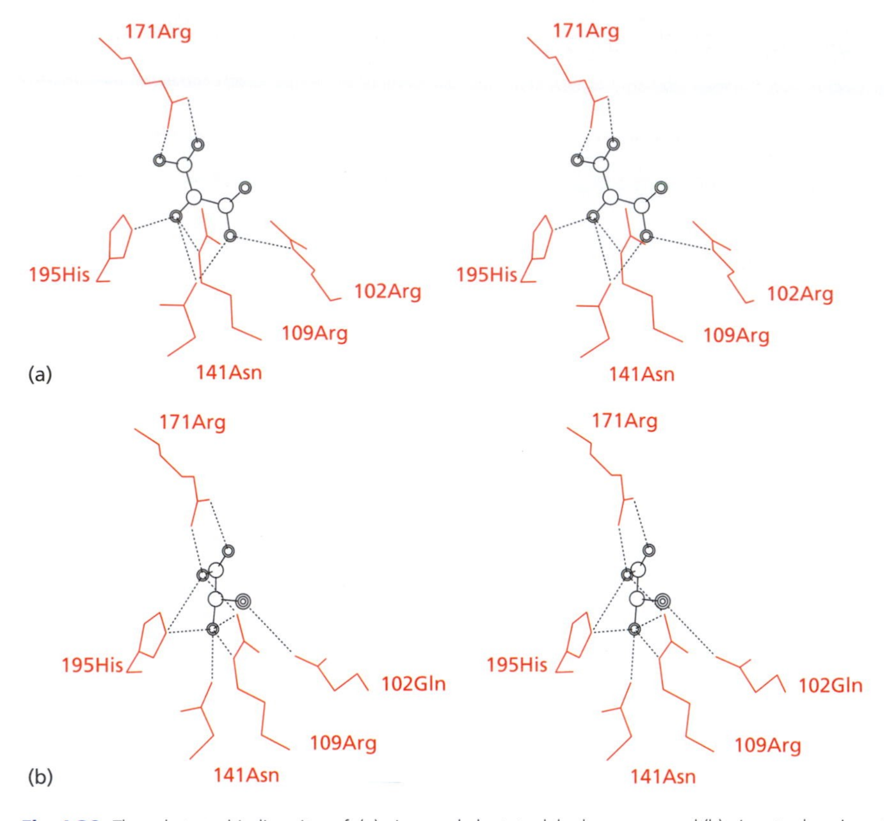

Fig. 4.38 The substrate-binding sites of: (a) pig muscle lactate dehydrogenase, and (b) pig cytoplasmic malate dehydrogenase. The lactate dehydrogenase binds the inhibitor oxamate, CH_3CONH_2, instead of the natural pyruvate product CH_3COCOO^-. The malate dehydrogenase binds the inhibitor α-ketomalonate, $^-OOCCOCOO^-$, instead of the natural oxaloacetate product $^-OOCCH_2COCOO^-$.

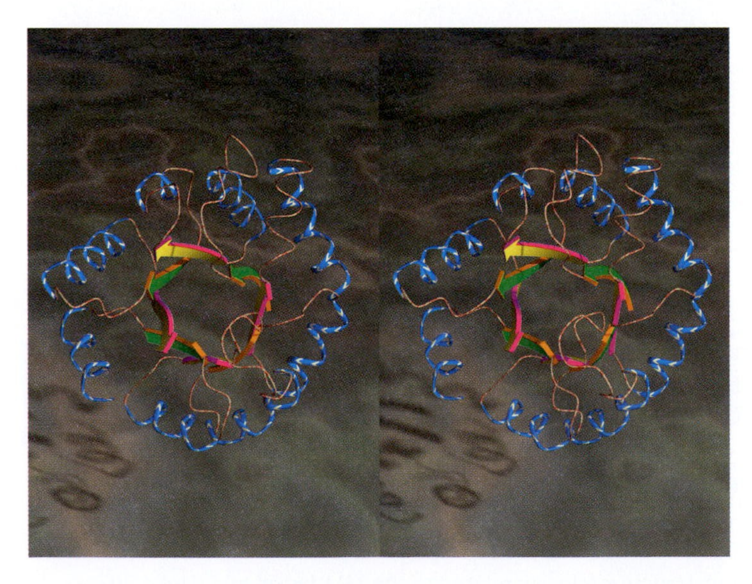

Fig. 4.39 The TIM-barrel fold, a cylindrically closed, eight-stranded β-sheet, surrounded by helices. This folding pattern has appeared in over 100 enzymes.

CASE STUDY Enolase, mandelate racemase, and muconate lactonizing enzyme catalyse different reactions but have related mechanisms

The enolase superfamily contains several enzymes that catalyse different reactions with shared features of their mechanisms. These include enolase itself, mandelate racemase, and muconate lactonizing enzyme I. From the point of view of sequence similarity, these enzymes are reasonably close relatives. Mandelate racemase and muconate lactonizing enzyme I have 25% sequence identity.

However, looking only at sequence and structure runs the risk of overlooking a more subtle similarity. What these enzymes share is a common feature of their *mechanism*: Each acts by abstracting a proton adjacent to a carboxylic acid to form an enolate intermediate (Fig. 4.40). The stabilization of a negatively charged transition state is conserved. In contrast, the subsequent reaction pathway, and the nature of the product, vary from enzyme to enzyme. These enzymes have not only a similar overall structure, a variant of the TIM-barrel fold, but each requires a divalent metal ion, bound by structurally equivalent ligands. Different residues in the active site produce enzymes that catalyse different reactions.

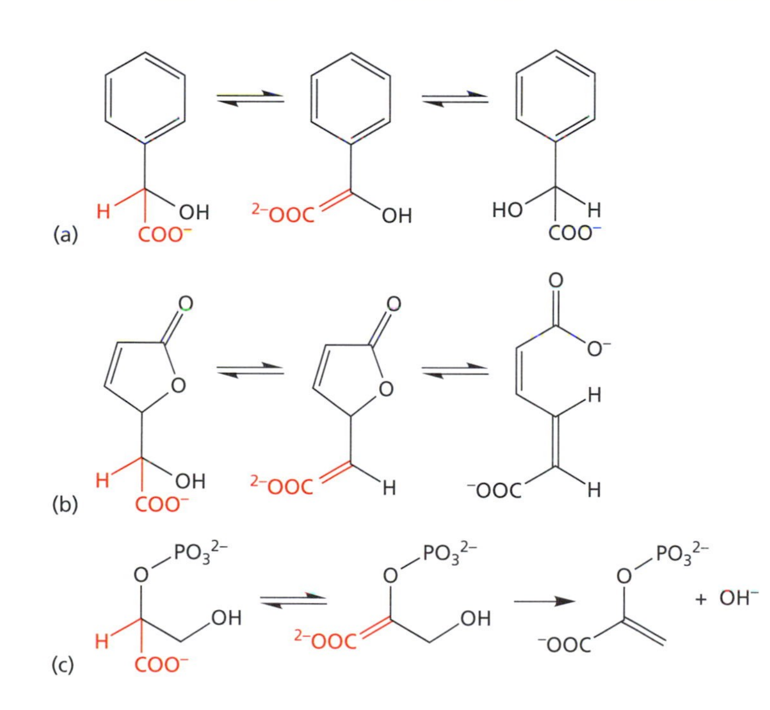

Fig. 4.40 Common mechanism in the enolase family of enzymes, (a) mandelate racemase, (b) muconate lactonizing enzyme, (c) enolase.

Such families of enzymes illustrate the kinds of structural features that change and those that stay the same. In some cases, the catalytic atoms occupy the same positions in molecular space, although the residues that present them are located at different points in the fold. In other cases, the positions in space of the catalytic residues are conserved even though the identities and functions of the catalytic residues vary. In these cases, there appears to be a set of conserved functional *locations* within the space of the molecule.

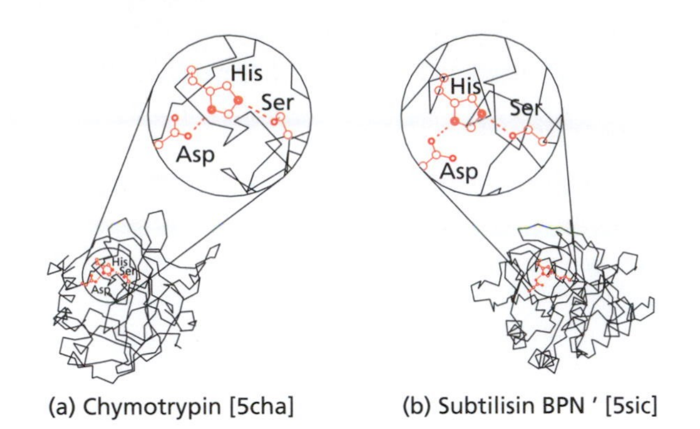

Fig. 4.41 (a) Chymotrypsin and (b) subtilisin, two proteinases that even share a common Ser–His–Asp catalytic triad, are not homologous, and show entirely different folding patterns. The Ser–His–Asp triad also appears in other proteins, including lipases and a natural catalytic antibody.

Several enzyme families show an even greater degree of divergence. The apurinic/apyrimidic endonuclease superfamily, a large diverse group of phosphoesterases, includes members that cleave DNA and RNA, and lipid phosphatases. Even catalytic residues vary between different subfamilies. For example, a His essential for the function of the DNA repair enzyme DNaseI is not conserved in exonuclease III.

Conversely, many functions are provided by unrelated proteins. Chymotrypsin and subtilisin have produced the same catalytic mechanism for proteolysis by convergent evolution (Fig. 4.41).

2. Recruitment

Many people ask how much a protein must change its sequence before its function changes. The answer is: Not at all! There are numerous examples of proteins with multiple functions:

- Eye lens proteins in the duck are identical in sequence to active lactate dehydrogenase and enolase in other tissues, although they do not encounter the substrates in the eye. They have been recruited to provide a structural and optical function. Several other avian eye lens proteins are identical or similar to enzymes. In some cases residues essential for catalysis have mutated, proving that the function of these proteins in the eye is not enzymatic. In those species, the coexistence of mutated, inactive, enzymes in the eye and active enzymes in other tissues implies that the gene must have been duplicated.
- Some proteins interact with different partners to produce oligomers with different functions. In *E. coli*, a protein that functions on its own as lipoate dehydrogenase is also an essential subunit of pyruvate dehydrogenase, 2-oxoglutarate dehydrogenase, and the glycine cleavage complex.
- Proteinase Do functions as a chaperone at low temperatures and as a proteinase at high temperatures. The logic, apparently, is that under conditions of moderate stress it attempts to salvage misfolded proteins; under conditions of higher stress it gives up and recycles them.

- Phosphoglucose isomerase (= neuroleukin = autocrine motility factor = differentiation and maturation mediator) functions as a glycolytic enzyme in the cytoplasm, but as a nerve growth factor and cytokine outside the cell.

Divergence and recruitment are at the ends of a broad spectrum of changes in sequence and function. Aside from cases of 'pure' recruitment—such as the duck eye lens proteins or phosphoglucose isomerase, in which a protein adopts a new function with no sequence change at all—there are examples of relatively small sequence changes correlated with very small function changes (which most people would think of as relatively pure divergence), relatively small sequence changes with quite large changes in function (which most people would think of as recruitment), and also many cases in which there are large changes in both sequence and function.

3. 'Mixing and matching' of domains, including duplication/oligomerization, and domain swapping or fusion (see Fig. 2.24)
Many large proteins contain tandem assemblies of domains which appear in different contexts and orders in different proteins. Censuses of genomes suggest that many proteins are multimodular. Of 4401 genes in *E. coli*, 287 correspond to proteins containing 2, 3, or 4 modules. The structural patterns of 510 *E. coli* enzymes involved in the metabolism of small molecules can be accounted for in total, or in part, by 213 families of domains. Of the 399 which can be entirely divided into known domains, 68% are single-domain proteins, 24% comprise two domains, and 7% three domains. Only 4 of the 399 had 4, 5, or 6 domains. There are marked preferences for the pairing of different families of domains.

Multidomain proteins present particular problems for the assignment of function in genome annotation, because the domains may possess independent functions, modulate one another's function, or act in concert to provide a single function. On the other hand, in some cases the presence of a particular domain or combination of domains is associated with a specific function. For example, NAD-binding domains appear almost exclusively in dehydrogenases.

Directed evolution

One strand of Darwin's thinking that led to the theory of evolution was the observation that farmers could improve the qualities of their livestock by selective breeding. He drew an analogy between this *artificial selection* and the idea of *natural selection* that he was proposing as the mechanism of evolution. We now recognize that evolution by natural selection takes place at the molecular level. Why not artificial selection also?

Natural proteins do many things, but not everything we'd like them to. For applications in technology, it would be useful to have proteins that would:

- have activities unknown in Nature;
- show activity towards unnatural substrates, or altered specificity profiles;
- be more robust than natural proteins, retaining their activity at higher temperatures or in organic solvents; and
- show different regulatory responses, enhanced expression, or reduced turnover.

See Chapter 5.

Someday, we shall be able to design amino acid sequences *a priori* that will fold into proteins with desired functions. As this is not yet possible, scientists have used *directed evolution*—or artificial selection—to generate molecules with novel properties starting from natural proteins.

Evolution requires the generation of variants, and differential propagation of those with favourable features. Molecular biologists dealing with microbial evolution have advantages over the farmers whom Darwin observed. We can generate large numbers of variants artificially. Screening and selection can, in many cases, be done efficiently, using stringent growth conditions. And there are virtually no limits on the size of the 'flock' or 'litter'. Darwin might well have been envious. He wrote:

> ... as variations manifestly useful or pleasing to man appear only occasionally, the chance of their appearance will be much increased by a large number of individuals being kept. Hence, number is of the highest importance for success. On this principle Marshall formerly remarked, with respect to the sheep of parts of Yorkshire, 'as they generally belong to poor people, and are mostly *in small lots*, they never can be improved.' (*Origin of species*, Chapter 1.)

The procedure of directed evolution comprises these steps:

1. Create variant genes by mutagenesis or genetic recombination.
2. Create a library of variants by transfecting the genes into individual bacterial cells.
3. Grow colonies from the cells, and screen for desirable properties.
4. Isolate the genes from the selected colonies, and use them as input to Step 1 of the next cycle.

Strategies for generating variants include: (1) single and multiple amino acid substitutions, (2) recombination, and (3) formation of chimeric molecules by mixing and matching segments from several homologous proteins. Each method has its advantages and disadvantages. The smaller the change in sequence, the more likely that the result will be functional. On the other hand, multiple substitutions or recombinations give a greater chance of generating novel features. The choice depends, in part, on the nature of the goal. For instance, it is easier to lose a function than to gain one. (Why would you want to *lose* a function? Removal of product inhibition to enhance throughput in an enzymatically catalysed process is an example.)

Classification of protein functions

Of the goals of organizing data on protein sequences, structures, and functions, the classification of function is the most difficult. Given two or more related *sequences*, it is possible to measure their similarity through aligning them. Given two or more related *structures*, it is possible to measure their similarity by superposing them and seeing what fraction of the residues can be superposed and how well they fit. But in many cases, given two or more protein *functions*, it would be difficult to measure how similar their functions are.

CASE STUDY Directed evolution of subtilisin E

Subtilisins are a family of bacterial proteolytic enzymes. Subtilisin E, from the mesophilic bacterium *Bacillus subtilis*, is a 275-residue monomer. It becomes inactive within minutes at 60 °C. Directed evolution has produced interesting variants, with features including enhanced thermal stability, and activity in organic solvents.

1. Enhancement of thermal stability

Thermitase, a subtilisin homologue from *Thermoactinomyces vulgaris*, remains stable up to 80 °C. The existence of thermitase is reassuring, because it shows that the evolution of subtilisin to a thermostable protein is possible. On the other hand, subtilisin E and thermitase differ in 157 amino acids. Do we have to go this far? Are all the changes essential for thermostability, or has there been considerable neutral drift as well?

A thermostable variant of subtilisin E, produced by directed evolution, differs from the wild type by only eight amino acid substitutions. The variant is identical to thermitase in its temperature of optimum activity: 76 °C (17 °C higher than the original molecule) and stability at 83 °C (a 200-fold increase relative to the wild type).

The procedure involved successive rounds of generation of variants, and screening and selection of those showing favourable properties. Mutations, via error-prone PCR (polymerase chain reaction) to produce an average of 2 or 3 base changes per gene, alternated with *in vitro* recombination to find the best combinations of substitutions at individual sites (Fig. 4.42). At each step several thousand clones were screened for activity and thermostability.

The optimal variant differed from the wild type at eight positions: N188S, S161C, P14L, N76D, G166R, N181D, S194P, and N218S. Figure 4.43 shows their distribution in the structure. Most of the substitutions are far from the active site, which is not surprising as the wild type and variant do not differ in function. Most of the sites of substitution are in loops between regions of secondary structure. These regions are the most variable in the natural evolution of the subtilisin family. However, only two of the substitutions produce the amino acids that appear, in those positions, in thermitase. Two of the substitutions are in α-helices, including P14L. P14L has a certain logic: proline tends to destabilize an α-helix because it costs a hydrogen bond (see Chapter 3).

2. Activity in organic solvents

For applications to industrial processes, it is useful to have enzymes that work under conditions other than those prevailing *in vivo*. Directed evolution produced another variant of subtilisin E that is active in an organic solvent, 60% dimethylformamide. In this case, no known homologue with the desired properties was known.

By random mutagenesis and screening, 12 amino acid substitutions were identified that produced a protein with 471 times the proteolytic activity in the organic medium as the wild type, measured in terms of k_{cat}/K_M. The changes were: D60N, D97G, Q103R, I107V, N218S, G131D, E156G, N181S, S182G, S188P, Q206L, and T255A. In this case 8 of the 12 mutations cluster around the active site (Fig. 4.44).

In addition to the enhancement of proteolytic activity in the organic medium, the variant can also synthesize polypeptides from an amino acid ester, a novel and unexpected activity not characteristic of the wild type.

These examples show some of the functional variation possible among molecules differing at only a few positions from a natural protein, and the power of directed evolution to find them.

continues ...

We can explore only a tiny fraction of possible sequences

Neither natural nor directed evolution can possibly explore more than a very small fraction of possible polypeptide sequences. The number of sequences of n residues is 20^n. To give such numbers some tangible meaning, the total mass of all possible sequences of even a small 37-amino acid polypeptide is larger than the mass of the Earth. And the numbers increase very steeply with sequence length. The total mass of all possible 50-residue polypeptides is roughly equal to the mass of the Milky Way

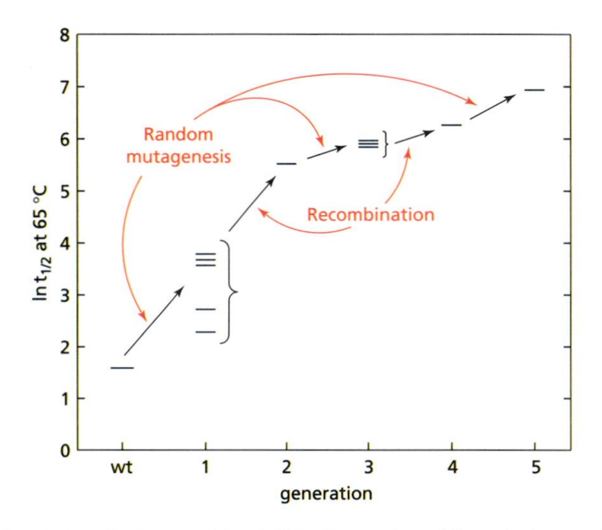

Fig. 4.42 Directed evolution of a thermostable subtilisin. The starting, wild type (wt) was subtilisin E from the mesophilic bacterium *B. subtilis*. Steps of random mutagenesis alternated with recombination. At each step screening for improved properties and artificial selection chose candidates for the next round. (After Zhao, H. and Arnold, F.H. (1999). Directed evolution converts subtilisin E into a functional equivalent of thermitase. *Protein Eng.*, **12**, 47–53.)

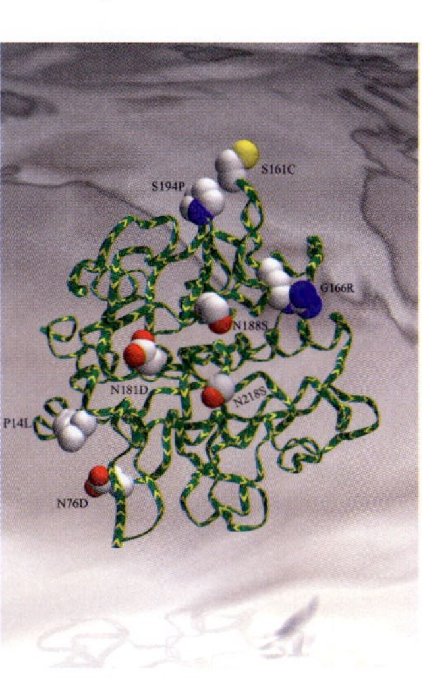

Fig. 4.43 The sites of mutation in *B. subtilis* subtilisin E, that produced a thermostable variant by directed evolution. Side chains shown are those of the final product.

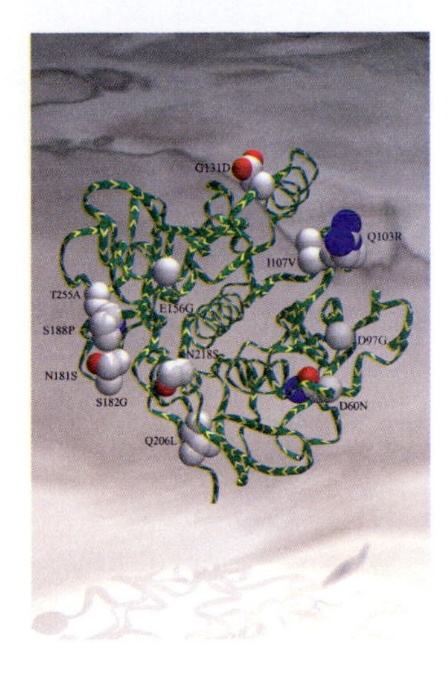

Fig. 4.44 The sites of mutation in *B. subtilis* subtilisin E that produced a variant active in organic solvent. Side chains shown are those of the final product.

Also, mapping sequence to structure is robust and conservative, as many small changes in sequence produce only small changes in structure. Protein function may depend on factors other than sequence and structure, including, for example, location.

Comparisons of functions could be based on suggested classifications of functions. One widely known classification is the Enzyme Commission scheme, limited, of course, to that class of functions. Other protein function classification schemes have been proposed, many in connection with individual organisms or individual families of proteins. However, a scheme appropriate for one organism is not necessarily appropriate for others, and until recently there has been no noticeable attempt at consistency.

Indeed, even for very well understood proteins, there are different legitimate points of view about what aspects of function to focus on. The biochemist looks for the process mediated by the isolated protein in dilute solution. The molecular biologist looks for the significance, in the overall scheme of the life of the cell, of the process or processes in which the protein participates. Is it possible to reconcile the different points of view? The Gene Ontology Consortium™ offers an attractive approach.

The Enzyme Commission (EC) classification

The origin of the EC classification was the action taken by the General Assembly of the International Union of Biochemistry (IUB), in consultation with the International Union of Pure and Applied Chemistry (IUPAC), in 1955, to establish an International Commission on Enzymes (see: **www.chem.qmul.ac.uk/iubmb/enzyme/**).

EC numbers (looking suspiciously like IP numbers) contain four fields, corresponding to a four-level hierarchy. For example, EC 1.1.1.1 corresponds to alcohol dehydrogenase, catalysing the general reaction:

$$\text{an alcohol} + \text{NAD} = \text{the corresponding aldehyde or ketone} + \text{NADH}_2.$$

Several reactions, involving different alcohols, would share this number; but dehydrogenation of one of these alcohols by an enzyme using the alternative cofactor NADP would be assigned EC 1.1.1.2. The Commission has emphasized that: *'It is perhaps worth noting, as it has been a matter of long-standing confusion, that enzyme nomenclature is primarily a matter of naming reactions catalyzed, not the structures of the proteins that catalyze them.'*

The first number in the EC hierarchy assigns the enzyme to one of the six main classes:

Class 1. Oxidoreductases
Class 2. Transferases
Class 3. Hydrolases
Class 4. Lyases
Class 5. Isomerases
Class 6. Ligases

The significance of the second and third numbers depends on the class. For Oxidoreductases, the second number describes the substrate and the third number the acceptor. For Transferases, the second number describes the class of item transferred, and the third number either more specifically what they transfer or in some cases the acceptor. For Hydrolases, the second number signifies the kind of bond cleaved (e.g., an ester bond) and the third number the kind of bond (e.g., a carboxylic ester or a thiolester). (Proteinases are treated slightly differently, with the third number including the mechanism: serine proteinases, thiol proteinases, and acid proteinases are classified separately.) For Lyases, the second number signifies the kind of bond formed (e.g., C–C or C–O), and the third number the specific molecular context. For Isomerases, the second number indicates the type of reaction and the third number the specific class of reaction. For Ligases, the second number indicates the type of bond formed and the third number the type of molecule in which it appears. For example, EC 6.1 for C–O bonds (enzymes acylating tRNA), EC 6.2 for C–S bonds (acyl-CoA derivatives), etc. The fourth number gives the specific enzymatic activity.

Specialized classifications are available for some families of enzymes. For instance, the MEROPS database by N.D. Rawlings and A.J. Barrett provides a structure-based classification of peptidases and proteinases (see **www.merops.sanger.ac.uk/**).

A problem with the EC classification is that the traditional biochemist's view of function arises from the study of isolated proteins in dilute solution, in the presence of carefully controlled concentrations of substrates. The molecular biologist knows that an adequate definition of function must recognize the biological role of a molecule in the living context of a cell (or intracellular compartment) or the complete organism on the one hand, and its role in a network of metabolic or control processes on the other. As a result, there is a generic problem with all attempts to force functional classifications into a hierarchical format.

Gerlt and Babbitt, among the most thoughtful writers on the subject, point out that 'no structurally contextual definitions of enzyme function exist'. They propose a general classification of function better integrated with sequence and structure. For enzymes they define:

- **Family**: homologous (often orthologous) enzymes that catalyse the same reaction (same mechanism, same substrate specificity). These can be hard to detect at the sequence level if the sequence similarity becomes very low.

- **Superfamily**: homologous enzymes catalysing either (1) a similar reaction with different specificity or (2) different overall reactions with a common mechanistic attribute (partial reaction, transition state, intermediate) that share conserved active-site residues.
- **Suprafamilies**: different reactions with no common feature. Distant members of the same suprafamily would not be expected to be detectable from sequence information alone.

The Gene Ontology Consortium™

A more general approach to the *logical* structure of a functional classification of proteins and other gene products has been adopted by The Gene Ontology Consortium™ (2000) (www.geneontology.org). Its goal is a systematic attempt to classify function, by creating a dictionary of terms and their relationships. By an *ontology* they mean a set of well-defined terms with well-defined interrelationships; that is, a dictionary and rules of syntax.

Organizing concepts of the Gene Ontology project include:

- **Molecular function**: a function associated with what an individual protein or RNA molecule does in itself; either a general description such as *enzyme*, or a specific one such as *alcohol dehydrogenase*. This is function from the biochemists' point of view.
- **Biological process**: a component of the activities of a living system, mediated by a protein or RNA, possibly in concert with other proteins or RNA molecules; either a general term such as *signal transduction*, or a particular one such as *cyclic AMP synthesis*. This is function from the cell's point of view.

Because many processes are dependent on location, Gene Ontology also classifies:

- **Cellular component**: the assignment of site of activity or partners; this can be a general term such as *nucleus* or a specific one such as *ribosome*.

An example of the GO classifiction is shown in Fig. 4.45. Note that it is more general than a hierarchy.

USEFUL WEB SITES

See www.bork.embl–heidelberg.de/Modules/ for pictures of modular proteins.
Classifications of protein structures:
SCOP: http://scop.mrc–lmb.cam.ac.uk/scop
FSSP: www.ebi.ac.uk/dali/fssp
DALI: www.ebi.ac.uk/dali
DALI Domain Dictionary: www.ebi.ac.uk/dali/domain
Peptidases: www.merops.sanger.ac.uk
Others: www.bioscience.org/urllists/protdb.htm

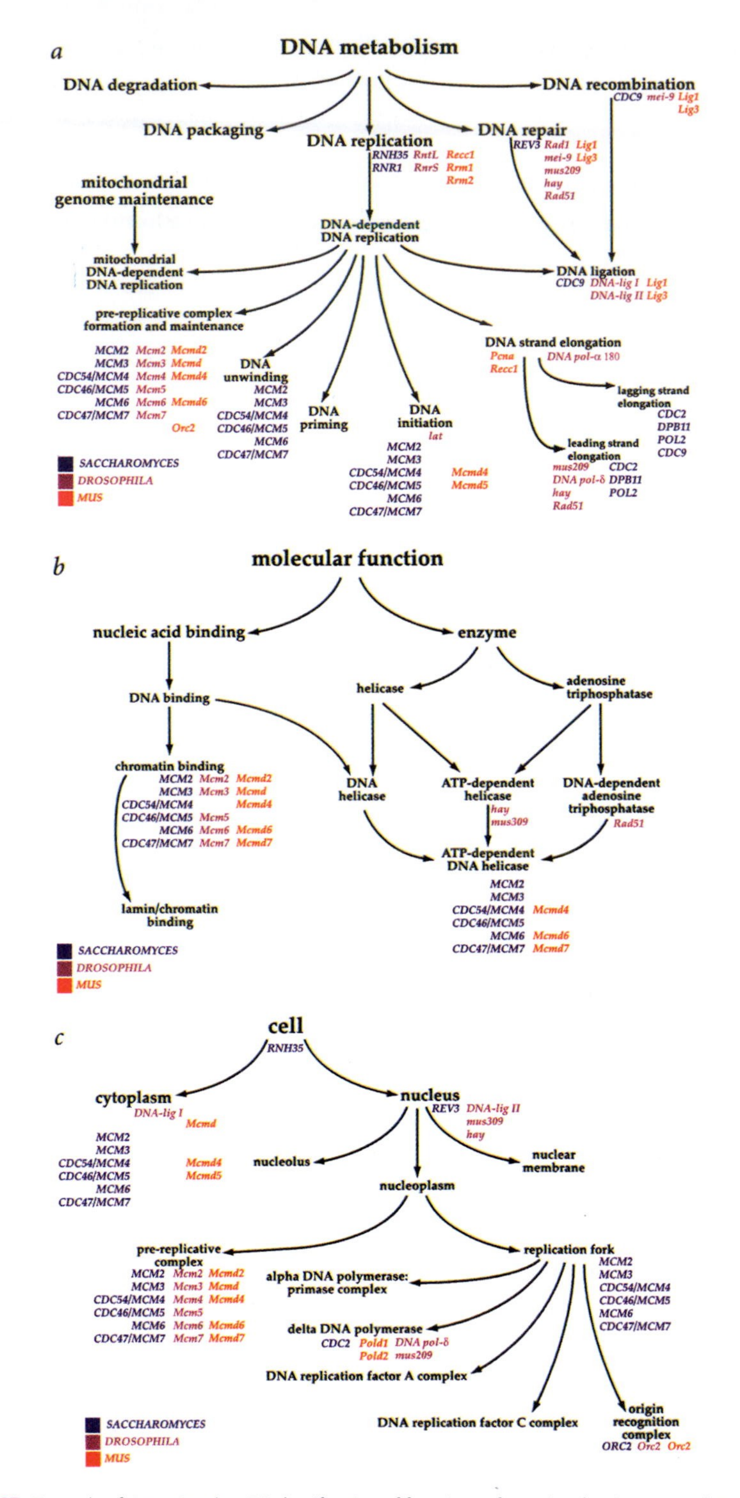

Fig. 4.45 Example of Gene Ontology™ classification of functions of proteins that interact with DNA. (From The Gene Ontology Consortium (2000). Gene ontology: tool for the unification of biology. *Nat. Genet.*, **25**, 25–8. © Nature Publishing Group, Reproduced with permission.)

RECOMMENDED READING

Fitch, W.M. (2000). Homology: a personal view on some of the problems. *Trends Genet.*, **16**, 227–31.

Golding, G.B. and Dean, A.M. (1998). The structural basis of molecular adaptation. *Mol. Biol. Evol.*, **15**, 355–69.

Liu, Y. and Eisenberg, D. (2002). 3D domain swapping: as domains continue to swap. *Prot. Sci.*, **11**, 1285–99.

Nathans, J. (1999). The evolution and physiology of human colour vision: insights from molecular genetic studies of visual pigments. *Neuron*, **24**, 299–312.

Suzuki, T. and Imai, K. (1998). Evolution of myoglobin. *Cell. Mol. Life Sci.*, **54**, 979–1004.

Tao, H. and Cornish, V.W. (2002). Milestones in directed enzyme evolution. *Curr. Opin. Chem. Biol.*, **6**, 858–64.

EXERCISES, PROBLEMS, AND WEBLEMS

Exercises

1. Which of the following pairs are orthologues? Which are paralogues? Which are neither? (a) Human trypsin, horse trypsin. (b) Human trypsin, horse chymotrypsin. (c) Human trypsin, human elastase. (d) *B. subtilis* subtilisin, horse chymotrypsin.
2. On a photocopy of the structure of thioredoxin (Fig. 4.8) circle (a) a β-hairpin, (b) a β–α–β-unit.
3. On a photocopy of the picture of cow odorant binding protein (Fig. 4.6(c)), circle the hinge region involved in the domain swapping.
4. Based on Fig. 4.5, assuming that the connection between A and B domains is flexible, sketch a possible structure of a polymer of A–B proteins. How might an insertion of several residues into the linker region between A and B promote polymerization?
5. What letter corresponds to the common core of the letters: (a) E and F, (b) V and A, (c) F and L?
6. In Fig. 4.2(a), the loop going across the top is one of the few regions showing a different conformation in papain and actinidin. Using Fig. 4.2(e) as the key, highlight the residues in this loop on a photocopy of the sequence alignment in Fig. 4.2(a).
7. In Fig. 4.2(c), human liver cathepsin B has two very long insertions relative to papain. (a) Circle them on a photocopy of the superposition of the structures. (b) Using Fig. 4.2(e), highlight the inserted residues in the sequence alignment. Indicate which regions in the sequence correspond to which parts of the structure.
8. In Fig. 4.2(d), the hairpin loop protruding down at the lower right is much longer in staphopain than in papain. Using Fig. 4.2(e), circle these residues on a photocopy of the sequence alignment in Fig. 4.2(d).
9. Two homologous proteins have 75% identical residues in an optimal alignment of their sequences. Estimate the root-mean-square deviation of an optimal superposition of their structures.
10. For all pairs of globin sequences that appear in Fig. 4.15(b), what is the percent amino acid identity of the pair?

11. (a) Find other positions in the alignment table of mammalian globins (Fig. 4.15(a)) at which human and horse haemoglobin α-chains contain the same residue, and human and horse haemoglobin β-chains contain the same residue, but at which the residues in the haemoglobin α-chains, the haemoglobin β-chains, and myoglobin are all different.

(b) In Fig. 4.15(b), find all positions that are conserved in mammalian globins but differ in at least one non-mammalian globin.

12. In the alignment table of Fig. 4.15(a), find sets of conserved residues separated by 3, 4, or 7 in the sequence. Referring to Fig. 4.20, are these indeed in helices? If so, which helix?

13. (a) From Fig. 4.16, describe a structural feature present in sperm-whale myoglobin but not in the α- or β-chains of human haemoglobin.

(b) From Fig. 4.18, describe a structural feature present in lupin leghaemoglobin but not in sperm-whale myoglobin or *Chironomus* erythrocruorin.

14. Using the alignment shown in Fig. 4.20, which parts of which helices in sperm-whale myoglobin are deleted in *Mycobacterium tuberculosis* globin? How many residues in *Mycobacterium tuberculosis* globin correspond to residues that are not in helices in sperm whole myoglobin?

15. On a photocopy of the NAD-binding domain of horse liver alcohol dehydrogenase, Fig. 4.22, label the helices αA, αB, ... according to their order of appearance along the chain, and label the strands of sheet βA, βB, ... according to their order of appearance along the chain.

16. On a photocopy of Fig. 4.22, circle the region shown in Fig. 4.32.

17. (a) On a photocopy of Fig. 4.25(a), circle the position of residue 90Asp.

(b) On a photocopy of Fig. 4.28(a), circle the position of residue 231Gly.

(c) On a photocopy of Fig. 4.30(a), circle the position of residue 306Ala.

18. On a photocopy of the structure of malate dehydrogenase in Fig. 4.38(b), sketch in the natural substrate, oxaloacetate.

19. The resting state of the chromophore in bacteriorhodopsin is all-*trans* retinal, with the methyl group at position 5 *cis* to the tail instead of *trans* as in Fig. 4.34. Excitation results in isomerization at the 13–14 double bond rather than at the 11–12 double bond. Draw the structure of the excited state of the bacteriorhodopsin chromophore.

20. (a) A visual pigment of the goat has residues 180A, 197H, 277Y, 285T, 308A, and a λ_{max} of 553 nm. What is the error of the prediction of λ_{max} by the method described on p. 163?

(b) A visual pigment of the rabbit has residues 180A, 197Y, 277Y, 285T, 308S, and a λ_{max} of 508 nm. What is the error of the prediction of λ_{max} by the method described on p. 163?

21. Mutations in the five critical residues in vertebrate MWS and LWS opsins influence the spectra almost independently, except for the double mutant S180A/H197Y from which one would predict an additional 11-nm shift in λ_{max} in addition to the individual separate contributions of the mutants. This observation might suggest the hypothesis that residues at positions 180 and 197 are in contact in the structure. Is this true? (See Fig. 4.37.)

22. How many ways are there to make eight mutations in subtilisin E, a protein containing 275 amino acids?

23. In the directed evolution of thermostable subtilisin E, two of the substitutions are in α-helices. One of these is P14L. Which is the other? Circle it on a photocopy of Fig. 4.43.

24. A student conjectured that a variant protein active in an organic medium would differ from the wild type by substitution of uncharged residues for charged or polar ones. How many (charged or polar) → uncharged substitutions are there among the 10 mutations that produced a variant of subtilisin E active in dimethylformamide? How many uncharged → (charged or polar) substitutions were there?

Problems

1. On photocopies of Fig. 4.46(a), (b), (c) identify the supersecondary structures in these proteins.

2. What parts of the folding pattern of *E. coli* thioredoxin are similar in topology to (a) acylphosphatase (Fig. 2.20(a)) and (b) the viral toxin from corn smut fungus (Fig. 2.20(b))?

3. From a visual inspection of Figs 4.2 and 4.18, for which homologue of papain is the extent of its structural divergence from papain comparable to the extent of structural divergence of sperm-whale myoglobin and lupin leghaemoglobin?

4. Pig odorant-binding protein, like most mammalian orthologues, is a monomeric domain approximately 160 residues long (Fig. 4.6(a)). Cow odorant-binding protein, exceptionally, is a domain-swapped dimer (Fig. 4.6(b)).

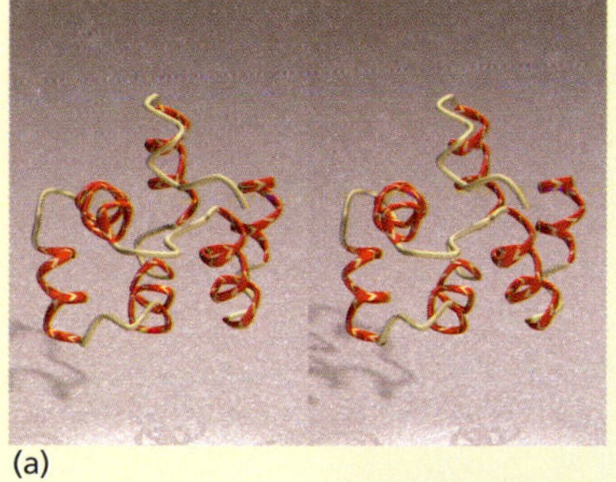
(a)

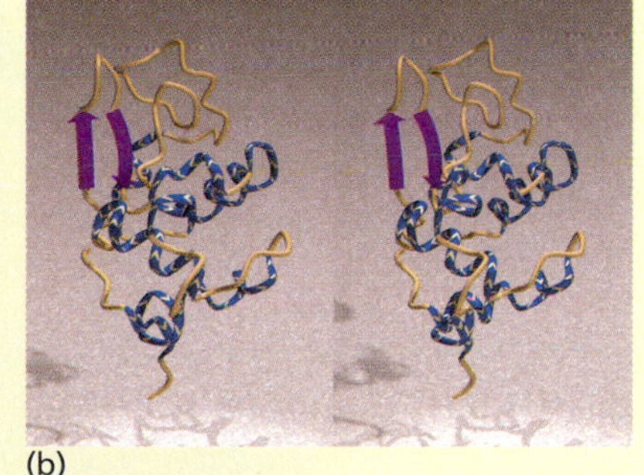
(b)

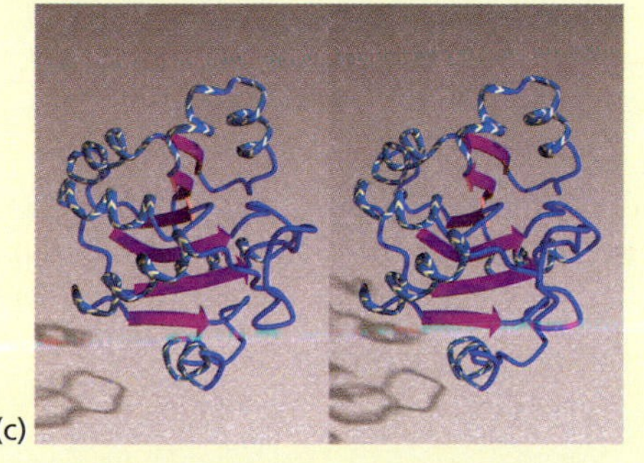
(c)

Fig. 4.46 (a) Death domain of p75 low-affinity neurotropin receptor (rat) [1NGR]. (b) Baboon α-lactalbumin [1ALC]. (c) Fructose permease subunit b [1BLE].

To identify the critical differences in the following sequence that account for this difference, consider the following observations:

(1) The hinge region in the pig sequence has a glycine inserted at *, which might give this region extra flexibility:

```
cow  QEEEAEQNLSELSGPWRTVYIGSTNPEKIQENGPFRTYFRELVFDDEKGT
               |||| | | ||||   ||| || ||    |   |||
pig         PF-ELSGKWITSYIGSSDLEKIGENAPFQVFMRSIEFDDKESK

cow  VDFYFSVKRDGKWKNVHVKATKQDDGTYVADYEGQNVFKIVSLSRTHLVA
     |   |  |  |         |||   ||   | | | |     | | |
pig  VYLNFFSKENGICEEFSLIGTKQEGNTYDVNYAGNNKFVVSYASETALII

cow  HNINVDKHGQTTELTGLFVK-LNVEDEDLEKFWKLTEDKGIDKKNVVNFL
      |||||  |  |  |||       || |||||   |   ||   | ||
pig  SNINVDEEGDKTIMTGLLGKGTDIEDQDLEKFKEVTRENGIPEENIVNII
                         *
cow  ENEDHPHPE
     |  | |
pig  ERDDCPA
```

(2) A disulphide bridge in the pig protein links residues 63–155 (Find these cysteine residues in the sequence alignment.). The cow protein contains no cysteines.

(3) Modelling a domain-swapped dimer with the pig sequence, based on the cow structure, shows unrelievable clashes at the interface, between the C-termini of the two monomers. In the cow structure, the C-terminal region of one of the monomers rearranges its conformation to avoid the collision.

(4) A study of mutants showed that deleting the inserted glycine from the pig protein produced a molecule that remained monomer. Inserting a glycine at this position in the cow protein produced a molecule that did not form a domain-swapped dimer but formed a functional monomer.

(a) Give an explanation for the effects of the mutations.

(b) Suggest minimal modifications to the pig odorant-binding protein sequence that would produce a domain-swapped dimer.

5. What root-mean-square deviation from the correct structure would you expect from a homology model, based on papaya papain as a parent structure, of (a) kiwi fruit actinidin, (b) human procathepsin L, (c) human cathepsin B, and (d) *Staphylococcus aureus* staphopain (see Fig. 4.2)?

6. Correlating the structural superposition in Fig. 4.21 with the sequence alignment in Fig. 4.20, indicate on a photocopy of Fig. 4.20 the residues in sperm-whale myoglobin that have structure in common with the truncated globin from *P. caudatum*.

7. The genes encoding most vertebrate globins contain three exons. The amino acid sequence encoded by the central exon of the gene for human myoglobin is:

RLFKSHPETLEKFDRFKHLKTEAEMKASED
LKKHGVTVLTALGAILKKKGHHEAELKPLA
QSHATKHKIPIKYLE

A synthetic molecule related to the central exon corresponds to residues 32–139 of horse heart myoglobin:

LFTGHPETLEKFDKFKHLKTEAEMKASEDL
KKHGTVVLTALGGILKKKGHHEAELKPLAQ
SHATKHKIPIKYLE

On photocopies of Figs 4.20 and 4.21, compare the sequences of *P. caudatum* globin, the amino acid sequence encoded by the second exon of the human myoglobin gene, and the synthetic minimyoglobin.

8. Estimate the photon flux in quanta m^2 s^{-1} at a point 2 m from a 100-watt green light bulb. Assume that the bulb is a perfect sphere (ignore the mounting), delivering energy with perfect efficiency at a monochromatic wavelength of 550 nm.

9. On the basis of comparing Figs 4.26(b) and 4.27(b), what appears to be the common part of the structure? Indicate the common part of the structure on photocopies of the pictures of both structures.

10. Humans achieve colour vision through the action of three visual pigments, with absorption maxima in the violet, green, and yellow–green regions of the spectrum. These pigments are homologous proteins, opsins, conjugated to a common ligand, 11-*cis*-retinal. They are distinguished according to their spectral properties as S, M, and L, for short, medium, and long wavelength absorbance maxima.

In humans and other Old-World primates, opsin S is encoded by an autosomal gene, and M and L by genes on the X chromosome. This is why certain kinds of colour-blindness in humans are sex-linked traits. In contrast, most species of New-World monkeys have only two loci, an autosomal gene for S opsin, and a polymorphic X-linked gene for M and L. In these species, females heterozygous at the opsin locus on the X chromosome can express three opsins (including the autosomal S), but males only two. As a result, monkeys of different genders have different colour vision: female heterozygotes have full colour vision, but all males are partially colour-blind.

Because the chromophore is common to all the proteins, the spectral shifts must arise from amino acid sequence differences. The table shows that the difference between M and L is the result of substitutions at three sites, residues 180, 277, and 285.

Species	L opsin residue number			M opsin residue number		
	180	277	285	180	277	285
Old-World species:						
Human	S	Y	T	A	F	A
Chimpanzee	S	Y	T	A	F	A
Gorilla	S	Y	T	A	F	A
Diana monkey	S	Y	T	A	F	A
Macaque	S	Y	T	A	F	A
Talapoin monkey	S	Y	T	A	F	A
New-World species:						
Capuchin monkey	S	Y	T	A	F	A
Marmoset	S	Y	T	A	Y	A
Howler monkey	S	Y	T	A	F	A

These results show that spectral tuning is the result, in most cases, of the same set of mutations. They suggest the hypothesis (1) that the divergence between L and M opsins *preceded* the divergence of Old-World and New-World primate species. This would require there to have been gene duplication and translocation of the gene on the X chromosome of Old-World primates. The alternative hypothesis (2) is that the proteins are showing convergent evolution, under the same selective pressure.

(a) For each hypothesis, sketch the appearance of the implied evolutionary tree relating the following four sets of proteins: Old-World primate L opsins, Old-World primate M opsins, New-World primate L opsins, New-World primate M opsins.

(b) According to each hypothesis, which would be expected to resemble Old-World L opsin more closely: the Old-World M opsin or the New-World L opsin?

(c) To choose between these alternatives, consider the following alignment table of partial sequences of opsins. (The asterisks mark the three sites at which the L/M divergence is the result of selective pressure on spectral properties. A '.' indicates that the residue at that position is the same as that of the human L sequence. Humans are Old-World and Howler monkeys New-World.)

```
Exon 3:
                                                 *
Human L       AIISWERWLVVCKPFGNVRFDAKLAIVGIAFSWIWSAVWTAPPI
Human M       ......M......................I.....A........
Howler L      ......R......................V.....S........
Howler M      ......R......................V.....A........

Exon 4:

Human L       GPDVFSGSSYPGVQSYMIVLMVTCCIIPLAIIMLCYLQVWLAIRA
Human M       ..........................T..S..V............
Howler L      .....................I...FL..G..E............
Howler M      ....................VI...IL..S..V............

Exon 5:
                           *       *
Human L       KEVTRMVVVMIFAYCVCWGPYTFFACFAAANPGYAFHPLMAALPAYFAKS
Human M       ..........VL.F.F.....A............P...............
Howler L      ...........M.Y.......T............................
Howler M      ...........I.F.......A............................
```

Excluding the three sites at which selection was operating, determine the number of differences between each pair of partial sequences. Which hypothesis is supported by the results?

(d) What additional data and analysis would you want to adduce to confirm or refute your conclusion?

Weblems

1. Find proteins that bind NAD that have structures in: (a) the all-β and (b) the $\alpha + \beta$ classes.
2. What are the EC numbers of: (a) malate dehydrogenase, (b) lactate dehydrogenase?
3. Which of the mutations selected in the directed evolution of the thermostable variant of subtilisin appear in thermitase?
4. Find two animations of the myosin power cycle viewable over the Web, and compare them.

Chapter 5

PROTEIN ENGINEERING, FOLDING, PREDICTION, AND DESIGN

LEARNING GOALS

1 **To appreciate that the mechanism of protein folding is complex and delicate.** The very large number of non-native conformations requires the protein to have evolved one or more folding pathways, the nature of which are difficult to measure or predict.

2 **To be familiar with the basic principles of thermodynamics and kinetics.**

3 **To recognize the effects of mutants on the rates of folding and unfolding, and the use of mutants to probe the structures of intermediates.** To be able to interpret 'chevron plots', and to determine and interpret **Φ-values**.

4 **To understand, and see the relationships among, several concepts that have emerged from the investigations of protein folding**, including the **molten globule** and the **folding funnel**.

5 **To be able to define and calculate contact order**, and to follow the implications about the folding process of the correlation between contact order and folding rate.

6 **To appreciate the dangers of misfolding, leading to protein aggregation in the concentrated intracellular milieu, and the mechanism of action of the GroEL–GroES system to rescue misfolded proteins.**

7 **To be able to distinguish the major approaches to protein structure prediction,** when they can be applied, and what confidence in the results is appropriate.

8 **To appreciate the state of the art in protein design.**

The significance of protein engineering

Protein scientists used to be like astronomers. We could observe our subjects but not influence them. Protein science has now become more invasive. We can modify amino acid sequences by genetic engineering, including the introduction of point mutations, insertions and deletions of specific peptides, and recombinations. Investigations of protein structure and folding have applied engineered mutants to studies aimed at clarifying different contributions to the stability of native states, or identifying intermediates or transition states in protein folding, and identifying or even modifying residues responsible for function. These scientific developments have spawned the biotechnology industry.

Protein folding

See Box: the Levinthal paradox.

The spontaneous folding of proteins to form native states solves the problem of how one-dimensional genetic information in DNA sequences is translated into three dimensions. But solving one problem, it raises another. *How* do proteins find their native states? Cyrus Levinthal pointed out that this is a serious problem.

Proteins have evolved mechanisms of folding. Might these involve a series of stable intermediates, that lead the proteins from the unfolded state to the native? If so, perhaps we could observe these intermediates by gradually destabilizing the native structure, through heating, pH changes, or the addition of denaturants such as urea or guanidinium chloride (GdmCl). In such experiments, however, many proteins behave as shown in Fig. 5.1.

This curve is expected for a simple two-state equilibrium between native (N) and denatured (D) states with no stable intermediate:

$$\text{denatured} \rightleftharpoons \text{native}$$

$$\text{or } D \underset{k_u}{\overset{k_f}{\rightleftharpoons}} N;$$

where k_f and k_u are the rate constants for folding and unfolding, respectively. For a two-state equilibrium, the thermodynamically measured equilibrium constant, $K = [N]/[D]$, should be equal to the ratio of the rate constants k_f for folding and k_u for unfolding:

$$K = \frac{[N]}{[D]} = \frac{k_f}{k_u}.$$

This is usually observed, and is confirming evidence for a simple two-state equilibrium.

For proteins that fold without stable intermediates, the lack of identifiable stages along the trajectory of folding makes it difficult to visualize the pathway. Another

The Levinthal paradox

Even if every residue had only two possible states—α_R and β, perhaps—a 100-residue peptide would have $2^{100} \sim 10^{30}$ possible conformations. If the rate of interconversion between conformations is $\sim 10^{-13}$ seconds, then the 100-residue polypeptide would require $10^{30} \times 10^{-13}$ s $= 10^{17}$ s $\sim 10^9$ *years* to explore its conformation space. Clearly proteins cannot fold by randomly groping in the dark for the native state.

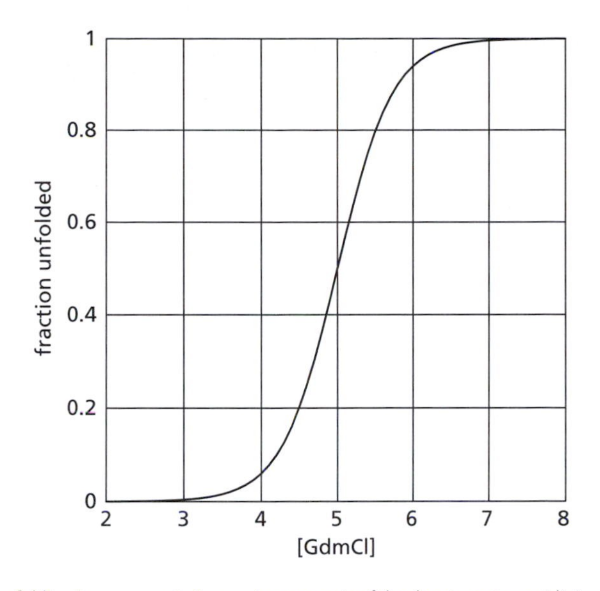

Fig. 5.1 Protein unfolding in response to increasing amounts of the denaturant guanidinium chloride (GdmCl).The amounts of native and denatured states are detected by some spectroscopic property such as circular dichroism. For low denaturant concentrations the molecules are all in their native states (fraction unfolded = 0). Over a fairly narrow range of denaturant concentration they unfold. Different proteins differ in the value of the GdmCl concentration at their mid-transition points, and in the breadths of their transitions

Many, but by no means all, proteins show this behaviour, which is expected for a simple two-state native $\rightleftharpoons$ denatured transition without intermediates.

difficulty is that the denatured state has no definite structure, so that there must be *many* folding trajectories converging to the same final native state. (Don't confuse the thermodynamic sense of the word 'state' with the idea of a definite structure. We prepare a thermodynamic state by specifying *macroscopic* variables, for example the temperature and pressure. Within a sample at a given temperature and pressure, there may be very large numbers of *microscopic* conformations, as in the denatured state of a protein, in which the polypeptide chain has great freedom and no two molecules in the sample are likely to have the same detailed structure. Alternatively, there can be a *single* microscopic conformation consistent with the macroscopic conditions, as in the native state of a protein.)

Not all proteins follow simple two-state folding schemes. Disulphide bridge formation can complicate the folding process—especially if formation of non-native disulphide bridges is possible—as can *cis* $\rightleftharpoons$ *trans* isomerization of proline residues.

Insulin takes steps to avoid such problems. Mature insulin contains two chains, one of 21 residues and the other of 30 residues, with one intrachain and two interchain disulphide bridges. Attempts to denature and renature insulin—breaking the disulphide bridges—give poor yields. Many incorrectly paired disulphide bridges form. Not good enough! *In vivo*, the precursor proinsulin is a single 86-residue polypeptide that folds into a three-dimensional structure with the cysteines in proper relative positions to form the correct disulphide bridges. Excision of a central region by peptidases leaves the mature dimer. Proinsulin can be unfolded—including reduction of the disulphide bridges—and *will* refold correctly.

Thermodynamics and kinetics—key concepts

All natural processes at ordinary temperatures are subject to two laws of thermodynamics:

- The First Law of Thermodynamics: *Energy is conserved*.
- The Second Law of Thermodynamics: *Entropy is always increasing*.

Note that the second law, but not the first, is asymmetric with respect to the direction of a process. The second law, but not the first law, can tell us which processes are spontaneous, and thereby set criteria for equilibrium.

Entropy

Think of entropy as a measure of the amount of conformational freedom accessible to a system.

The original approach to entropy linked it to heat. If a hot object is brought into contact with a cold one, heat will flow spontaneously from the hotter object to the colder one. The amount of heat absorbed by the colder object is equal to the amount of heat given up by the hotter one; satisfying the first law. The observation that heat will never flow spontaneously from a cold object to a hot one illustrates the second law.

A second approach colloquially regards entropy as equivalent to disorder, and the second law as recognizing a natural tendency towards increasing disorder. The tendency of a drop of ink to diffuse until it fully occupies a glass of water illustrates this concept of entropy. Another example is the tendency of desks spontaneously to become messy. It is difficult to think of the diffusion of ink, or the spontaneous cluttering of desks, in terms of heat. However the common idea linking heat with disorder is that systems spontaneously seek states in which they have the greatest conformational freedom. Heat is *random* molecular motion—in this context ‘random’ implying the greatest dynamical freedom. The kinetic energy of a dropped shoe turns into heat when the shoe lands on the floor. From the point of view of conformations, the kinetic energy of all the molecules moving *in tandem* in the falling shoe is converted to *random and independent* thermal vibrations of molecules of the shoe and the floor.

The classical laws of thermodynamics govern the energy and entropy differences in processes in which a system goes from one stable state to another. They say nothing about the pathway between initial and final states, or about the rates of the processes. In fact, the laws of thermodynamics do not depend on the details of the structure of matter. The laws were discovered in the nineteenth century, before microscopic models of matter were even available.

Spontaneity and equilibrium

The first law does not distinguish between forward and reverse *directions* of a process. The second law does: the reverse of a spontaneous process is not a natural process. The flow of heat from a hot object to a colder one is a spontaneous process, but heat does *not* flow spontaneously from a cold object to a hot one. The tendency of desks spontaneously to become cluttered is a natural process, and desks do *not* spontaneously unclutter themselves. The conversion of kinetic energy of a dropped shoe to heat, upon landing on the floor, occurs spontaneously, and the reverse process is never observed. True, a steam locomotive converts *some* of the heat in its boiler into kinetic energy, but the second law insists that no engine could convert heat energy *completely* into kinetic energy.

The second law tells us what processes are spontaneous. It thereby also tells us about equilibrium. For a system to be in equilibrium means that it is in a state that is *not* the initial state of any spontaneous process.

Many biological activities involve driving a non-spontaneous process by coupling it to a spontaneous one. For instance, other things being equal, the concentrations of solutes will tend to become equal throughout a solution. Dissipation of a concentration gradient by diffusion is a spontaneous process. If it is possible to establish a proton gradient across a membrane, it is possible to use the spontaneity of proton translocation to drive the synthesis of ATP. This is what ATPase does.

Usually, attainment of equilibrium requires a compromise between the demands of energy and entropy. The equilibrium between liquid water and steam at the boiling point is a balance between the higher cohesive forces in the liquid (lower energy) and the greater conformational freedom in the vapour (higher entropy). Similarly, the equilibrium between the native and denatured states of a protein at the melting point is a balance between the higher cohesive forces in the compact native state and the greater conformational freedom in the ensemble of denatured conformations.

Equilibrium in a system held at constant temperature and pressure is governed by the **Gibbs free energy**: $G = H - TS$. Here, $H = E + PV$ is the enthalpy; for any process carried out at constant pressure H is equivalent to heat. A system held at constant temperature and pressure will be at equilibrium if the Gibbs free energy is at a minimum. At the minimum its derivatives are zero. Therefore $\Delta G = 0$ *for any infinitesimal process*

Dictionary of thermodynamic quantities

E	Energy
P	Pressure
V	Volume
T	Absolute or Kelvin temperature
$H = E + PV$	Enthalpy
S	Entropy
$G = E - TS$	Gibbs free energy
$\Delta G = 0$	Criterion for equilibrium at constant temperature and pressure

starting from the current state is the criterion for equilibrium at constant temperature and pressure.

For any reaction, such as:

$$\text{haemoglobin} + 4O_2 = \text{haemoglobin–}(O_2)_4,$$

we can write an *equilibrium constant expression, K*:

$$K = \frac{[\text{haemoglobin} - (O_2)_4]}{[\text{haemoglobin}] \times p_{O_2}^4};$$

in which [haemoglobin] is the ratio of the haemoglobin concentration to the standard concentration 1 mol l^{-1}, or, equivalently the numerical value of the haemoglobin concentration expressed in units of mol l^{-1}; p_{O_2} is the ratio of the partial pressure of oxygen over the solution to the standard pressure, 1 atm.* Here, K measures the difference between ΔG for an actual reaction and the value ΔG^θ that would be observed if the reactants and products were in their standard states (for example, for the O_2, the standard state is p_{O_2} = 1 atm):

$$\Delta G - \Delta G^\ominus = RT \ln K.$$

Setting $\Delta G = 0$ as the condition for equilibrium gives the well-known equation relating the equilibrium constant and standard Gibbs free energy change:

$$\Delta G^\ominus = -RT \ln K.$$

At equilibrium ΔG is zero but $\Delta G^\ominus$ need not be zero.

Kinetics

The interpretation of kinetic data in terms of thermodynamic properties of unstable intermediates is a heuristic extension of thermodynamics. Figure 5.2 shows a free-energy diagram of a reaction that proceeds from a well-defined initial state (the reactants), through a set of unstable and ill-defined configurations, to a well-defined final state (the products). Some point on the reaction trajectory has the highest free energy. The free energy of the system at this point is higher than that of the initial state, and represents a barrier that the system must overcome in order to pass to the final state. The difference in free energy between the initial state and the transition state is the energy of activation, $\Delta G^\ddagger$.

Kinetics is blind, in general, to all but the highest peak in the reaction pathway. Figure 5.2 is drawn as if the free energy along the pathway changes smoothly from the reactants up to the intermediate and then smoothly down to the product. However, the reaction rate would be the same if the energy surface had many bumps and ripples smaller than the highest peak. It is the highest point of the barrier that governs the rate.

* Why the laboured periphrasis? It is *essential* to recognize that *equilibrium constants are dimensionless numbers*, else we couldn't take their logarithms. (Have you ever tried to take the logarithm of an atmosphere, or a mol l^{-1}?) Even this definition of K is somewhat simplified.

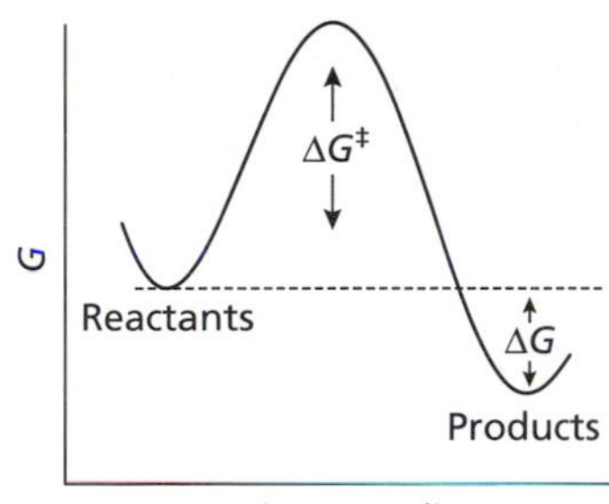

Fig. 5.2 Simplified diagram of a reaction. $\Delta G = G(\text{products}) - G(\text{reactants})$. Because $\Delta G < 0$ (G of the products is lower than G of reactants), the reaction is spontaneous. $\Delta G^{\ddagger}$ = free energy of activation = free energy at top of highest barrier − free energy of reactants. The activation energy of the reverse reaction is $\Delta G^{\ddagger} + |\Delta G|$.

The higher the energy of activation, the longer it will take the system to pass over the barrier. The higher the temperature, the more frequently the required activation energy will be available. This is why reaction rates generally increase with temperature.

The relationship between activation energy and reaction rate is the van't Hoff equation:

$$\text{reaction rate} = \text{constant} \times e^{-\frac{\Delta G^{\ddagger}}{RT}}.$$

It is thereby possible to determine $\Delta G^{\ddagger}$ from the straight line expected by plotting the logarithm of reaction rate against $1/T$:

$$\ln(\text{reaction rate}) = \ln(\text{constant}) - \frac{\Delta G^{\ddagger}}{R} \times \frac{1}{T}.$$

Thermodynamics of the protein folding transition

The stability of a native state is the free energy change of the folding transition, Denatured ⇌ Native or D ⇌ N: $\Delta G_{D \rightleftharpoons N} = G(N) - G(D)$. A stable native state corresponds to a *negative* value of $\Delta G_{D \rightleftharpoons N}$. Under physiological conditions the equilibrium D ⇌ N lies well over to the native side. In many cases, a mutation will destabilize the native state of a protein, shifting the equilibrium towards the denatured state.

The relatively low stability of native states of natural proteins is measured quantitatively by values of:

$$\Delta G^{\circ} = \Delta H^{\circ} - T\Delta G^{\circ}$$

(see Table p. 192). Typically $\Delta G^{\circ} \simeq 20\text{–}60\ \text{kJ mol}^{-1}$. The values of ΔH and $T\Delta S$ are much larger in magnitude than ΔG°. However, they largely cancel each other out, leaving as their difference only a small value of ΔG°. The high negative value of $T\Delta S^{\circ}$ arises from the loss of freedom in going from the unfolded state with many possible conformations to the native state with only one. To achieve a negative ΔG°—that is, stability of

Thermodynamics of protein folding at 298 K

Protein	ΔG°/(kJ mol^{-1})	ΔH°/(kJ mol^{-1})	TΔS°/(kJ mol^{-1})
CI2	−27.7	−135	−107.3
Eglin c	−36.9	−115	−78.1
RNase T1	−37.5	−281	−243.6
Cytochrome c	−37.1	−89	−51.9
Barnase	−48.9	−307	−258.2
Lysozyme	−57.8	−242	−184.3
Chymotrypsin	−45.7	−268	−222.4
Tendamistat	−37.5	−70	−32.5

From: Makhatadze, G.I. and Privalov, P.L. (1995). Energetics of protein structure. *Adv. Prot. Chem.*, **47**, 307–425.

the native state—the protein must find strong interresidue interactions to provide large negative ΔH°.

Thermodynamics of mutated proteins

Studies with both natural and engineered mutants show the effects of amino acid sequence changes on protein stability. Different mutations, at different positions, vary in importance. A small fraction of residues appear to be absolutely essential. These are resistant to substitution. Others—typically about one-third of the residues—make significant contributions to stability, in that substitution decreases but does not entirely destroy stability. The magnitude of the thermodynamic effects on protein stability correlates well with substitution patterns in aligned sequences of protein families.

The effect of denaturants on rates of folding and unfolding: chevron plots

The kinetics of folding of mutated proteins gives clues to the structure of the transition state for folding. The usual methods of structure determination are inapplicable to unstable transition states, because their concentrations are too low and they are too short-lived. However, comparison of the stabilities and rates of unfolding and refolding between wild type and mutated proteins reveal the importance of the mutated site in the structure of the transition state.

Reasonable structural assumptions allow us to interpret the effects of mutations on rates of folding and unfolding. Suppose that a mutated residue makes no intramolecular interactions in the denatured state, but has a destabilizing effect on the native state. If the mutated residue also makes no intramolecular interactions in the transition state, the activation energy for folding $\Delta G^{\ddagger}_{\text{folding}} = G(\text{transition state}) - G(\text{denatured})$ will be the same in the wild type and the mutant, but the activation energy for unfolding $\Delta G^{\ddagger}_{\text{unfolding}} = G(\text{transition state}) - G(\text{native})$ will be lower in the mutant (see Fig. 5.3(a)). Conversely, if the mutated residue makes the same interactions in the transition state that it does in the native state, the activation energy for unfolding $\Delta G^{\ddagger}_{\text{unfolding}} = G(\text{transition state}) - G(\text{native})$ will be the same in the wild type and the mutant, but the activation energy for folding $\Delta G^{\ddagger}_{\text{folding}} = G(\text{transition state}) - G(\text{denatured})$ will be higher in the mutant (see Fig. 5.3(b)). By applying this approach to many residues spread over the sequence, a fair job of mapping out the interactions in the transition state is possible.

In many cases, the interactions of a residue in the transition state are not identical either to those in the denatured state or to those in the native state. That is, many mutations affect the native and transition states to different extents. The relative amount is measured by a parameter Φ:

$$\Phi_{\text{f}} = \frac{\Delta G^{\ddagger}(\text{wild type}) - \Delta G^{\ddagger}(\text{mutant})}{\Delta G_{\text{N}\to\text{D}}(\text{wild type}) - \Delta G_{\text{N}\to\text{D}}(\text{mutant})}.$$

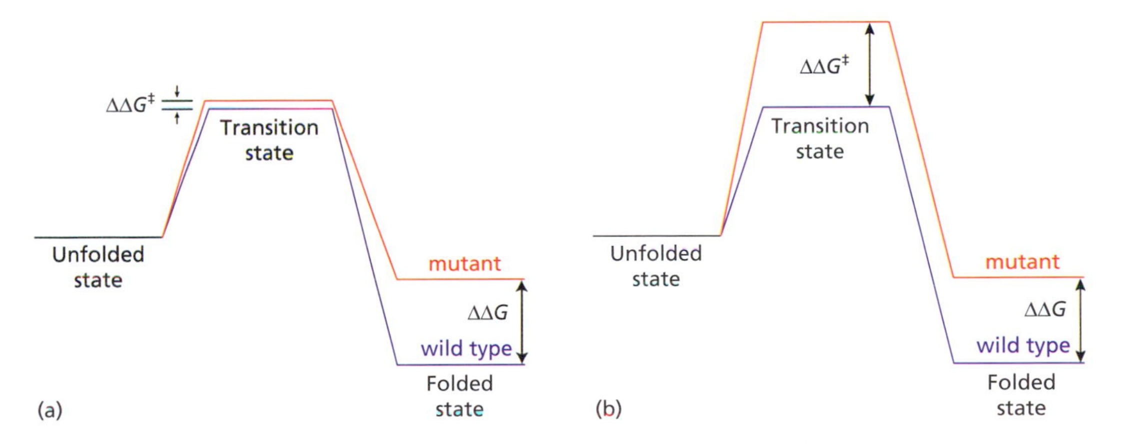

Fig. 5.3 Reaction diagrams showing two possible effects of mutants on the relative destabilization of transition states and folded states relative to the unfolded state, as measured by Φ-values = ratios of destabilization of the transition state to the destabilization of the folded state. The energies of the unfolded states for wild type and mutant are set at the same level, but it is not true that the unfolded states have identical energies. (a) This mutant destabilizes the native state but does *not* destabilize the transition state. In this case $\Phi = 0$. The implication is that the residue at the mutated position is *not* making the interactions in the transition state that it does in the native state, or that the mutated position is *not* surrounded by native-like structure. (b) This mutant destabilizes the transition state by the same amount that it destabilizes the folded state. In this case $\Phi = 1$. The implication is that the residue at the mutated position is making the same interactions in the transition state as in the native state, or that the structure of the transition state in the neighbourhood of the mutated position is native-like.

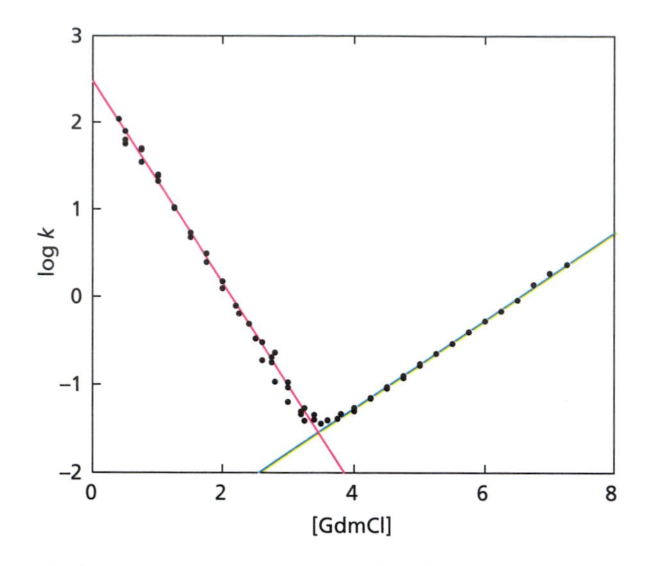

Fig. 5.4 Folding and unfolding rates of ribosomal protein S6 from *Thermus thermophilus* at different concentrations of denaturant guanidinium chloride. Points at [GdmCl]< 3.34 represent folding rates; points at [Gdm Cl]> 3.34 represent unfolding rates. This figure shows a **chevron plot** of folding and unfolding rate constants as a function of denaturant concentration. Upon denaturation, residues buried inside the protein in the native state are exposed to solvent. The reason that high concentrations of denaturant unfold the protein is that there is a more favourable interaction of peptides with denaturant (in the unfolded state) than with other peptides (buried in the folded state) or even with water.

On this basis we can explain the form of the chevron plot. Assuming that the transition state is partially unfolded, the effect of the denaturant is (see Fig. 5.5):

the lowering of energy of native state
< the lowering of energy of transition state
< the lowering of energy of unfolded state

Don't you agree that a diagram would make this clearer? See Problem 2.

Therefore, as the concentration of denaturant increases, the activation energy for unfolding is reduced—and the unfolding rate increased—and the activation energy for folding is raised—and the folding rate decreased.

(Data from: Otzen, D.E., Kristensen, O., Proctor, M., and Oliveberg, M. (1999). Structural changes in the transition state of protein folding: alternative interpretations of curved chevron plots. *Biochemistry*, **38**, 6499–511.)

To measure Φ, determine $\Delta G_{N\rightarrow D}$ for wild type and mutant either from equilibrium or rate measurements, and $\Delta\Delta G^{\ddagger}$ from the kinetics of folding and unfolding (see Fig. 5.4).

Different proteins show different distributions of Φ-values. In barnase, the Φ-values are almost all nearly 0 or 1, suggesting that the stage of the folding process between the transition state and the native state involves the docking of preformed structural domains.

In CI2 the values of Φ are broadly distributed, suggesting that folding forms around an extended structural nucleus.

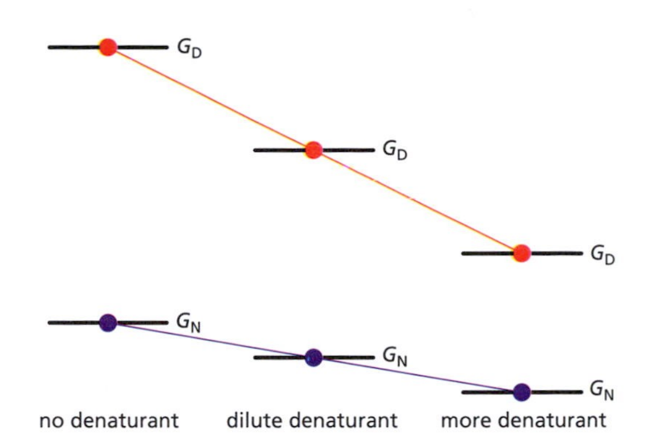

Fig. 5.5 Dependence of free energy of native and denatured states on denaturant concentration. The energy of interaction of peptides with denaturant is greater than their energy of interaction with water. Therefore, increasing concentrations of denaturants stabilize both the denatured state and the native state. However, the effect on the denatured state is greater, because more peptide is exposed to solvent in the denatured state than in the native state, as a result of the compact packing in the native state. This is seen in the figure as a greater change in the level of the denatured state (G_D, red line) than the level of the native state (G_N, blue line) with increasing denaturant concentration.

Increasing concentrations of denaturant reduce the stabilization of the native state with respect to the denatured state. The quantity:

$$G_{\text{denatured state}} - G_{\text{native state}} = \Delta G_{N \to D}$$

is a measure of the stability of the native state of a protein. In the absence of denaturant, it is typically in the range 20–60 kJ mol^{-1}. A *linear* dependence of stability on denaturant concentration is often observed:

$$\Delta G_{N \to D} = m \times \text{denaturant concentration.}$$

This will certainly be true if both G_N and G_D individually depend linearly on the denaturant concentration as in this figure.

Eventually the red and blue lines will cross. At this point, $G_N = G_D$ and the concentrations of denatured and native molecules will be equal. At higher denaturant concentrations, the sample will become more completely denatured (see Fig. 5.1).

CASE STUDY Comparison of folding pathways of a natural protein and a circular permutant

S6, a 101-residue protein isolated from the small ribosomal subunit of *Thermus thermophilus*, undergoes a sharp Native $\rightleftharpoons$ Denatured transition. Lindberg, Tångrot, and Oliveberg* studied the effects of mutations on the folding of S6, and of circular permutants synthesized by inserting linking peptides at the N- and C-termini, and cutting individual peptide bonds.

Figure 5.6 shows the rate constants for folding and unfolding as a function of guanidinium chloride concentration [GdmCl] for the wild type and an A88V mutant, for the original sequence (Fig. 5.6(a)), and for a circular permutant cleaved between residues 13 and 14 (Fig. 5.6(b)). Figure 5.7 shows the structures, and indicates the sites of mutation and cleavage.

* Lindberg, M., Tångrot, J., and Oliveberg, M. (2002). Complete change of the protein folding transition state upon circular permutation. *Nat. Struct. Biol.*, **9**, 818–22.

continues...

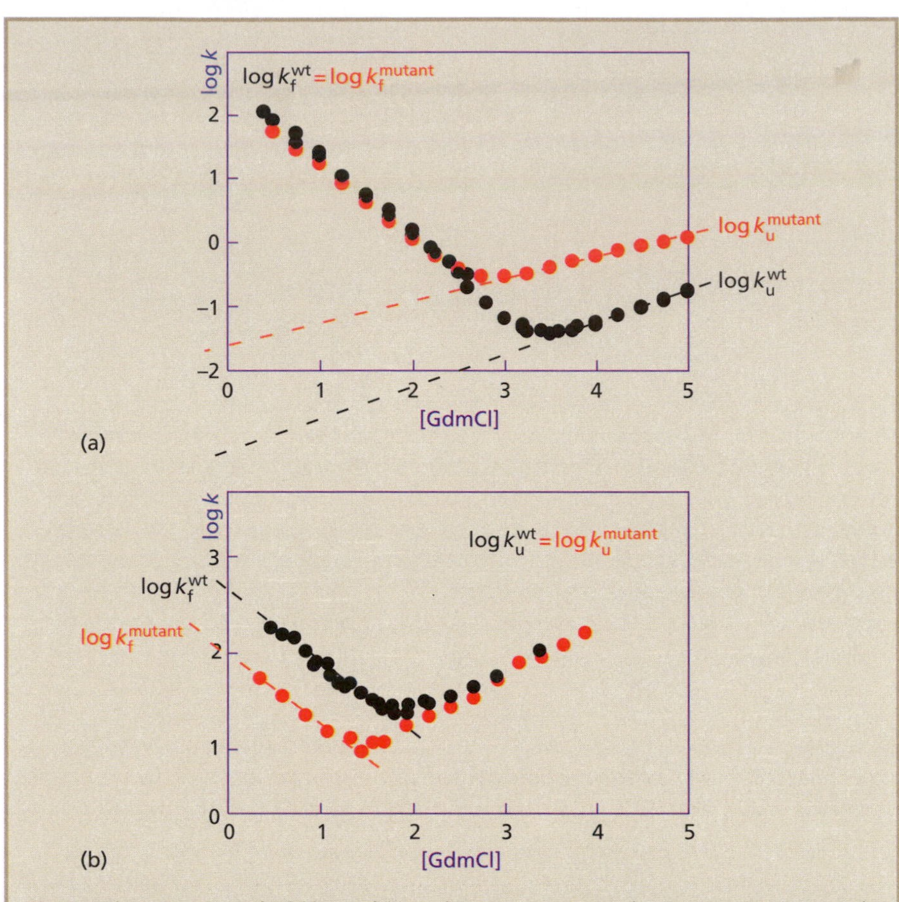

Fig. 5.6 The rate constants for folding/unfolding of S6 and mutant V88A, for the original sequence and for a circular permutant cleaved between residue 13 and 14; (a) wild type, (b) circular permutant.

In the wild type, the V88A mutation has no effect on the rate of folding, $\Delta\Delta G_{D-\ddagger} \sim 0$, but a large effect on the rate of unfolding (Fig. 5.6(a)). $\Delta\Delta G_{D-N} > \Delta\Delta G_{D-\ddagger}$, and $\Phi = 0.1–0.2$. For the circular permutant, in contrast, the V88A mutation has a large effect on the rate of folding and little effect on the rate of unfolding (Fig. 5.6 (b)): $\Phi \sim 1$.

These data show that the transition states for folding of the wild type and circular permutant are different, at least as far as residue 88 is concerned. In the transition state of the circular permutant, this residue makes native-like contacts. In the transition state of the wild type, it remains unconstrained.

See p. 198 contact order.

The topologies of the structures make these results reasonable. S6 contains a four-stranded β-sheet, and two helices packed against it. *In the wild type*, the order of the strands from the left in Fig. 5.7 is 2–3–1–4, with position 88 in the C-terminal, rightmost strand, and strands 2 and 3 forming an internal hairpin (light blue in Fig. 5.7). The idea that residues nearby in the sequence tend to participate in early folding events suggests that the hairpin formed by the two strands central in the sequence in the wild type might fold early, and the N- and C-terminal strands (the two rightmost strands in Fig. 5.7, containing the mutated position), might fold late, *after* passing the transition state.

continues ...

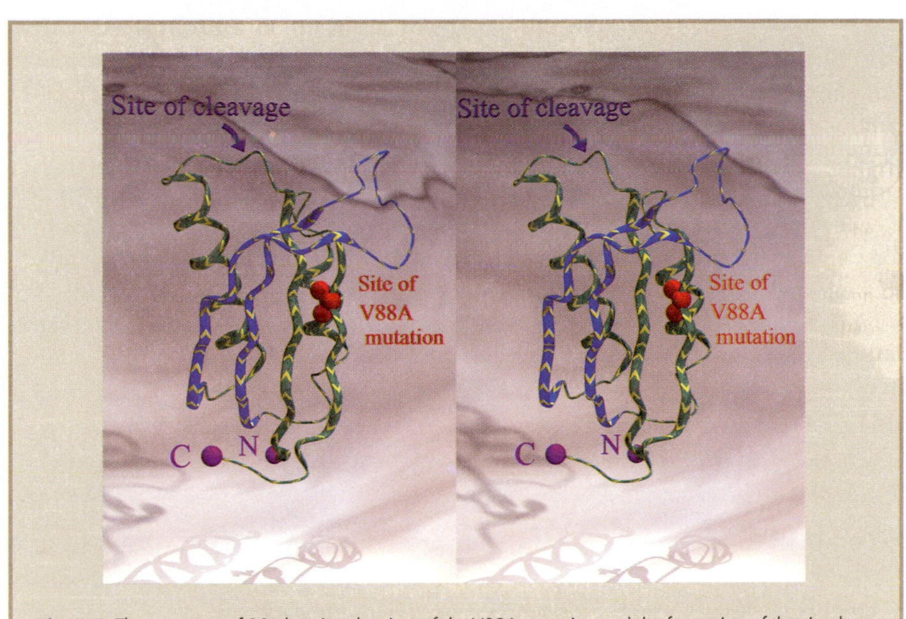

Fig. 5.7 The structure of S6, showing the sites of the V88A mutation and the formation of the circular permutant. To make the circular permutant, the original N- and C-termini, shown here with large purple spheres, are connected by a linking peptide, and the chain cleaved between residues 13 and 14. Residues shown in blue form an internal hairpin.

In the circular permutant, the mutated position is in a hairpin, between two strands consecutive in the sequence, and it is reasonable that they fold before the transition state.

A more complete picture of the transition state requires surveying a broad distribution of mutations. Lindberg, Tångrot, and Oliveberg measured Φ-values for additional mutants deployed throughout the molecule. For the circular permutant, the distribution of Φ-values is bimodal, most values being either close to 0 or close to 1. Residues with high Φ-values appeared in the newly linked strands 1 and 4; residue 88, with $\Phi \sim 1$, is an example. For the wild type, in contrast, Φ-values were generally low, and distributed fairly uniformly throughout the sequence.

The results suggest that the wild type has evolved a diffuse folding pathway, by creating stronger interactions between residues distant in the sequence, which, in principle, have a harder time finding their correct partners. Forming the circular permutant disturbs the folding pathway. Instead, in the circular permutant, the mutated residue interacts with residues nearby in *both* the sequence and the structure, creating a localized nucleus for folding.

These S6 data show that both decentralized and locally condensed transition states are possible. It has been suggested that decentralized states, with highly cooperative folding transitions, have evolved to reduce problems caused by aggregation of partially folded proteins, and to compensate for low thermodynamic stability of native states. Given that interactions between residues close in the sequence have a natural kinetic advantage and hence a tendency to form early, *how* do proteins avoid local folding nuclei? Data on S6 mutants show that, in the wild type, contacts between residues distant in the sequence are relatively more stabilizing than contacts between residues close in the sequence. Destabilizing the interactions between residues close in the sequence gives an energetic advantage to interactions between residues far apart in the sequence, compensating for their kinetic disadvantage—'levelling the playing field'.

The molten globule

En route to the native state, many folding proteins pass through a partially structured state with the intriguing name of the **'molten globule'**. This state can be observed under mildly denaturing conditions. Energetically, proteins differ in the stability of their molten-globule states, and in the structural relationship between the molten-globule and the native states. For instance, the molten-globule state of chicken lysozyme is only a transient intermediate in the pathway. In contrast, the molten-globule state of the homologue α-lactalbumin is a long-lived intermediate in the unfolding. The folding of α-lactalbumin is not a simple native–denatured equilibrium. For some proteins, observation of the molten–globule state is difficult unless its formation is promoted by the addition of moderate amounts of denaturant.

Structurally, the molten–globule state has:

- a relatively compact conformation, perhaps 10–30% larger than the native state, as shown from measurements of molecular size by hydrodynamic measurements, small-angle X-ray scattering, and gel-filtration chromatography;
- a substantial amount of intact secondary structure, as shown from circular dichroism measurements;
- loose packing of the hydrophobic core, without formation of the native-like tertiary structure; residues that are buried in the native structure are accessible to water, as revealed by interactions with the hydrophobic dye ANS (8-anilino-1-naphthalene-sulphonic acid), or by NMR studies;
- more structural flexibility than the native state, corresponding to a spectrum of rapidly interconverting conformations.

The picture of the molten-globule state is that it has at least some of the correct secondary structure and folding pattern. However, the helices and sheets have not come together with their side chains interlocking tightly as in the native state.

The relationship between the structure and kinetics of folding: contact order

Contact order is a measure of the separation in the sequence of residues that are close by in space:

$$\text{Contact order} = \frac{1}{LN}\Sigma(i - j);$$

where L is the number of residues in the protein, N is the number of residues in contact, $i - j$ is the distance in the sequence between residues i and j, and the sum is taken over all pairs of residues in contact. It is observed that for proteins with simple two-state behaviour, the folding rate is strongly correlated with the contact order: the greater the tendency of residues nearby in the sequence to interact in the structure, the faster the folding tends to be:

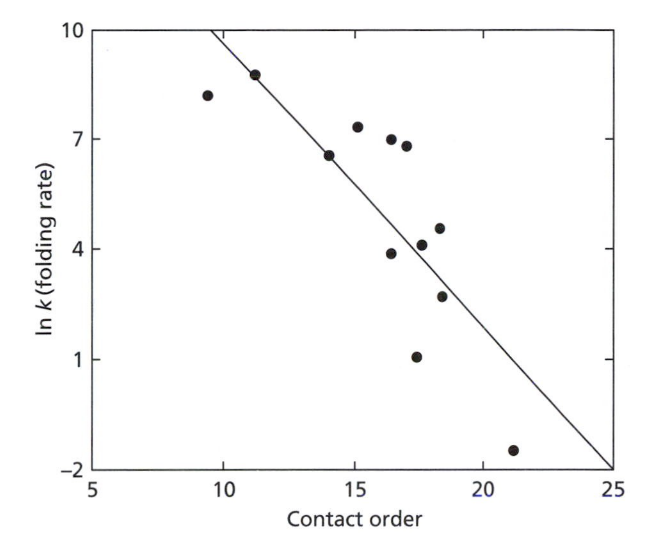

Unfolded proteins can most quickly explore interactions between residues nearby in the sequence. The greater the extent that interactions between residues nearby in the sequence are native-like, the faster the folding rate.

Folding funnels

This chapter contains many conventional diagrams in which the Gibbs free energy of the system is plotted against a single, not very precisely defined, variable called the '**reaction coordinate**'. Even for the simplest chemical systems, this is a simplification. For protein folding the simplification is so severe as to be downright misleading.

Two problems with the conventional reaction coordinate diagram are:

1. The diagram suggests that the system follows a unique trajectory. In fact, the denatured state comprises many different conformations, and, in principle, there should be a different reaction coordinate describing how each denatured conformation folds to the common native state.
2. Although it would be easy to show a folding pathway that passed through several intermediates, by producing a reaction coordinate diagram with several successive minima, it is not possible in the one-dimensional diagram to show 'blind alleys'—off-path or non-productive intermediates.

The current view is that an appropriate schematic representation of the folding process is a **folding funnel** (Fig. 5.8). The funnel is wide at the top, to encompass all the different denatured conformations. It is possible to get from different unfolded conformations to the native state by many routes, each passing through different sets

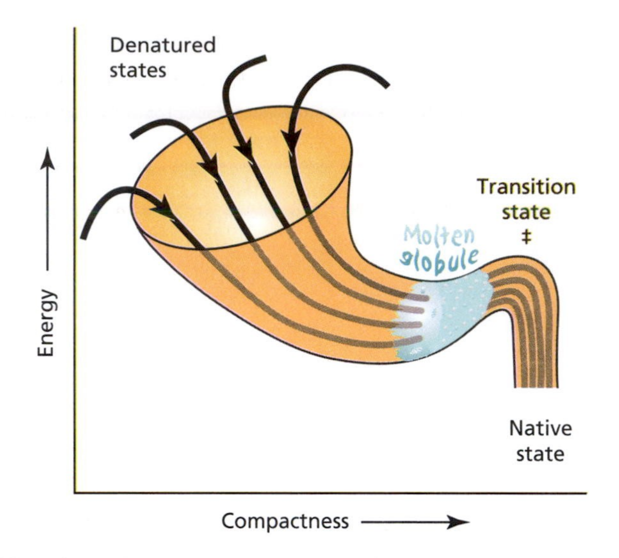

Fig. 5.8 The **folding funnel,** a schematic representation of the progress from high-energy, spatially diffuse denatured conformations to the low-energy, compact native state. The process can be thought of as a kind of intramolecular crystallization. Folding proceeds by collapse of the chain to a relatively compact molten globule state, in which much of the secondary structure is formed but the side chain packing that permits the very high compactness of the native state is not locked in.

This figure generalizes the simple reaction-coordinate picture (see, e.g., Fig. 5.2). In recognition of the conformational heterogeneity of the unfolded state, a multiplicity of trajectories links the denatured and native states. The trajectories converge in the molten globule and transition states.

The simple reaction-coordinate diagram (to speak in its defence) is intended as a description of macroscopic, or thermodynamic, states. In contrast, the folding-funnel idea alludes to microscopic states.

Of course even this diagram is a simplification. To make it easier to understand, there is no indication of local minima, non-productive intermediates, multiple transition states, competition from aggregation, etc. In Problem 6 the reader is invited to sketch them in.

of conformations with different partial native-like features. But these different trajectories eventually coalesce, shown by the narrowing of the funnel.

Protein misfolding and the GroEL–GroES chaperone protein

Proteins fold spontaneously to their native states, based only on information contained in the amino acid sequence . . . but sometimes they need a little help. After all, the dilute salt solution of the physical chemistry laboratory is one thing. The intracellular medium is quite another, with concentrations of macromolecules in the range of 300–400 mg ml^{-1}. The threat is that partially folded or misfolded proteins may form aggregates.

Cells have therefore developed molecules that catalyse protein folding, called '**chaperones**'. The name is apt: molecular chaperones supervise the states of nascent proteins, hold them to the proper pathway of folding, and keep them away from improper influences that might lead to incorrect assembly or non-specific aggregation. Heat-shock or viral infection enhances the danger. These conditions induce the over-expression of chaperone proteins, which is how some of them were originally discovered and why they are often called 'heat-shock proteins'.

Chaperones do not contradict the basic tenet that amino acid sequences dictate the three-dimensional structures of proteins. Chaperones act *catalytically* to speed up the process of protein folding, by lowering the activation barrier between misfolded and native states. They do not alter the result. Indeed, the observation that a chaperone can catalyse the folding of many proteins, with very different secondary and tertiary structures, is consistent with the idea that chaperones themselves contain no information about particular folding patterns. They anneal misfolded proteins, and allow them, rather than direct them, to find the native state.

The chaperonin system GroEL–GroES of *E. coli* contains two products of the *GroE* operon, GroEL (L for large, r.m.m. = 58 000) and GroES (S for small, r.m.m. = 10 000). The active complex contains 14 copies of GroEL and 7 copies of GroES, for a total r.m.m. of almost 10^6.

In the absence of GroES, 14 GroEL molecules form two 7-fold rings packed back-to-back (Fig. 5.9). Each ring surrounds a cavity open at one end to receive substrate (i.e. misfolded protein). The cavity is closed at the bottom (the wall between the two rings), so that bound protein cannot pass internally from one ring to the other. Nevertheless, the two rings are not independent; they communicate via allosteric structural changes that are an essential component of the mechanism. Mutants containing only a single GroEL ring form complexes with substrate, ATP and GroES, but cannot release them.

With substrate in the cavity, binding of ATP and GroES enlarges the cavity, closes it off, and changes its structure. The GroES subunits form another 7-membered ring that

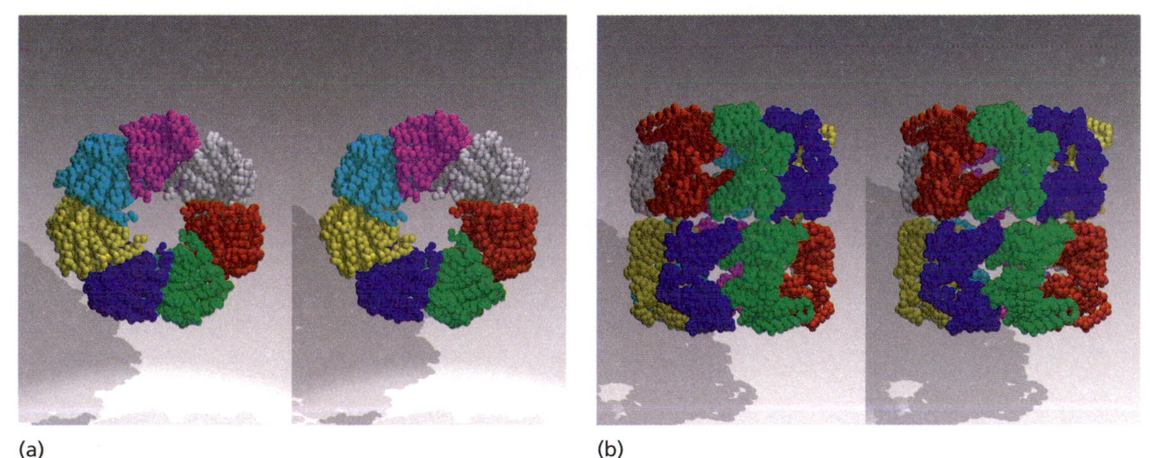

(a) (b)

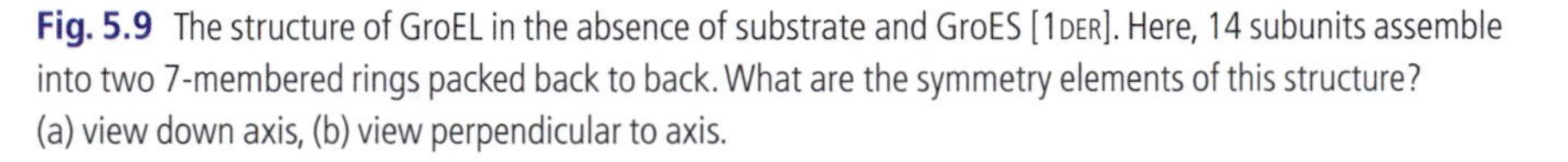

Fig. 5.9 The structure of GroEL in the absence of substrate and GroES [1DER]. Here, 14 subunits assemble into two 7-membered rings packed back to back. What are the symmetry elements of this structure? (a) view down axis, (b) view perpendicular to axis.

caps the GroEL ring (Fig. 5.10). The GroEL ring capped by GroES is called the *cis* ring, and the non-capped ring is the *trans* ring. Formation of the GroEL–GroES complex requires a large and remarkable conformational change in the *cis* GroEL ring, changing the interior surface of the cavity from hydrophobic to hydrophilic, and breaking the symmetry between the two GroEL rings (Fig. 5.11). An allosteric change mediates a negative cooperativity between the rings, precluding *both* rings from forming the GroES-capped structure simultaneously.

The enclosed cavity is the site of protein folding. Misfolded proteins in the cavity are given a chance to refold. After ~20 seconds they are expelled, either folded successfully, or, if not, with the chance to enter the same or another complex to try again. Misfolded proteins are treated as juvenile offenders: they are caught, incarcerated, kept in solitary confinement, and—having been given an opportunity to reform—they are released.

The GroEL–GroES conformational change

The GroEL monomer contains three domains (Fig. 5.11), an apical domain (red), a hinge domain (green), and an equatorial domain (blue). Figure 5.11 compares the

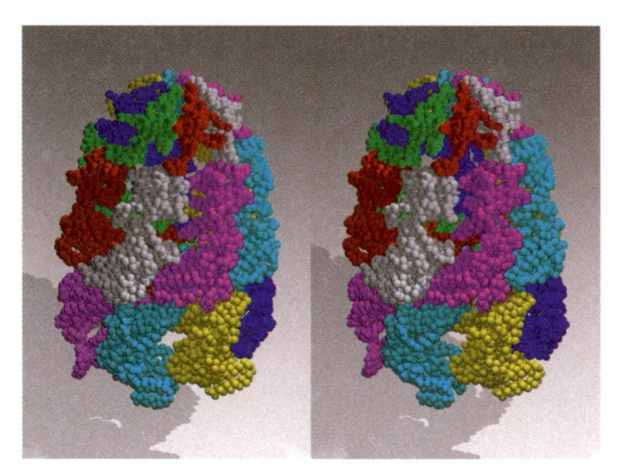

Fig. 5.10 The GroEL-GroES complex [1AON]. One of the GroEL rings is 'capped' by a third 7-membered ring of GroES subunits.

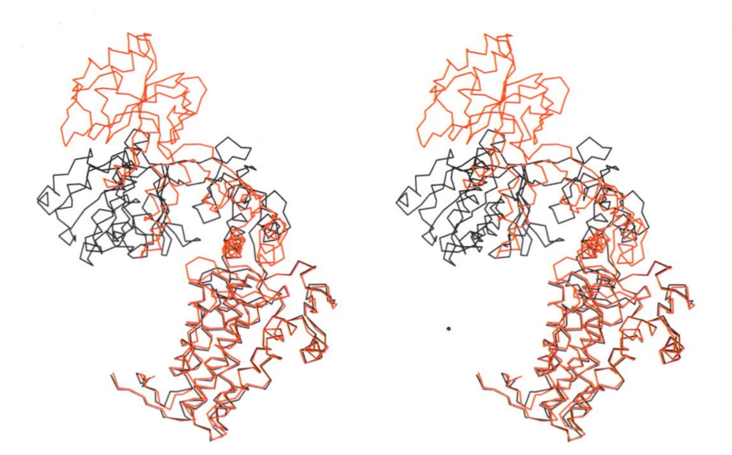

Fig. 5.11 The conformational change in one GroEL subunit. The apical domain hinges up, and rotates by ~90° to change the interior surface of the cavity. A second hinge motion locks in the ATP [1DER] and [1AON].

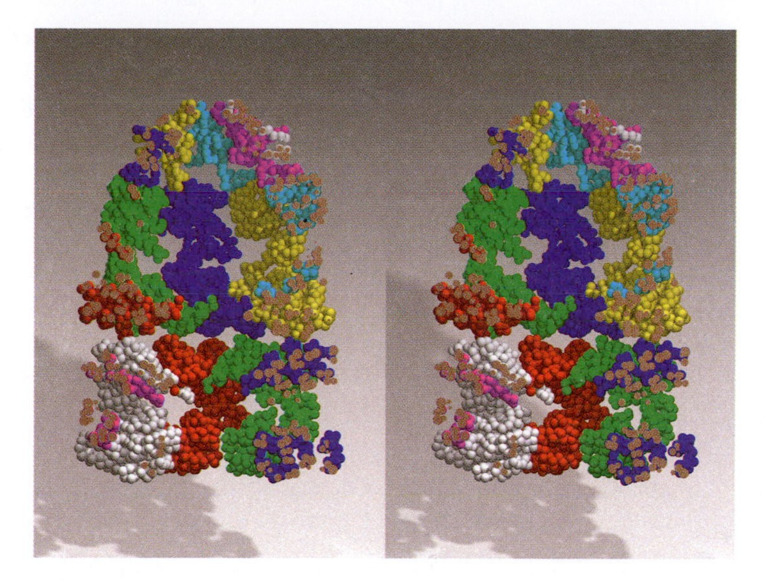

Fig. 5.12 A cutaway view of the GroEL-GroES complex, showing the cavity [1AON].

structures of (a) the unbound form of GroEL, with (b) the bound form, with one subunit of GroES bound. The viewpoint is perpendicular to the 7-fold axis, that is, tangential to the ring. Figure 5.12 shows the complex, cut away to expose the cavity.

The most obvious conformational change in GroEL is the swinging up of the apical domain by an ~60° hinge motion. A second hinge, near the ATP-binding site in the equatorial domain, rotates to lock in the ATP, and to expose a residue critical for its hydrolysis. The most remarkable feature of the conformational change is the ~90° rotation of the apical domain to expose a different surface to the interior. In the unbound form, the residues lining the GroEL cavity are hydrophobic. In the bound form, they are hydrophilic, with the hydrophobic residues that formed the lining now taking part in intersubunit contacts.

This is inspired, but of course perfectly logical. Recall that typical globular proteins have preferentially hydrophobic interiors and charged/polar exteriors. The characteristic of misfolded proteins, that renders them subject to non-specific aggregation, is the accessibility of hydrophobic residues that are buried in the native state. Proteins in such states bind to the open form of GroEL, with its channel lined with hydrophobic residues. What would one want a chaperone to do to such a misfolded protein, once it has it in its clutches? Altering the interior surface from hydrophobic to hydrophilic encourages the protein to turn itself right side out.

Operational cycle

The assembly functions like a two-state motor. Each of the 7-member GroEL rings may be in one of two states—open, ready to receive misfolded proteins, or closed, containing misfolded proteins and capped by the GroES ring.

1. In the unbound state the GroEL ring is open to allow protein to enter. The interior interface presents a flexible hydrophobic lining, suitable to bind misfolded proteins

by non-specific hydrophobic and Van der Waals interactions. Indeed, the binding process may even partially unfold proteins in incorrect states.

2. The binding of ATP and GroES, and the conformational change in the *cis* GroEL ring, creates the closed cavity in which the substrate protein, once released from the apical domains, can refold, sequestered away from potential aggregation partners. The conformational change more than doubles the volume of the cavity, to accommodate non-compact unfolding/refolding transition states. The interior surface changes from hydrophobic to hydrophilic, peeling the bound misfolded protein off the surface and unfolding it even further. The burial of the original interior GroEL surface in intersubunit contacts within the GroEL–GroES complex itself breaks the binding of the protein to the original hydrophobic internal surface, leaving a macroclathrate complex.
3. Hydrolysis of ATP in the *cis* ring weakens the structure of the *cis*-ring/GroES complex. Binding of ATP (but not necessarily its hydrolysis) in the *trans* ring triggers the disassembly of the *cis* assembly and the release of GroES and substrate protein, restoring the ring to its original state.

Each cycle of the engine requires hydrolysis of 7 or even 14 ATP molecules. The energetic cost is much larger than the energy of unfolding of a protein, but small compared to synthesis of the polypeptide chain, and very small compared to the death of the cell.

Protein structure prediction and modelling

If proteins fold spontaneously into unique three-dimensional native conformations, nature must have a rule, or algorithm, for determining protein structure from amino acid sequence. It should be possible for us to reproduce this algorithm in a computer program.

There are two approaches to a prediction of protein structures. One is to learn as much as possible about the natural process of protein folding, and try to simulate it in a computer, either in detail or in simplified ways. The other is to ignore the natural folding process, and to consider any possible approach that takes in the sequence and prints out the structure, including methods that make use of known protein sequences and structures in databanks.

Some attempts to predict protein structure by following the natural mechanism try to represent interatomic interactions in proteins explicitly, to define a computable energy associated with any conformation. Then the problem of protein structure prediction becomes a computational task of finding the global minimum of this conformational energy function. So far this approach has not succeeded consistently, partly because of the imprecision of the energy function and partly because, even with the power of modern computers, the calculations cannot be run for long enough.

Other approaches, that ignore the natural folding process, are based on assembling clues to a structure from similarities to known structures. As of now, these empirical methods are the most powerful we have. They include:

- **attempts to predict secondary structure without attempting to assemble these regions in three dimensions**. The results are lists of regions of the sequence predicted to form α-helices and regions predicted to form strands of β-sheet. The most powerful methods of secondary structure prediction are based on a computational technique called *neural networks*.
- **homology modelling**: prediction of the three-dimensional structure of a protein from the known structures of one or more related proteins. The results are a complete coordinate set for main chain and side chains, intended to be comparable to at least a low-resolution experimental structure.
- **fold recognition**: given a library of known structures, determination of which of them shares a folding pattern with a query protein of known sequence but unknown structure. The results are selections of a known structure predicted to have the same fold as the query protein, or a statement that no protein in the library has the same fold as the query protein.
- **prediction of structures with novel folding patterns**. The results are a complete coordinate set for at least the main chain and sometimes the side chains also. The goal is a model with the correct folding pattern, but the results would not be expected to be comparable in quality to an experimental structure. D. Jones, a thoughtful observer of the field, likened the distinction between *a priori* modelling and fold recognition to the difference between an essay and a multiple-choice question on an exam.

Critical Assessment of Structure Prediction (CASP)

CASP organizes blind tests of protein structure predictions, in which participating crystallographers and NMR spectroscopists reveal the amino acid sequences of proteins they are investigating, and agree to keep the experimental structure secret until predictors have submitted their models. The first CASP programme took place in 1994, and it has run on a two-year cycle since then. The latest, CASP5, ran in 2002. Sequences are published in the spring, and predictions are due in the autumn.

A panel of judges compares the predicted and experimental structures, and grades the predictions. At the end of the year a gala meeting brings the predictors together to discuss the current results and to gauge progress. Speakers include the organizers, the assessors, and selected predictors: including those who have been particularly successful, or who have an interesting novel method to present.

Many predictions are prepared by groups of researchers who inspect the results generated by their computer programs, and select and edit them before submission. In addition, the target sequences are sent to Web servers, that return predictions without human intervention. This activity is called CAFASP: Critical Assessment of Fully Automated Structure Prediction. It is thereby possible to determine to what extent successful procedures could be made fully automatic. CASP thus comprises three challenges:

- human against protein: CASP
- computer against protein: CAFASP
- human against computer: CASP vs. CAFASP.

A separate program of blind tests of prediction evaluates methods for predicting protein–protein interactions, or 'docking'. This is called CAPRI—Critical Assessment of PRedicted Interactions.

Predictions in CASP fall into three main categories: (1) homology modelling, (2) fold recognition, and (3) prediction of novel folds (see Table).

CASP category	Nature of target
Comparative modelling	Close homologues of known structure are available; homology modelling methods are applicable.
Fold recognition	Structures with similar folds are available, but no relative sufficiently close for homology modelling; the challenge is to identify known structures with similar topology.
New Fold	No structure with same folding pattern known; requires either a genuine *a priori* method or a knowledge-based method that can correctly combine features of several structures.

In the latest CASP programme, there were initially 67 targets, predictions for 55 of which could be assessed. (The main reason for attrition of targets was that they were not solved in time.) There were 265 groups participating, submitting a total of 28 728 predictions.

Homology modelling

Model-building by homology is a useful technique when one wants to predict the structure of a target protein of known sequence, that is related to at least one other protein of known sequence *and* structure. If the proteins are closely related, the known protein structures—called the 'parents'—can serve as the basis for a model of the target. Although the quality of the model will depend on the degree of similarity of the sequences, it is possible to specify this quality before experimental testing (see Fig. 4.12). In consequence, knowing how good a model is necessary for an intended application permits intelligent prediction of the probable success of the exercise.

Steps in homology modelling are:

1. Align the amino acid sequences of the target and the homologous protein or proteins of known structure. It will generally be observed that insertions and deletions lie in the loop regions between helices and sheets.
2. Determine main chain segments to represent the regions containing insertions or deletions. Stitching these regions into the main chain of the known protein creates a model for the complete main chain of the target protein.

3. Replace the side chains of residues that have been mutated. For residues that have not mutated, retain the side chain conformation. Residues that have mutated tend to keep the same side chain conformational angles, and could be modelled on this basis. However, computational methods are now available to search over possible combinations of side chain conformations.
4. Refine the model by limited energy-minimization. The role of this step is to fix up the exact geometrical relationships at places where regions of the main chain have been joined, and to allow the side chains to wriggle around a bit to place themselves in comfortable positions.

In most families of proteins the structures contain some relatively constant regions and other more variable ones. The core of the structure of the family retains the folding topology, although it may be distorted, but the periphery can entirely refold. A single parent structure will permit reasonable modelling of the conserved portion of the target protein, but will often fail to produce a satisfactory model of the variable portion. Moreover, it will not be easy to predict which are the variable and constant regions. A more favourable situation occurs when several related proteins of known structure can serve as parents for modelling a target protein. These reveal the regions of constant and variable structure in the family. The observed distribution of structural variability among the parents dictates an appropriate distribution of constraints to be applied to the model.

Mature software for homology modelling is available. SWISS-MODEL is a Web site that will accept the amino acid sequence of a target protein, determine whether a suitable parent or parents for homology modelling exist, and, if so, deliver a set of coordinates for the target. SWISS-MODEL was developed by T. Schwede, M.C. Peitsch, and N. Guex, now at The Geneva Biomedical Research Institute. Another effective program is MODELLER, by A. Šali.

Comparisons between CASP and CAFASP suggest that, although programs for automatic homology modelling work well, manual intervention and editing can improve the results, if done carefully by an expert.

In a sense, this procedure produces 'what you get for free,' in that it defines the model of the protein of unknown structure by making minimal changes to its known relative. Unfortunately it is not easy to make substantial improvements. A rule of thumb (referring again to Fig. 4.12) is that if the two sequences have at least 30–40% identical amino acids in an optimal alignment of their sequences, the procedure described will produce a model of sufficient accuracy to be useful for many applications. If the sequences are very distantly related, neither the procedure described nor any other currently available method will produce a model, correct in detail, of the target protein from the structure of its relative.

Threading

Threading is a method for fold recognition. Given a library of known structures, and a sequence of a query protein of unknown structure, does the query protein share a folding pattern with any of the known structures? The fold library could include some or all of the Protein Data Bank, or even hypothetical folds.

The basic idea of threading is to build many rough models of the query protein, based on each of the known structures and using different possible alignments of the sequences of the known and unknown proteins. This systematic exploration of the many possible alignments gives threading its name: imagine trying out all alignments by pulling the query sequence gently through the three-dimensional framework of any known structure. Gaps must be allowed in the alignments, but if the thread is thought of as being sufficiently elastic the metaphor of threading survives.

Both threading and homology modelling deal with the three-dimensional structure induced by an alignment of the query sequence with known structures of homologues. Homology modelling focuses on one set of alignments, and the goal is a very detailed model. Threading explores many alignments and deals with only rough models, usually not even constructed explicitly (see the Table)

Homology modelling	**Threading**
First, identify homologues	Try all possible parents
Then, determine optimal alignment	Try many possible alignments
Optimize one model	Evaluate many rough models

Successful fold recognition by threading requires:

(1) a method to score the models, so that we can select the best one; and

(2) a method for calibrating the scores, so that we can decide whether the best-scoring model is likely to be correct.

Several approaches to scoring have been tried. One of the most effective is based on empirical patterns of residue neighbours, derived from known structures. Observe the distribution of interresidue distances in known protein structures, for all 20×20 pairs of residue types. For each pair, derive a probability distribution, as a function of the separation in space and in the amino acid sequence. For instance for the pair Leu–Ile, consider every Leu and Ile residue in known structures, and, for each Leu–Ile pair, record the distance between their $C\beta$ atoms, and the difference in their positions in the sequence. Collecting these statistics permits an estimation of how well the distributions observed in a model agree with the distributions in known structures, and how good this agreement needs to be to regard the identification as reliable.

Prediction of novel folds

Conformational energy calculations and molecular dynamics

The interactions between atoms in a molecule can be divided into:

(1) primary chemical bonds—strong interactions between atoms that must be close together in space. These interactions are not broken when the conformation of a protein changes, but are equally consistent with a large number of conformations.

(2) weaker interactions that depend on the conformation of the chain. These can be significant in some conformations and not in others—they affect sets of atoms that are brought into proximity by different folds of the chain.

The conformation of a protein can be specified by giving the list of atoms in the structure, their coordinates, and the set of primary chemical bonds between them (this can be inferred with only slight ambiguity, from the amino acid sequence). Terms used in the evaluation of the energy of a conformation typically include:

- **Bond stretching**: $\Sigma_{\text{bonds}} K_r(r - r_0)^2$. Here r_0 is the equilibrium interatomic separation and K_r is the force constant for stretching the bond. K_r and r_0 depend on the type of chemical bond.
- **Bond angle bend**: $\Sigma_{\text{angles}} K_\theta(\theta - \theta_0)^2$. For any atom *i* that is chemically bonded to two other atoms *j* and *k*, the angle *i–j–k* has an equilibrium value θ_0 and a force-constant for bending K_θ.
- **Other terms to enforce proper stereochemistry** penalize deviations from planarity of certain groups, or enforce correct chirality (handedness) at certain centres.
- **Torsion angle**: $\Sigma_{\text{dihedrals}} \frac{1}{2}V_n[1 + \cos n\phi]$. For any four connected atoms: *i* bonded to *j* bonded to *k* bonded to *l*, the energy barrier to rotation of atom *l* with respect to atom *i* around the *j*–*k* bond is given by a periodic potential. V_n is the height of the barrier to internal rotation; *n* barriers are encountered during a full 360° rotation. The main chain conformational angles ϕ, ψ, and ω are examples of torsional rotations.
- **Van der Waals interactions:** $\Sigma_i\Sigma_{j<i}(A_{ij}R_{ij}^{-12} - B_{ij}R_{ij}^{-6})$. For each pair of non-bonded atoms i and j, the first term accounts for a short-range repulsion and the second term for a long-range attraction between them. (See Fig. 7.3.) The parameters A and B depend on atom type.
- **Hydrogen bond:** $\Sigma_i\Sigma_{j<i}(C_{ij}R_{ij}^{-12} - D_{ij}R_{ij}^{-10})$. The hydrogen bond is a weak chemical/electrostatic interaction between two polar atoms, mediated by a hydrogen atom. Its strength depends on distance, and also on the bond angle. The approximate hydrogen bond potential given here does not explicitly reflect the angular dependence of hydrogen bond strength; other potentials attempt to account for hydrogen bond geometry more accurately.
- **Electrostatics**: $\Sigma_i \Sigma_{j<i} Q_i Q_j/(\epsilon R_{ij})$. Q_j are the effective charges on the atoms, R_{ij} is the distance between them, and ϵ is the dielectric 'constant'. This formula applies only approximately to media that are not infinite and isotropic, including proteins.
- **Solvent**: Interactions with the solvent, water, and co-solutes such as salts and sugars, are crucial for the thermodynamics of protein structures. Attempts to model the solvent as a continuous medium, characterized primarily by a dielectric constant, are approximations. With the increase in available computer power, it is now possible to include solvent explicitly, simulating the motion of a protein in a box of water molecules.

The energy of a conformation is computed by summing these terms over all appropriate sets of interacting atoms. There are numerous sets of conformational energy potentials of this or closely related forms, and a great deal of effort has gone into the tuning of parameter sets.

Many potential functions satisfy necessary, but not sufficient, conditions for successful structure prediction. One test is to take the right answer—an experimentally determined protein structure—as a starting conformation, and minimize the energy starting from there. In general, most energy functions produce a minimized conformation that is about 1 Å (root-mean-square deviation of the main chain) away from the starting model. This can be thought of as a measure of the resolution of the force field. Another test has been to take deliberately misfolded proteins and minimize their conformational energies, to see whether the energy value of the local minimum in the vicinity of the correct fold is significantly lower than that of the local minimum in the vicinity of an incorrect fold. Such tests reveal that multiple local minima cannot be reliably distinguished from the correct one on the basis of calculated conformational energies.

Rosetta

ROSETTA is a program by D. Baker and colleagues that predicts protein structure from an amino acid sequence by assimilating information from known structures. At the 2002 CASP programme, ROSETTA showed the most consistent success on targets in both the 'Novel Fold' and 'Fold Recognition' categories.

ROSETTA predicts a protein structure by first generating the structures of fragments using known structures, and then combining them. For each contiguous region of three and nine residues, instances of that sequence and related sequences are identified in proteins of known structure. For fragments this small, there is no assumption of homology to the target protein. The distribution of conformations of the fragments in the proteins of known structure thus models the distribution of possible conformations of the corresponding fragments of the target structure.

ROSETTA explores the possible combinations of fragment conformations, evaluating compactness, paired β-sheets, and burial of hydrophobic residues. The structures that result from these simulations are clustered, and the centres of the largest clusters presented as predictions of the target structure. The idea is that a structure that emerges many times from independent simulations is likely to have favourable features.

Linus

LINUS (= Local Independently Nucleated Units of Structure) is a program for the prediction of protein structure from an amino acid sequence, designed by G.D. Rose and R. Srinivasan. It is one of the few methods that is a completely *a priori* procedure, making no explicit reference to any known structures or sequence–structure relationships. LINUS folds the polypeptide chain in a *hierarchical* fashion, first producing structures of short segments, and then assembling them into progressively larger fragments.

An insight underlying LINUS is that the structures of local regions of a protein—short segments of residues consecutive in the sequence—are controlled by local interactions within these segments. During natural protein folding, each segment will preferentially sample its most favourable conformations. However, these preferred conformations of local regions, even the one that will ultimately be adopted in the native state, are below the threshold of stability. Local structure will form transiently and break up many times before a suitable interacting partner stabilizes it. But in the computer one is free to anticipate the results. In a LINUS simulation, favourable structures of local fragments, as determined by their frequent recurrence during the simulation, transmit their preferred conformations as biases that influence subsequent steps. The procedure applies the principle of a rachet to direct the calculation along productive lines.

Currently LINUS is generally successful in getting the correct structures of small fragments (size between supersecondary structure and domain), and in some cases can assemble them into the right global structure.

Prediction of protein function

The problem of predicting protein function arises in two contexts. (1) A research group has a committed interest in a gene and its protein product. They may identify cofactors and post-translational modifications, perform a structure determination, and check the phenotypic effect of a knockout. They assign function from a thick dossier of information. This was the paradigm. (2) Now, complete genomes challenge us to annotate their coding sequences with protein functions, on the basis of information often limited to amino acid sequences alone. The goal of providing protein structures for their implications about function is an important motivation of structural genomics projects.

A common way to try to assign function to a protein is to identify a putative homologue of known function, and assume that both share a common function. Even for very closely related proteins this does not always work, and the assumption grows less and less safe as the sequences diverge. Multidomain proteins present another level of difficulty.

Often it is impossible to predict function precisely, but only to provide hints. For instance, a protein known to be a TIM barrel is likely to be a hydrolytic enzyme. Multiple sequence alignments can be informative, in pinpointing conserved residues. The literature may provide crucial information about the family as a whole and the role of conserved residues. Phylogenetic trees can provide information as to whether an unknown clusters with a particular functional grouping.

A protein of unknown function from structural genomics

The crystal structure of protein HI1679 was determined in the course of the *Haemophilus influenzae* structural genomics project. It has an α/β-hydrolase fold, with putative remote homology to members of the L-2-haloacid dehydrogenase family, the P-domain of Ca^{2+} ATPase, and phosphoserine phosphatase. It was the first structure of a protein in the L-2-haloacid dehydrogenase family to be determined, and one of the motives for selecting it for crystallographic attention was the desire to learn about the structure and the mechanism of function of this family. The structure was consistent with a phosphatase, and this was confirmed by trying a variety of potential substrates: the protein cleaved 6-phosphogluconate and phosphotyrosine. Modelling of these observed substrates into the binding pocket revealed how sequence variation in the active site might affect specificity.

If an unknown protein shares significant sequence similarity with a well-characterized family, and possesses the right essential conserved residues, then a prediction of its general functional class (proteinase, exonuclease, etc.) may be reasonable. If the unknown also forms part of a well-characterized functional cluster within a phylogenetic tree, then a more detailed level of functional prediction may be possible. Such hints are useful in guiding experimental investigations.

Sometimes specific constellations of residues provide signature patterns in the sequence that identify an active site

1. A **motif** describes a single consecutive set of residues.
2. **Multiple motifs** combine several motifs involving separate consecutive sets of residues.
3. **Profile methods** are based on entire sequences and weighting different residue positions according to their variability.

Motifs may be expressed in terms of uniquely defined sequences, such as:

GWTLNSAGYLLGP

which characterizes the neuropeptide galanin. Or, they may contain alternatives; for instance [LIVMF]–T–S–P–P–[FY], the signature of N-4 cytosine-specific DNA methylases. Here [LIVMF] means that that first position may contain *any* of the amino acids L, I, V, M, or F, followed by the unique sequence TTPP, followed by a position that may contain either F or Y. The database PROSITE contains a collection of motifs covering a wide range of proteins.

Single-motif patterns have the weakness that an active site of a protein may be defined by regions that are distant in the sequence, although nearby in space. BLOCKS and PRINTS are databases containing *multiple motifs*, typically ~20 residues long, presented as ungapped multiple sequence alignments.

If different sets of sequences match different individual regions in a multiple motif pattern, it is possible to classify subsets of a family. G-protein-coupled receptors (GPCRs) form a large family of cell-surface proteins that detect and signal hormones and growth factors, and mediate the senses of sight and smell. A motivation for classifying subtypes is that GPCRs are common drug targets. The PRINTS database contains a seven-motif general fingerprint for GPCRs—each motif corresponding to one of the seven transmembrane helices. Additional sets of motifs identify subfamilies of GPCRs and receptor subtypes.

Overall sequence similarity is sensitive to the general folding patterns of proteins, sometimes at the expense of focus on specific active-site residues. Conversely, some motifs are sensitive to active-site residues, but in their insensitivity to the sequence as a whole may pick up non-homologous proteins as false positives.

Structural data provide additional routes to function prediction

Some successes

- Identification of structural relationships unanticipated from a sequence can suggest similarity of function. The crystal structure of AdipoQ, a protein secreted from adipocytes, showed a similarity of folding pattern to that of tumour necrosis factor. The inference that AdipoQ is a cell signalling protein was subsequently verified.

- The histidine triad proteins are a broad family with no known function. Analysis of their structures indicated a catalytic centre and nucleotide-binding site, identifying them as a nucleotide hydrolase. This did *not* depend on detection of a distant homology.
- Structural similarity of a gene product of unknown function from *Methanococcus jannischii* and other proteins containing nucleotide-binding domains led to experiments showing it to be a xanthine or inosine triphosphatase.

Interaction patterns, contextual information, and intergenomic comparisons are useful indicators of function

- **Interaction patterns.** As part of the development of full-organism methods of investigation, data are becoming available on patterns of protein interactions. The network of interactions often reveals the function of a protein.
- **Gene fusion.** A composite gene in one genome may correspond to separate genes in other genomes. There is likely to be a relationship between the functions of these genes.
- **Local gene context.** It makes sense to co-regulate and co-transcribe components of a pathway. In bacteria, genes in a single operon are usually functionally linked.
- **Phylogenetic profiles.** Proteins in a common structural complex or pathway are functionally linked and expected to co-evolve. For each protein, a phylogenetic profile indicates which organisms contain homologues. Clustering the profiles identifies sets of proteins that co-occur in the same groups of organisms. Some relationship between their functions is expected.

  There need be no sequence or structural similarity between those proteins that share a phylogenetic distribution pattern. A welcome feature of this method is that it derives information about the function of a protein from features it shares with *non-homologous* proteins.

Protein design

If we understood how amino acid sequence detemines protein structure, we could create proteins unknown in nature. The *inverse folding problem* is the challenge of starting with a structure, and designing sequences that fold into that structure. A simplified version of the problem: select a few residues in a protein interior, delete them and forget what they were, then check combinations of side chains that might pack the space vacated. Even with high-powered computers, subtle algorithms are required.

CASE STUDY *Ab initio* design of a hyperstable variant of streptococcal protein G, β1-domain

In the parent 56-residue domain an α-helix packs against a four-stranded β-sheet (see Fig. 5.13). The midpoint of its thermal unfolding transition (its 'melting point') is 83 °C, with a ΔG° of stabilization of 11.7 kJ mol^{-1} at 50 °C.

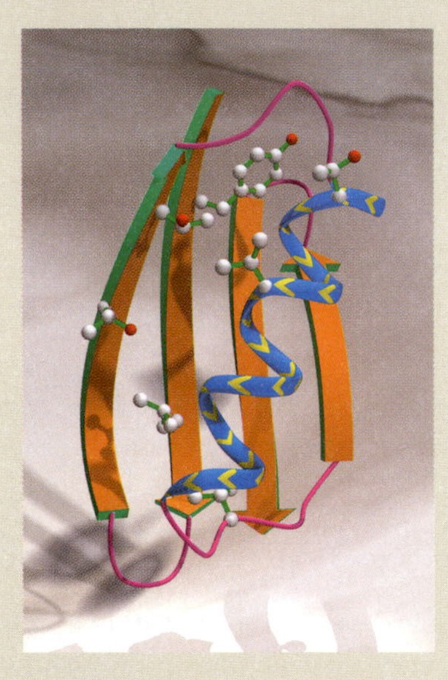

Fig. 5.13 Streptococcal protein G, β1 domain [1PGA]. Side chains shown are those mutated to form a hyperstable variant.

Mayo and colleagues have redesigned its sequence to achieve remarkable thermostability.*

A total of 11 residues buried in the core were allowed to assume all combinations of A, V, L, I, F, Y, and W, with conformations chosen from a rotamer library. The conformations of the main chain and of other side chains were held fixed. The seven side chains considered have 217 possible rotamers, giving a very large number of possible combinations of side chain conformations. Optimization of an energy function reflecting Van der Waals interactions and buried surface area reproduced the wild-type residue at eight positions, and suggested the mutations Y3F, L7I, and V39I. This variant melted at 91 °C. Optimization of five additional residues at the periphery of the core, allowing all residues *except* G, P, C, M, R, suggested a sequence with four additional positions changed: T16I, T18I, T25E, V29I; position 43 stayed W.

The final result differed at 7 sites from the wild type, out of 16 variable positions. Its melting temperature could not be determined as the transition was incomplete at 100 °C, comparable

* Dahiyat, B.I. and Mayo, S.L. (1997). Probing the role of packing specificity in protein design. *Proc. Natl. Acad. Sci.* (USA), **94**, 10172–77; Malakauskas, S.M. and Mayo, S.L. (1998). Design, structure and stability or a hyperthermophile protein variant. *Nat. Struct. Biol.*, **5**, 470–5.

continues...

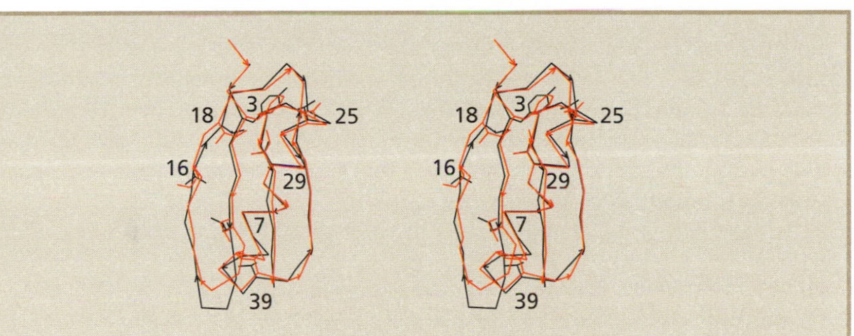

Fig. 5.14 Streptococcal protein G, β1 domain; comparison of wild type (black) and a hyperstable variant (red).

to proteins from hyperthermophiles. At 50 °C, ΔG° of stabilization was 30 kJ mol^{-1}, 18 kJ mol^{-1} higher than the original.

The great change in stability did *not* require substantial structural change (see Fig. 5.14). Indeed, if it did, the calculations based on fixed main chain conformation might not be adequately realistic.

```
wt      MTYKLILNGKTLKGETTTEAVDAATAEKVFKQYANDNGVDGEWTYDDATKTFTVTE
          *   *         * *      *   *          *
design  MTFKLIINGKTLKGEITIEAVDAAEAEKIFKQYANDNGIDGEWTYDDATKTFTVTE
```

It is difficult to pinpoint the source of the enhanced stability, because the structural changes are so small. The largest effects appear to be enhanced packing, reduction in side chain strain and additional buried surface area. Residue T25E can make additional electrostatic and hydrogen-bond interactions.

The difficulty of relating energetic effects to structural changes is a general one. There is a mismatch, unfortunate from the interpretative point of view, between the very large energies that can arise from very small structural changes. It is fortunate that computer programs can cope with these subtleties.

USEFUL WEB SITES

SWISS-MODEL: **www.expasy.org/swissmod/SWISS-MODEL.html**
MODBASE: **http://alto.compbio.ucsf.edu/modbase-cgi/index.cgi**
ROSETTA: **www.bioinfo.rpi.edu/ ̃bystrc/hmmstr/server.php**
LINUS: **www.roselab.jhu.edu**
CASP: **http://predictioncenter.llnl.gov/**
CAPRI: **http://capri.ecbi.ac.uk/capri.html**

RECOMMENDED READING

Baldwin, R.L. and Rose, G.D. (1999). Is protein folding hierarchic? I. Local structure and peptide folding; and II. Folding intermediates and transition states. *Trends Biochem. Sci.*, **24**, 26–33 and 77–83.

Bieri, O. and Kiefhaber, T. (1999). Elementary steps in protein folding. *Biol Chem.*, **380**, 923–9.

Chan, H.S. and Dill, K.A. (1998). Protein folding in the landscape perspective: chevron plots and non-Arrhenius kinetics. *Proteins*, **30**, 2–33.

Dill, K.A. and Chan, H.S. (1997). From Levinthal to pathways to funnels. *Nat. Struct. Biol.*, **4**, 10–19.

Finkelstein, A.V. and Ptitsyn, O.B. (2002). *Protein physics*. Academic Press, New York. ('The Feynman Lectures of Protein Science.')

Hellinga, H.W. (1997). Rational protein design: combining theory and experiment. *Proc. Natl. Acad. Sci. (USA)*, **94**, 10015–17.

King, J., Haase-Pettingell, C. and Gossard, D. (2002). Protein folding and misfolding. *Am. Sci.*, **90**, 445–53.

Nolting, B. (1999). *Protein folding kinetics: biophysical methods*. Springer-Verlag, Heidelberg.

Tramontano, A. (2003). Of men and machines. *Nat. Struct. Biol.*, **10**, 87–90.

Wolynes, P.G. (2001). Landscapes, funnels, glasses, and folding: from metaphor to software. *Proc. Am. Philos. Soc.*, **145**, 555–63.

EXERCISES, PROBLEMS, AND WEBLEM

Exercises

1. Consider a reaction A + B = C + D and the formula $\Delta G - \Delta G^{\circ} = RT \ln K$. At equilibrium, $\Delta G = 0$. What could be inferred if $\Delta G^{\circ} = 0$ also?
2. A protein folds in the simple way N=D, with no intermediates. A mutant was found to destabilize the native state relative to the wild type, and to leave the rate of folding unaffected. Explain why the rate of unfolding must be increased.
3. Calculate the Φ-value for the mutant shown in the following figure:

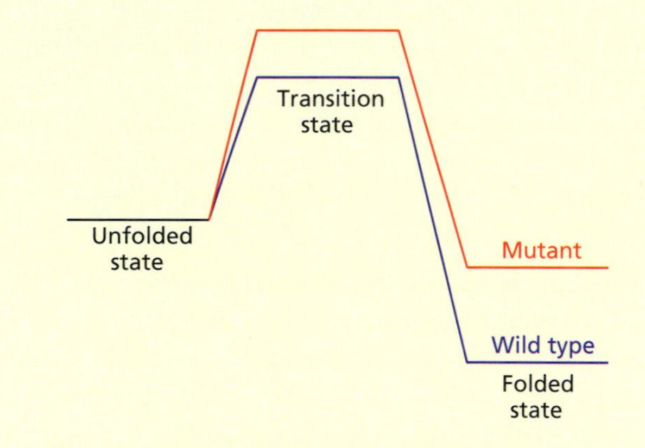

4. In Fig. 5.5, does the rightmost column, labelled *more denaturant*, represent conditions in which the protein is mostly unfolded? Explain your answer.
5. Consider these three sketches of enzyme-catalysed reactions. (E + S = enzyme + substrate, unbound; ES = enzyme–substrate complex = Michaelis complex; $ES^{\ddagger}$ = transition state; EP = enzyme–product complex; E + P = enzyme + product, unbound.)

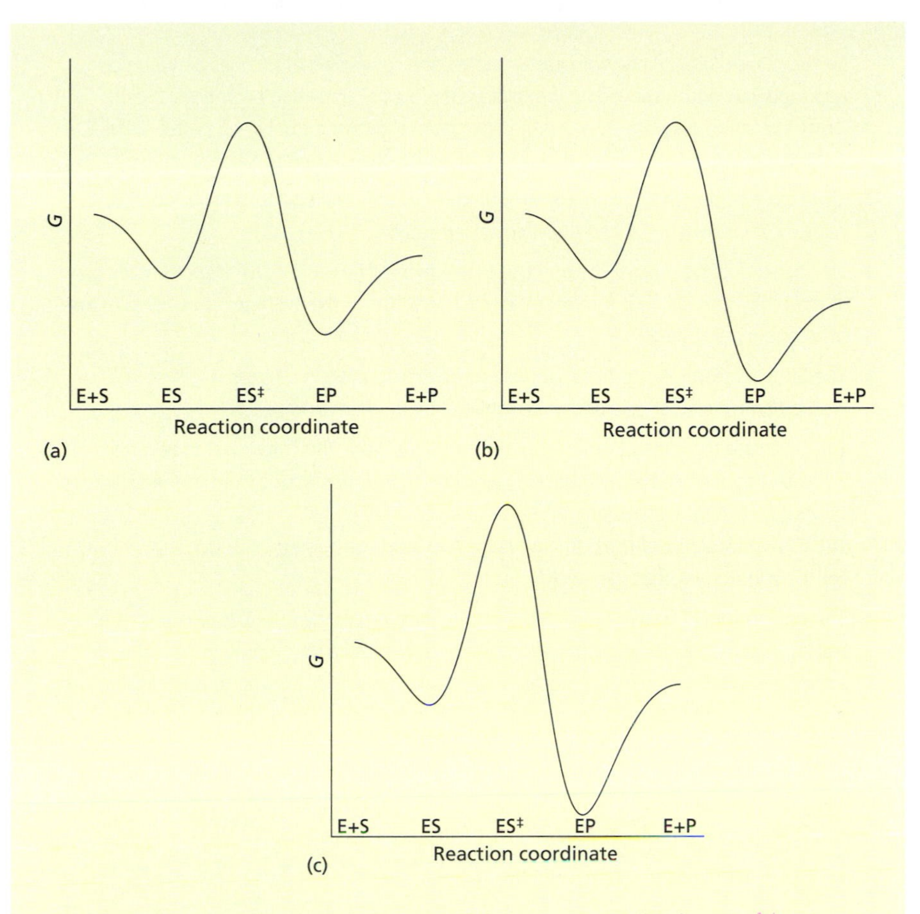

(a) Which reaction would have the highest equilibrium constant in the absence of the enzyme?

(b) Which reaction has the highest activation energy in the presence of the enzyme?

(c) Which reaction has the highest Michaelis constant = dissociation constant of enzyme–substrate complex = $[E][S]/[ES]$?.

(d) Suppose that the enzyme in frame (c) speeds up the reaction by a factor of 100 at 37 °C, relative to the uncatalysed rate. On a photocopy of part (c) of this figure, sketch the reaction diagram in the absence of the enzyme.

6. Calculate the contact order for the HP lattice protein shown in Fig. 2–31.

Problems

1. The energy difference between *cis*- and *trans*-proline conformations is about 84 kJ mol^{-1}. For an isolated proline residue in a denatured protein at 300 K, estimate the fraction that are in the *cis* conformation. (If two states differ by an energy ΔE, at thermal equilibrium their populations will be in the ratio $\exp(-\Delta E/kT)$, where k = Boltzmann's constant = 1.38×10^{-23} J K^{-1} and T is the absolute temperature.) For a denatured protein containing two isolated prolines, estimate the fraction of molecules for which both prolines are in the *trans* conformation. (Proline isomerization is one of the factors governing the kinetics of refolding. Prolyl isomerases catalyse the conversion of prolines between *cis* and *trans* isomers and speed up the folding of proteins containing proline residues.)

2. Draw a reaction diagram showing the energetic relationships between native state, transition state, and unfolded state as described in the caption to Fig. 5.4.
3. From examination of Fig. 5.7, draw topology diagrams analogous to those on page 42 for wild-type S6 and the circular permutant.
4. Plot $\Delta S^{\ominus}$ of folding against the number of residues in the protein, for the examples given in the Table on page 192. Interpret the result in terms of the average number of degrees of freedom per residue.
5. The 56-residue, amino-terminal domain of ribosomal protein L9 from *Bacillus stearothermophilus* shows two-state folding kinetics with $\ln k_f = 6.57$.
   (a) Calculate the contact order from the structure, 1DIV. The criterion that two residues are in contact is that at least one pair of non-hydrogen atoms from the residues are within 6 Å of each other.
   (b) On a photocopy of the figure on page 199, add the point for this domain.
   (c) Does it fit the other data well?
6. The folding funnel shown in Fig. 5.8 is a simplification, in the interests of clarity. It does not show the undoubted roughness of the energy surface that would disrupt the smoothness of the trajectories, local minima, and dead ends leading to non-productive intermediates. Modify a photocopy of Fig. 5.8 to make it more realistic.
7. Analyse the folding and unfolding kinetics of mutants of chicken brain α-spectrin (see Table).

Protein	$[urea]_{50\%}$ mol l^{-1}	k_f at $[urea] = 0$ s^{-1}	k_u at $[urea] = 5$ mol l^{-1} s^{-1}
Wild type (wt)	3.0	23.3	0.80
F117L	1.9	5.4	6.00
A119G	2.5	4.0	0.89
F157L	2.3	20.5	10.0
V171A	2.4	17.1	5.62
G198A	3.4	53.8	0.39
L203	1.2	3.1	181.0

$[urea]_{50\%}$ is the concentration of urea at which $[N]/[D] = 1$.
All measurements were carried out at 298.15 K.
Data from Moran, S., Scott, K., and Clarke, J. (unpublished data).

For each mutant:

(a) calculate

$$\Delta\Delta G_{D-N} = \Delta G^{wt}_{D-N} - \Delta G^{mutant}_{D-N} = m([urea]^{wt}_{50\%} - [urea]^{mutant}_{50\%})$$

(Take $m = 2$.)

(b) Calculate $\Delta\Delta G_{D-\ddagger} = RT \ln (k_f^{wt}/k_f^{mutant})$.

(c) Calculate $\Phi_f = \Delta\Delta G_{D-\ddagger}/\Delta\Delta G_{D-N}$.

(d) Calculate $\Delta\Delta G_{\ddagger-N} = -RT \ln (k_u^{wt}/k_u^{mutant})$. (Use the data for unfolding in 5 mol l^{-1} urea.)

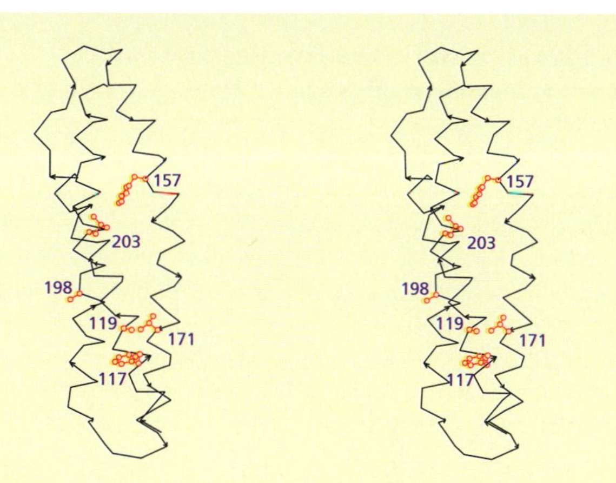

Fig. 5.15 Chicken brain α-spectrin, repeat 17, showing the positions of mutations for which kinetic data are given. Wild-type side chains are shown except for 198A.

(e) Calculate $\Phi_u = 1 - \Delta\Delta G_{\ddagger-N}/\Delta\Delta G_{D-N}$.

(f) To what average accuracy do the values of Φ_f and Φ_u for the same mutant agree?

(g) Divide the mutants into sets for which Φ is small ($\Phi < 0.33$), medium ($0.33 \le \Phi < 0.67$), and large ($0.67 \le \Phi$). Is there a correlation of Φ-values with the structure (see Fig. 5.15)? What can you say about the structure of the intermediate?

8. Eftink *et al.* measured the fluorescence intensity of staphylococcal nuclease as a function of urea concentration at 20 °C.

$[urea]/\text{mol l}^{-1}$	Relative fluorescence intensity
0.00	69.8
0.00	74.7
0.97	70.7
1.45	72.2
1.92	68.3
1.91	64.6
2.14	56.3
2.23	52.3
2.36	40.4
2.61	30.0
2.83	22.3
3.79	15.6
4.75	15.9

Data from: Eftink, M.R., Ghiron, C.A., Kautz, R.A., and Fox, R.O. (1991). Fluorescence and conformational stability studies of *Staphylococcus* nuclease and its mutants, including the less stable nuclease–concanavalin A hybrids. *Biochemistry*, **30**, 1193–9.

(a) Draw a graph of these data. Explain its general form.

(b) Fit the data to a function of the form:

$$F([\text{urea}]) = \{1 - f_U([\text{urea}])\}F_N + f_U([\text{urea}])\{F_U - S_U \times [\text{urea}]\};$$

where $F([\text{urea}])$ is the relative fluorescence intensity, and $f_U([\text{urea}])$ is the fraction of molecules unfolded, as functions of urea concentration. F_N, F_U, and S_U are constants (independent of [urea]), the values of the relative fluorescence of native and unfolded molecules *in the absence of urea*, and the slope of the relative fluorescence intensity of the unfolded state as a function of [urea] (assumed constant). The fraction of unfolded molecules as a function of urea concentration is:

$$f_U([\text{urea}]) = \exp(\Delta G_{UN}([\text{urea}])/RT)/\{[1 + \exp(G_{UN}([\text{urea}])/RT\};$$

and

$$\Delta G^\circ_{UN}([\text{urea}]) = \Delta G^\circ_{UN}([\text{urea} = 0]) - m \times [\text{urea.}]$$

Determine thereby the values of $\Delta G^\circ_{UN}([\text{urea} = 0])$, F_N, F_U, S_U, and m. (Solution of this problem will require access to a suitable curve-fitting program.)

(c) What fraction of molecules is unfolded in 2.1 mol l^{-1}urea?

Weblem

1. For purposes of testing structure prediction programs, the Protein Data Bank will, in some cases, release the sequences of proteins in advance of the release of the structures. At the Web site:

www.rcsb.org/pdb/status.html select:
Holding Status: release on a certain date
Sequence Availability: release immediately
and then Search.

From the information returned, select, if possible, up to three sequences for which the release date of the structure is no more than 4 weeks in the future. Call these the target sequences.

For each target sequence:

(a) Determine whether there is a homologue of known structure. If so, record the optimal alignment between the homologue and the target sequence, and the per cent identical residues in the optimal alignment.

(b) Submit the target sequence to a variety of structure-prediction Web sites, including sites that predict secondary structure, sites that create homology models, and sites that predict tertiary structures. You can identify these by a Web search for the terms Protein Structure Prediction Web Sites. (If this exercise is being done in a classroom context, it would be useful if different students try different sites, so that a comparison of the results will be possible.)

(c) Record the predictions returned, before the date of release of the structure.

(d) After the structure is released, evaluate the quality of the prediction.

Chapter 6

PROTEINS WITH PARTNERS

LEARNING GOALS

1. **To understand the wide variety of structures and functions associated with protein-protein and protein–DNA complexes.**
2. **To know the general features of a typical protein–protein interface**, including size, chemical composition, shape and charge complementarity, and the role of embedded water molecules.
3. **To understand accessible surface area**, how it is measured, and its significance in terms of the thermodynamics of protein folding and interaction.
4. **To recognize the very great variety of protein–nucleic acid interactions** and be familiar with a few of them in detail.
5. **To understand the basic principles of viral capsid assembly**, the purely geometric problems that viruses face in capsid design, and how different viruses solve these problems.

Introduction

Proteins rarely act alone.

Many proteins form parts of stable aggregates, including oligomeric proteins such as haemoglobin (Fig. 2.22) or phosphofructokinase (Fig. 6.1), many fibrous proteins such as keratins or actin (see Chapter 2), and viral coat proteins. Many metabolic and regulatory processes involve the successive activities of different proteins, which may associate only transiently. For example, arrival of a ligand at a cell surface may initiate a signal transduction cascade by inducing receptor dimerization (Fig. 2.16). Regulation of gene transcription depends on interaction between proteins and specific DNA sequences. Protein–protein interactions may modulate transcriptional regulation; for example, interactions with protein cofactors affect the DNA-binding specificity of homeodomain proteins.

Stable oligomeric proteins may contain many copies of one protein, or combine different ones. Some prokaryotic proteins containing identical subunits are homologous

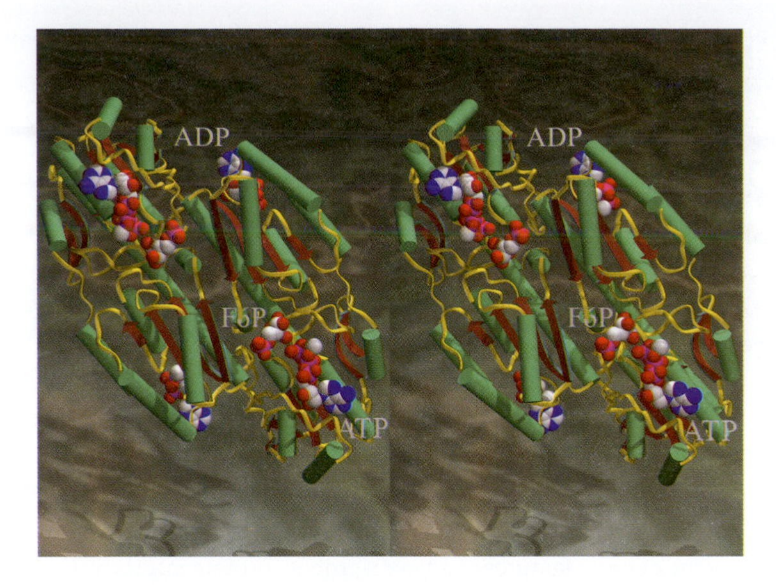

Fig. 6.1 Phosphofructokinase, R state, from *Bacillus stearothermophilus*, showing substrate fructose-6-phosphate (F6P), cofactor ATP, and allosteric effector ADP. The complete molecule is a tetramer, of which only two subunits are shown. Homologous enzymes from mammals are octamers.

to eukaryotic proteins containing related but non-identical subunits, arising by gene duplication and divergence. The proteasome is an example. Some viruses achieve diversity *without* gene duplication, by combining proteins with the same sequence but different conformations.

Interacting proteins span a range of structures and functions:

- **simple dimers** in which the two monomers appear to function independently;
- **oligomers with functional 'cross-talk'**, including ligand-induced dimerization of receptors, and allosteric proteins such as haemoglobin, phosphofructokinase, and asparate carbamoyltransferase;
- **large fibrous proteins** such as actin or keratin;
- **non-fibrous structural aggregates** such as viral capsids.
- **large aggregates with dynamic properties** such as F1-ATPase, pyruvate kinase, the GroEL–GroES chaperonin, and the proteasome;
- **protein–DNA complexes**—in many cases initial binding is followed by recruitment of additional proteins to form large complexes;
- **many proteins, whether monomeric or oligomeric, function by interacting with other proteins**—these include all enzymes with protein substrates, and many antibodies, inhibitors, and regulatory proteins;
- **protein interactions are frequently associated with disease**, as misfolded or mutant proteins are prone to aggregation. Amyloidoses are diseases characterized by extracellular fibrillar deposits, usually with a common crossed-β-sheet structure. They arise from a variety of causes, including overproduction of a protein, destabilizing

mutations, and inadequate clearance in renal failure. Alzheimer and Huntington diseases are also associated with protein aggregation.

Structural studies have elucidated several important features of the interactions between soluble proteins:

- **What holds the proteins together?** Burial of hydrophobic surface, hydrogen bonds, and salt bridges.
- **Do proteins change conformation upon formation of complexes?** In some cases they do. In these cases the interaction energy has to 'pay for' the conformational change, and the interface tends to be larger.
- **What determines specificity?** Complementarity of the occluding surfaces, in shape, hydrogen-bonding potential, and charge distribution. Prediction of protein complexes from the structures of the partners is the **docking** problem. The docking problem is *much* easier to solve if the components do not change main chain conformation upon association.

General properties of protein–protein interfaces

Sets of interacting proteins of known structure include many proteinase-inhibitor complexes, antibody–antigen complexes, and complexes involved in signalling. The interfaces can be characterized by: (1) the overall amount of surface buried, (2) the chemical composition of the buried surfaces, (3) the shape and charge complementarity of occluding surfaces, and (4) specific interactions, such as hydrogen bonds.

Burial of protein surface

The accessible surface area (ASA) of a protein is calculated by rolling a probe sphere of radius 1.4 Å over the protein and determining the area of the surface generated. The surface buried by the formation of a complex is computed by calculating the ASA of the complex and subtracting the ASA of the components separately. The results give the area buried in the interface, and identify the residues or atoms that change accessibility.

See Chapter 3.

If the structure of the complex is known but the structures of the free partners are not, a buried surface in the complex can be calculated by separating the partners artificially in the computer. However, this would not accurately measure the change in surface area, because it necessarily ignores conformational changes that might be taking place on association. In a number of cases, a region that is disordered in a component becomes ordered in the complex (and, in the very special case of the serpin–proteinase complex, vice versa!).

The interface is defined by the atoms that lose accessible surface area in forming the complex. Not all interface atoms are completely buried. A general picture of an interface would be a continuous, relatively flat surface patch, with some atoms at the centre buried completely, and others at the periphery buried only partially. A typical interface might involve 22 residues, including 90 atoms of which 20% would be main chain atoms, and an occasional water molecule.

A histogram of surface area buried in binary protein complexes shows a peak centred at 1600 Å^2, containing about 70% of the complexes. The minimum buried surface for stability under typical conditions is about 1200 Å^2. Complexes that bury >2000 Å^2 tend to involve conformational changes upon complex formation.

The composition of the interface

The chemical character of protein–protein interfaces is intermediate between that of the surfaces and interiors of monomeric globular proteins. Interfaces are enriched in neutral polar atoms at the expense of charged atoms:

	Atomic composition (%)		
	Non-polar	Polar (neutral)	Polar (charged)
Surfaces of globular proteins	57	24	19
Interfaces	56	29	15
Interiors of globular proteins	58	39	4

Main chain atoms constitute an average of 19% of interfaces.

The amino acid composition of interfaces are enriched in aromatic residues—His, Phe, Tyr, and Trp—relative to the remaining exposed surface. There is a lesser degree of enrichment in aliphatic side chains—Leu, Ile, Val, Met—and Arg (but, surprisingly, not Lys).

Complementarity

Antibody–antigen interfaces are discussed in Chapter 7.

Complementarity of interfaces is reponsible for specificity. It involves geometric complementarity; that is, good packing at the occluding surfaces; and also complementarity in hydrogen-bond and charge–charge interactions.

Most protein–protein interfaces are as well-packed as protein interiors. This optimizes the Van der Waals interactions that help stabilize the complex. In many cases, isolated water molecules occupy sites in the interface, and avoid leaving holes in places where the fit of the proteins is not exact.

Specific interactions at protein–protein interfaces

Atom–atom contacts include non-polar Van der Waals interactions, hydrogen bonds, and salt bridges. Typically there is one hydrogen bond per 170 Å^2 of area, one-third of the hydrogen bonds involving a charged side chain. There is, on average, one fixed water molecule per 100 Å^2 of interface, usually forming hydrogen bonds to both proteins. That is, a typical interface of 1700 Å^2 would be expected to have about 10 intermolecular hydrogen bonds and about 17 fixed water molecules.

CASE STUDY Phage M13 gene III protein and *E. coli* TolA

During infection of *E. coli* by phage M13, a complex forms between the N-terminal domain of the minor coat gene III protein of the phage and the C-terminal domain of a receptor protein in the bacterial cell membrane, TolA. Figure 6.2 shows different representations of this interface.

The complex is stabilized by the burial of 1765 Å^2 of surface area, by combination of β-sheets from both proteins to form an extended β-sheet (see Fig. 6.2(a)), and by several side chain–side chain hydrogen bonds and salt bridges. The area buried in the complex, 1765 Å^2, is divided almost evenly between the two partners.

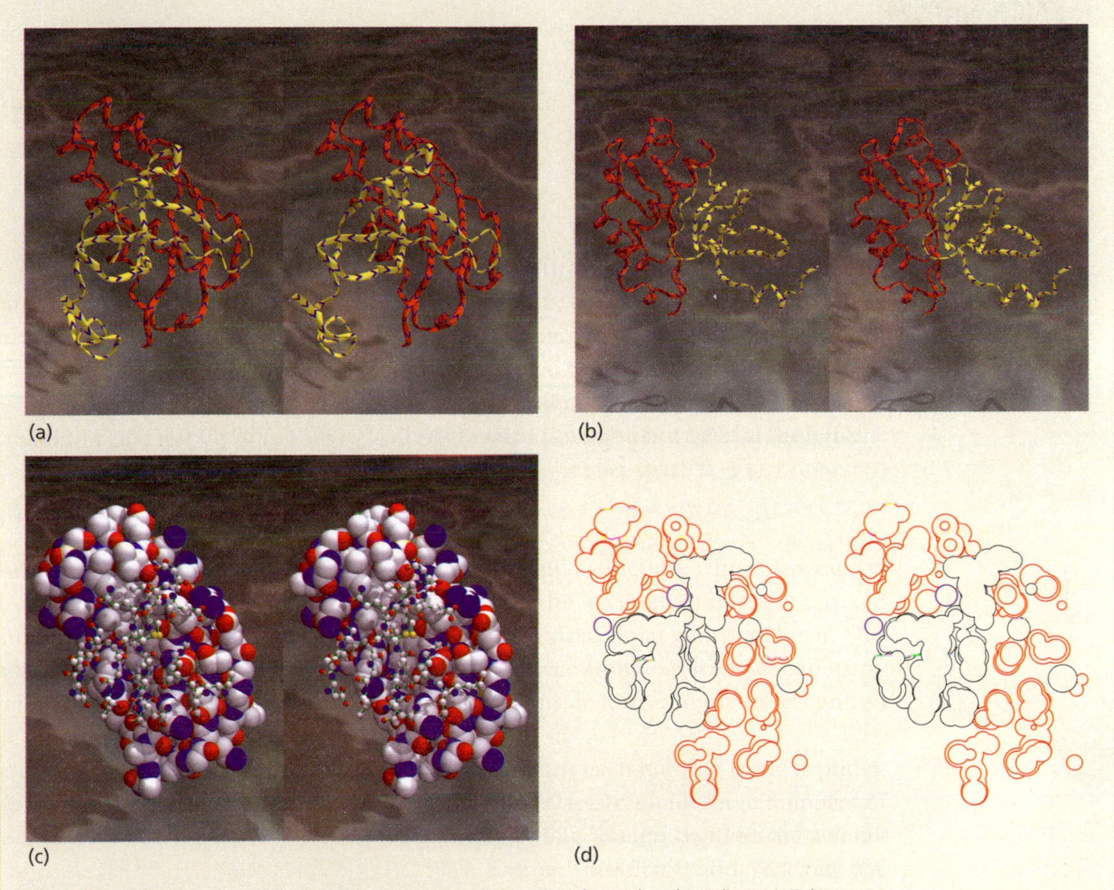

Fig. 6.2 Interface between phage M13 gene III protein (N-terminal domain) and *E. coli* protein TolA (C-terminal domain). (a, b) Folding patterns and relative orientation of domains, viewed approximately (a) perpendicular and (b) parallel to the interface. Note the β-sheet formed from strands contributed by both partners. (c) TolA domain shown as spheres, gene III protein shown in ball-and-stick representation. (d) Slice through the interface, TolA black, gene III protein red, water molecule blue. It is possible that another water molecule sits next to the one inside the structure.

continues . . .

N-terminal domain of phage M13 gene III protein/C-terminal domain of *E. coli* protein TolA

Partner	Accessible surface area		
	Separately	In the complex	Difference
Gene III protein	4270	3382	888
TolA	5311	4434	877
Total	9581	7816	1765

Multisubunit proteins

Proteins fit together in many ways.

It would be useful to revisit the pictures of multisubunit proteins in this book, with the following questions in mind:

- **What is the stoichiometry**? How many different types of subunits appear, and how many of each are present? Most proteins are homodimers or homotetramers. Monomers and hetero-oligomers are less common. The ribosome is an extreme example of a hetero-oligomer. Proteins containing odd numbers of subunits are rarer than those containing even numbers of subunits.
- **What is the relationship between the contributions of different subunits to the interface**? Consider a dimer of two identical subunits: In *isologous* binding, the interface is formed from the same sets of residues from both monomers. In *heterologous* binding, different monomers contribute different sets of residues to the binding site. A handshake is isologous.

An isologous open structure is not possible—why?

- **Is the structure open or closed**? In an open structure, at least one of the sites forming the binding surface is exposed in at least one of the subunits, so that additional subunits could be added on. In a closed structure, all binding surfaces are in contact with partners, and the assembly is saturated. Domain swapping often, but not always, produces closed isologous dimers.
- **What is the symmetry of the structure**? Symmetry is the rule, rather than the exception, in structures of oligomeric proteins. The subunits in most dimers are related by an axis of twofold symmetry. Yeast hexokinase is an exception. It forms an asymmetric dimer. In the human growth-hormone receptor, a nearly symmetric dimer binds an asymmetric ligand (Fig. 2.17).
- **Do any of the subunits undergo conformational changes on assembly**? Often we don't know. In cases of extensively interlocked interfaces, such as trp repressor (Fig. 6.3), the monomers could not adopt the same structure in the absence of their partners. In ATP synthase a three-fold symmetric complex of $\alpha\beta$-subunits is distorted by interaction with the γ-subunit (see Fig. 6.31).

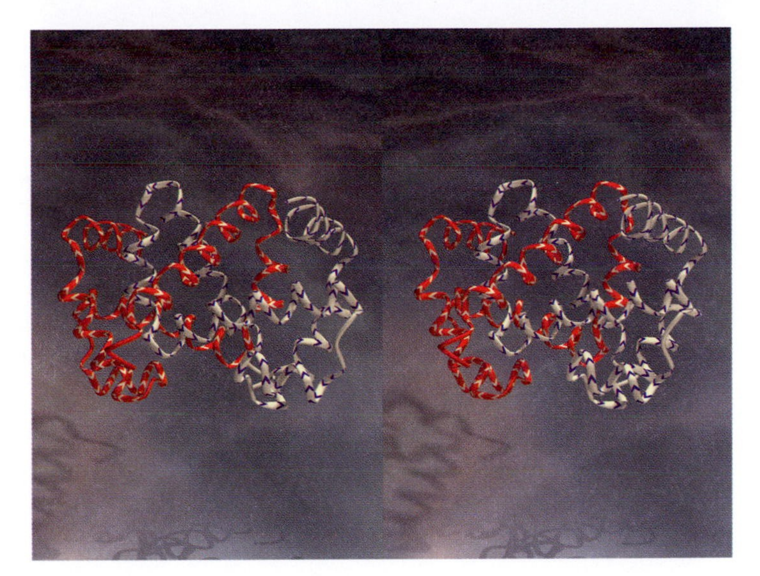

Fig. 6.3 Trp repressor, an intimately intertwined dimer [3WRP].

- **What features contribute to the stability of oligomeric proteins?** The usual suspects: burial of hydrophobic surface, hydrogen bonds, and salt bridges. Some oligomers, such as immunoglobulins, form interchain disulphide bridges (see Fig. 2.23). Others, such as insulin, share metal-binding sites between monomers (see Fig. 3.5).

Protein–DNA interactions

In this section we acknowledge that DNA is made of atoms, not character strings. ('One writes poems not with ideas, but with words.' (Mallarmé))

Protein–DNA complexes mediate several types of processes:

- replication, including repair and recombination;
- transcription;
- regulation of gene expression;
- DNA packaging, including nucleosomes and viral capsids.

Different processes require different degrees of DNA-sequence specificity. Differences in DNA-sequence specificity carry implications about the structures of the proteins that participate in different processes. For example, proteins involved in the regulation of gene expression show great variety, because of the complexity and diversity of the network of control mechanisms.

DNA-binding proteins show varying degrees of DNA-sequence specificity

- Some DNA-binding proteins are relatively non-specific with respect to nucleotide sequence, including DNA replication enzymes, and histones.
- Some, for instance *Eco*RV, *bind* to DNA with low specificity, but *cleave* only at GATATC. This combination permits a mechanism of finding the target sequence by initial non-specific binding followed by diffusion in one dimension along the DNA.
- Some recognize specific nucleotide sequences. For example, the *Eco*R1 restriction endonuclease binds specifically to GAATTC sequences with almost absolute specificity. It is a homodimer that recognizes palindromic sequences.
- Some DNA-binding proteins recognize consensus sequences. For example, the phage Mu transposase and repressor proteins bind 11 base-pair sequences of the form CTTT[T|A]PyNPu[A|T]A[A|T] (where: [A|T] = A or T; Py = either pyrimidine = C or T; Pu = either purine = A or G; and N = any of the four bases).
- Some recognize nucleotide sequences indirectly, via modulations of local DNA structure. For example, the TATA-box binding protein takes advantage of the greater flexibility of AT-rich sequences to form complexes in which the DNA is very strongly bent. The distinction between sequence specificity achieved through direct interaction with bases, or through recognition of local structure, has been termed 'digital versus analogue readout'.
- Some recognize general structural features of DNA, such as mismatched bases, or supercoiling.
- Some DNA-binding proteins form an initial complex with high DNA sequence specificity, followed by recruitment of other proteins of low specificity, to enhance overall binding affinity.

Structural themes in protein–DNA binding and sequence recognition

What does a protein looking at a stretch of DNA in the standard B conformation see? What could it hope to grab hold of? Prominent general features are the sugar-phosphate backbone, including charged phosphates suitable for salt bridges, and potential hydrogen-bond partners in the sugar hydroxyl groups (Fig. 6.4). Contact with the bases is accessible through the major and minor grooves; although, unless the DNA is distorted, the bases are visible only 'edge-on'. Hydrogen-bonding patterns between bases in the grooves and particular amino acids account for some of the DNA-sequence specificity in binding (Fig. 6.5). However, many protein–DNA hydrogen bonds are mediated by intervening water moleules, an effect that tends to *reduce* the specificity.

The idea that an α-helix has the right size and shape to fit into the major groove of DNA was noted in the 1950s. The structures of the first protein–DNA complexes confirmed this prediction. It became the paradigm for protein–DNA interactions. Indeed, when a student solving the structure of the Met repressor–DNA complex told his

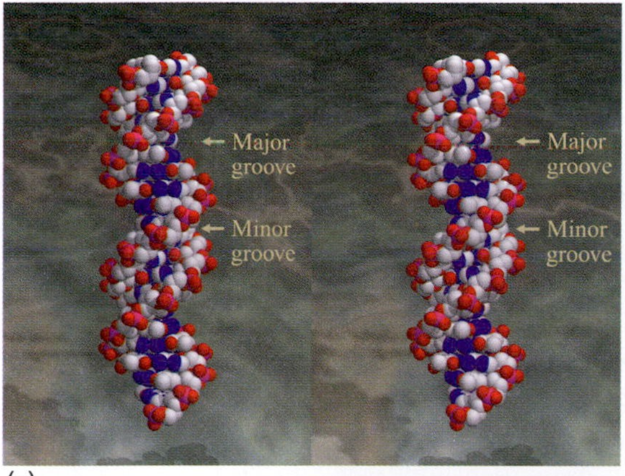

(a)

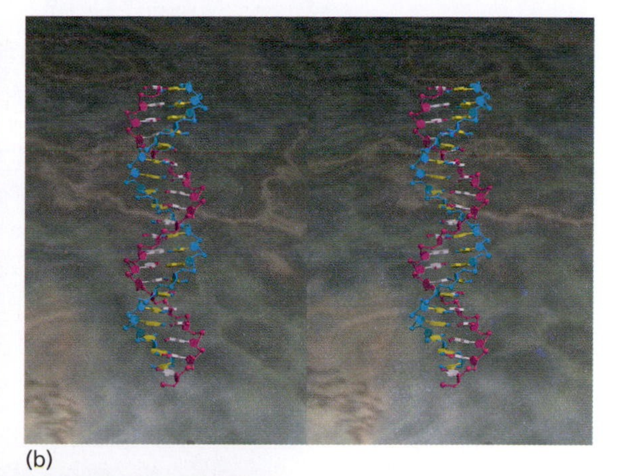

(b)

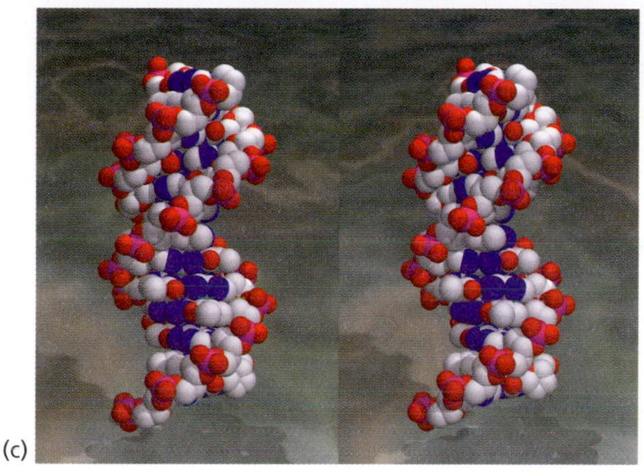

(c)

Fig. 6.4 DNA in the standard B conformation. (a) All-atom representation. (b) Schematic representation. This is the conformation of DNA under physiological conditions. At lower ionic strength it forms the A conformation, with a narrower major groove and a wider and shallower minor groove. Certain sequences, rich in GC repeats, can form a left-handed, double-helical Z conformation. (c) Detailed view *looking into* the major and minor grooves. The grooves are outlined by charged phosphate groups (P is in magenta). Bases are clearly more accessible through the major groove than through the minor groove. This figure contains a 12 base-pair segment of DNA, B conformation—slightly more than one turn.

supervisor that, in the electron-density map he was interpreting, it appeared that a β-sheet was binding in the major groove, he was advised, with patience tinged by condescension, to go back and look for the helix.

We now recognize great structural variety in protein–DNA interactions. A few examples include:

- **Helix-turn-helix domains.** These appear in prokaryotic proteins that regulate gene expression, eukaryotic homeodomains involved in developmental control, and histones that package DNA in chromosomes.
- **Zinc fingers** include eukaryotic transcription factors, and steroid and hormone receptors.

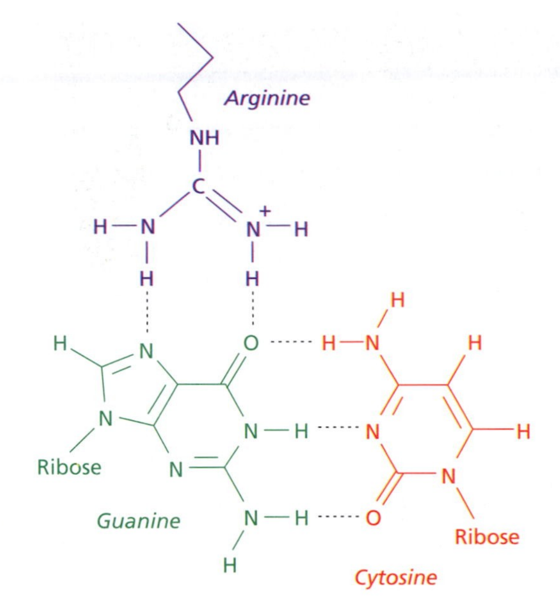

Fig. 6.5 Specific hydrogen-bonding pattern between an arginine side chain of a protein and a guanine of DNA.

- **Proteins with β-sheets that interact with DNA**, for instance the gene regulatory proteins, Met and Arc repressors; and the TATA-box binding protein.
- **Leucine zippers** that act as eukaryotic transcriptional regulators.
- **The high-mobility group in eukaryotes, and the prokaryotic protein HU**, which bind sequences non-specifically and bend DNA.
- **Enzymes that interact with DNA**, involved in replication, translation, repair, and uncoiling. Some are relatively small; others are large multiprotein complexes. They show many different types of folding patterns. Many distort the DNA structure, in order to get access to the bases that are the target of their activity.
- **Viral capsid proteins, and histones**, which package DNA into compact forms.

These examples form an anecdotal list, not a classification.

RNA-binding proteins have a separate variety. Some resemble DNA-binding proteins. Others bind to RNA molecules of defined structure; for instance, enzymes that interact with tRNA, including, but not limited to, amino acid tRNA synthetases and the ribosome itself.

Some protein–DNA complexes that regulate gene transcription

λ cro

Bacteriophage λ is a virus containing a double-stranded DNA genome of 48 502 base pairs. A λ phage infecting an *E. coli* cell chooses—depending on which genes are active—between **lysis** or **lysogeny**. λ can replicate, and *lyse* the cell, releasing ~100 progeny. Alternatively, it can integrate its DNA into the host genome. The phage in

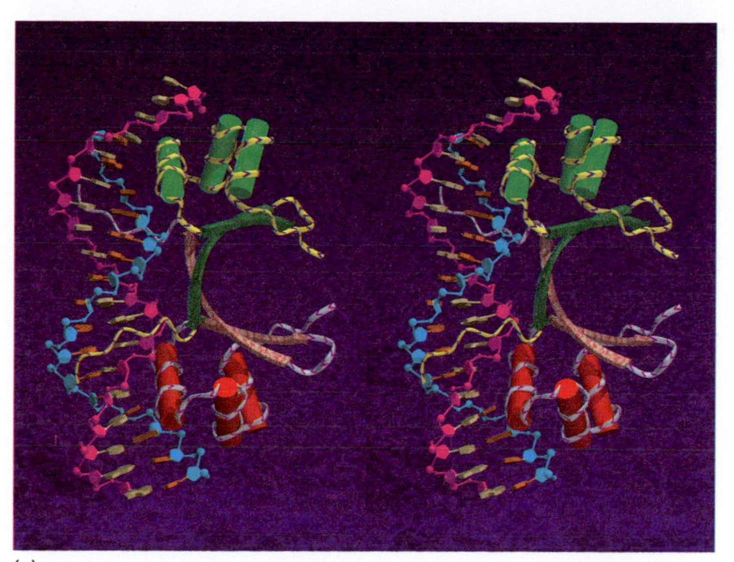

(a)

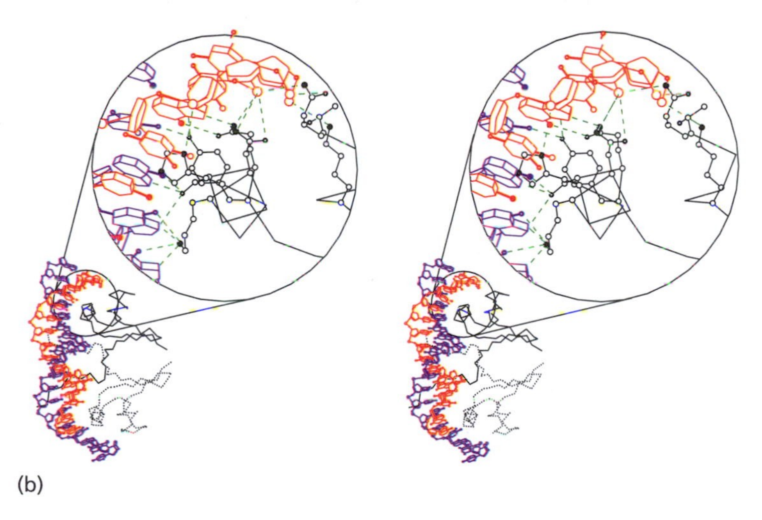

(b)

Fig. 6.6 Phage λ cro repressor–DNA complex. (a) A dimer binds an approximately palindromic sequence in DNA. (b) Pattern of protein–DNA hydrogen bonds.

such a *lysogenized* cell is dormant, and can be released by stimuli that switch from the lysogenic to the lytic state.

λ cro binds to DNA as a symmetrical dimer (Fig. 6.6) Its target sequence is approximately palindromic:

C T A T C A C C **G C** A **A G G G A T** A A
G A T **A G T G G** C **G T** T C C C **T A T T**

The bases to which the protein makes contact are shown in bold-face. The protein interacts with both strands. The DNA is slightly bent.

λ cro is an example of the 'helix-turn-helix' structural motif. Following along the chain in Fig. 6.6(a), the first secondary structure is a helix, followed by two more helices that frame the motif. The second of these two helices (the third helix in the molecule)—called the *recognition helix*—lies in the major groove and makes extensive contacts with the DNA (Fig. 6.6(b)). The hairpin that follows is involved in dimerization of the protein. It interacts with the corresponding hairpin of the other monomer to form a four-stranded β-sheet. A long C-terminal tail wraps around the DNA, following the minor groove.

The eukaryotic homeodomain antennapedia

Homeodomains are highly conserved eukaryotic proteins, active in the control of animal development. They regulate **homeotic** genes; that is, genes that specify the locations of body parts. Antennapedia is a *Drosophila* protein responsible for initiating leg development. The earliest mutations found in antennapedia produced ectopic legs at the positions of, and instead of, antennae. Loss-of-function mutations convert legs into antennae.

The structure of the antennapedia–DNA complex (Fig. 6.7(a)) resembles, in some respects, prokaryotic helix-turn-helix proteins such as λ cro. However, the tail that wraps around into the minor groove is N-terminal to the helix-turn-helix motif in antennapedia, instead of C-terminal as in λ cro.

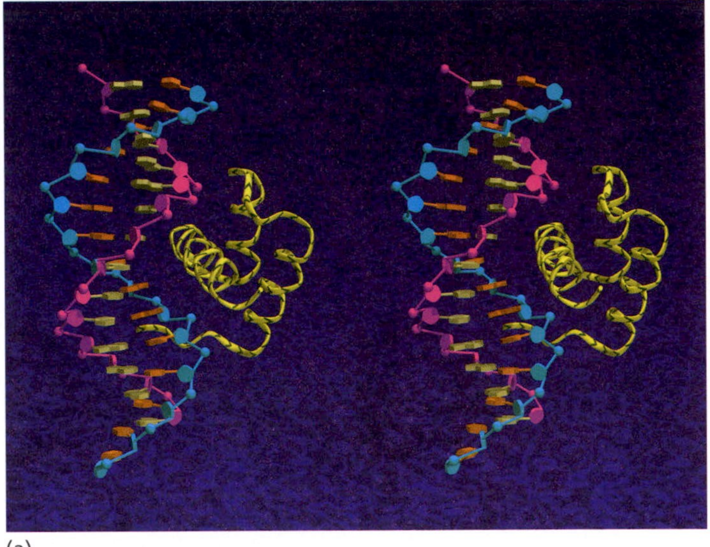

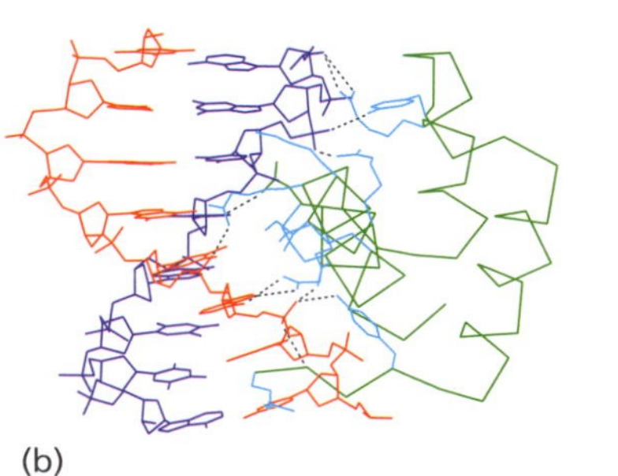

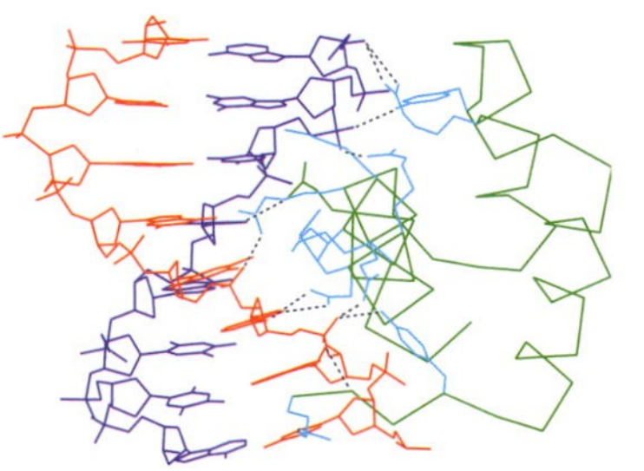

(b)

Fig. 6.7
(a) Antennapedia homeodomain–DNA complex [9ANT].
(b) Details of protein–DNA interactions in the antennapedia complex.

A comparison of the protein–DNA hydrogen bonds in the antennapedia complex (Fig. 6.7(b)) with that of λ cro shows that in the antennapedia complex a greater fraction of DNA-protein hydrogen bonds involve phosphate oxygens rather than bases.

Leucine zippers as transcriptional regulators

Leucine zippers, which we have already met in connection with keratin, form part of another type of dimeric transcriptional regulator. The Jun protein forms a homodimer consisting of an N-terminal domain with many positively charged side chains that binds to DNA, and a C-terminal leucine zipper domain involved in dimerization.

The proteins grip the DNA as if they were picking it up with chopsticks. The α-helices bind in major grooves on opposite sides of the double helix (Fig. 6.8). This structure shares with λ cro, and many other DNA-binding proteins, the *symmetry* of the complex, which mimics the dyad symmetry of the DNA double helix. This requires, on the part of the protein, formation of symmetrical dimers; and on the part of the DNA, an approximately palindromic (=complementary to its reverse) target sequence. For Jun dimers the target sequence is: ATGACGTCAT.

Jun can dimerize not only with itself but with other related proteins, notably Fos. Different dimers have different DNA-sequence specificities, and different affinities, affording subtle patterns of control.

Zinc fingers

Zinc fingers are small modules found in eukaryotic transcription regulators. Each finger recognizes a triplet of bases in DNA. Tandem arrays of fingers recognize an extended region (Fig. 6.9). Understanding the relationship between the amino acid sequences of

Fig. 6.8 bZIP (basic DNA-binding domain)–leucine zipper transcriptional regulator Jun homodimer binding to DNA [1JNM].

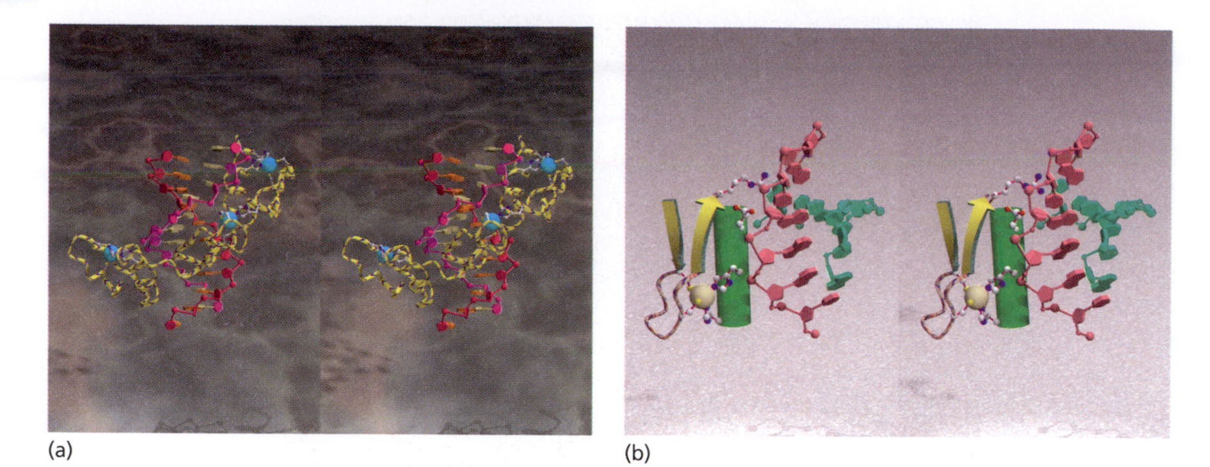

(a) (b)

Fig. 6.9 (a) Zif268, a tandem three-finger structure binding the sequence GCGTGGGCG. [1AAY]. (b) Structure of a single module of a Class 1 (Cys_2–His_2) zinc-finger, taken from the Zif268–DNA complex. Each finger interacts with three consecutive bases. Three positions along the α-helix, non-consecutive in the amino acid sequence, contain primary determinants of the DNA-sequence specificity. One of them, a histidine, binds to both the Zn^{2+} and a phosphate of the DNA. The other three side chains that bind the Zn^{2+} are also shown.

Zn fingers and the DNA sequences they bind would permit the design of gene-specific repressors.

Three types of Zinc fingers differ in structure: in Class 1 one Zn^{2+} ion binds two Cys and two His residues to stabilize the packing of a β-hairpin against an α-helix (Fig. 6.9(b)). Residues in the helix interact with three consecutive nucleotides. (In Class 2 zinc fingers, one Zn^{2+} ion binds four Cys side chains. In Class 3, two Zn^{2+} ions bind six cysteines.)

The *E. coli* Met repressor

Like many other DNA-binding proteins, the Met repressor binds as a symmetrical dimer (Fig. 6.10). In the complex, each monomer contributes one strand of a 2-stranded β-sheet, which sits in the major groove (the student was right! (see p. 288)), with side chains making hydrogen bonds to bases. The co-repressor, *S*-adenosyl methionine, increases the affinity of the complex. *S*-adenosyl methionine is a product of methionine metabolism, and its effect is an example of feedback inhibition—not on the activity of the enzyme directly but on its *expression*.

The TATA-box binding protein

A TATA box is a sequence (consensus TATA[A|T]A[A|T]) upstream of the transcriptional start site of bacterial genes. Recognition of this sequence by the TATA-box binding protein (Fig. 6.11) initiates the formation of the basal transcription complex, a large

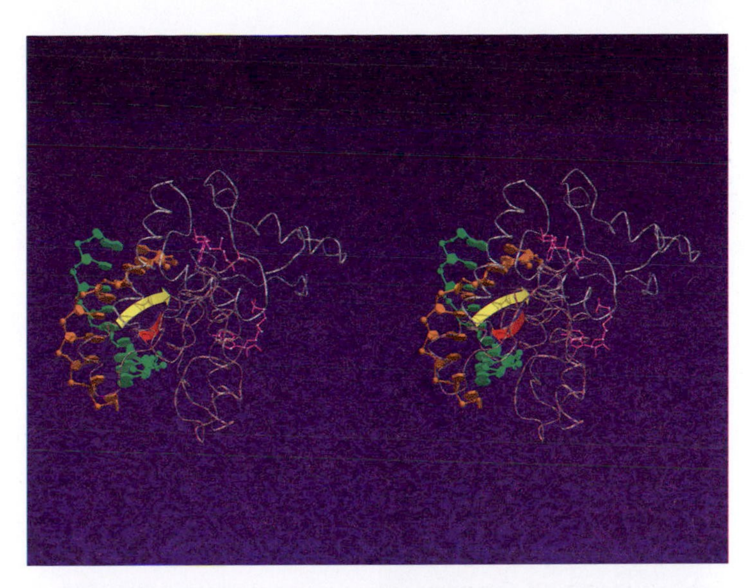

Fig. 6.10 *E. coli* Met repressor–DNA complex [1CMA].

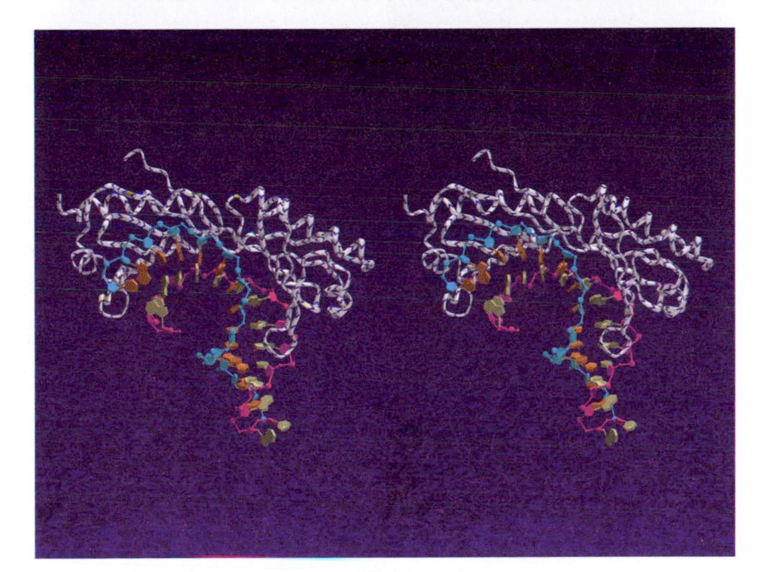

Fig. 6.11 TATA-box binding protein YTBP [1YTB].

multiprotein particle. This is an example of initial binding of a protein to DNA followed by recruitment of other proteins to form an active complex.

The most obvious feature of the complex is the very strong bending and unwinding induced in the DNA. A long curved β-sheet sits against an unusually flat surface on the DNA, the result of prying open the minor groove. Phe side chains intercalate between the bases (Fig. 6.12).

p53 is a tumour suppressor

p53 is a DNA-binding transcriptional activator. It is of great clinical importance because mutations in the *p53* gene are very common in tumours.

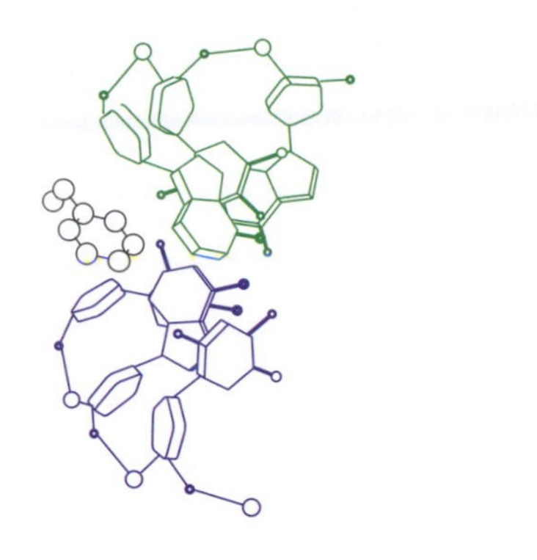
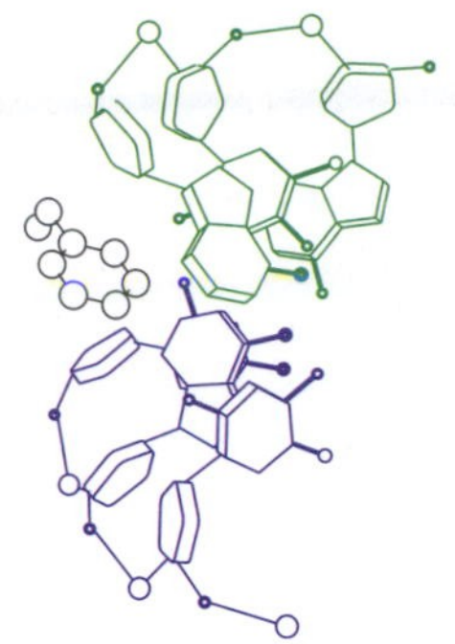

Fig. 6.12 Intercalation of a Phe side chain between bases in the distorted double helix in TATA-box binding protein–DNA complex [1YTB].

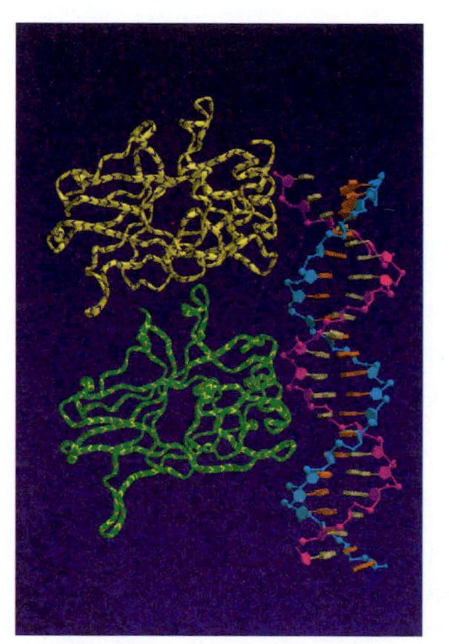

Fig. 6.13 p53 core domain in complex with DNA [1TSR].

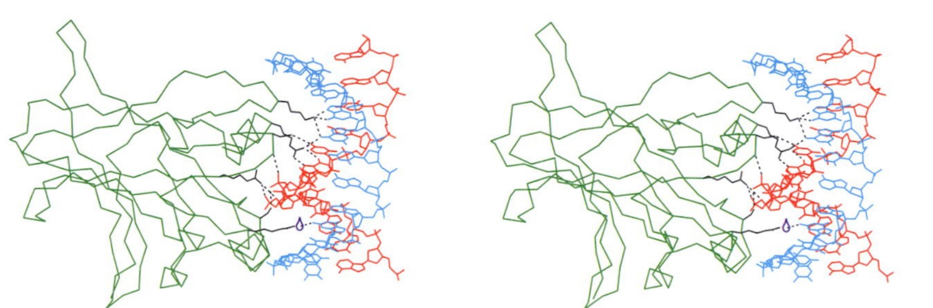

Fig. 6.14 Details of protein–DNA interaction, in p53 core domain–DNA complex [1TSR]. One protein–DNA contact is water-mediated.

p53 acts by surveilling for genome integrity. Damage to DNA induces enhanced expression of p53, which stalls cell-cycle progression. This gives time for DNA repair; if repair is unsuccessful, the 'fail-safe' mechanism is **apoptosis**.

The structure of the DNA-binding subunit of p53 shows a double-β-sheet fold (Fig. 6.13). A helix sits in the major groove, and side chains from loops connecting strands of the β-sheet insert into the minor groove (Fig. 6.14).

Virus structures

Viruses are small particles containing DNA or RNA enclosed in a protein **capsid**, that enter cells and make use of cellular enzymes to reproduce. Some viruses acquire a lipid envelope derived from a host membrane. Viral genomes encode a minimal set of proteins unavailable from the host, including the capsid protein. Some viruses integrate their genomes into host DNA. This makes clinical management more difficult, because even after killing active virus with drugs, more can re-emerge. This is a very serious problem in the treatment of people with AIDS.

Typically over half the weight of a virus is protein. In principle, a nucleic acid can encode one-tenth of its molecular weight in protein. This creates a packaging problem: a typical virus capsid with an outer radius of 125 Å and an inner radius of 100 Å can enclose only a ~3000-nucleotide, single-stranded RNA molecule.

How do viruses cope? The efficient way to encode a capsid is to synthesize many copies of a relatively small protein, and assemble them in identical environments, as far as possible. Compatible with this general approach, viruses show diverse morphologies. The two basic architectures are:

- **Helical symmetry**. Tobacco mosaic virus (TMV) is a cylinder of radius 180 Å and 3000 Å long, with a central hole 20 Å in radius. It contains 2130 subunits of a 158-amino-acid coat protein, arranged in a right-handed helical array with 16.3 subunits per turn. They enclose a molecule of single-stranded RNA 6395 nucleotides long. The RNA follows a helical path through the molecule, three nucleotides binding to each protein subunit (see Figs 6.15 and 6.16)
- **Capsids roughly spherical in overall shape**. For identical subunits to pack in identical environments, they must observe the constraints on possible symmetry patterns. Satellite tobacco necrosis virus exemplifies the simplest possible structure of this type. It contains 60 copies of a 195-residue coat protein, forming an approximately spherical

Viral infections pose severe threats to humans, animals, and plants

- The influenza epidemic of 1918 killed 25–40 million people, more than died in the war that preceded it.
- Today, AIDS is a major killer, accounting for 5% of deaths worldwide and a higher proportion in the most severely affected regions.
- An epidemic of foot and mouth disease in Britain in 2001 cost the lives of millions of animals and devastated the countryside.
- Plant viruses can kill crops over wide geographical areas and have caused famines. On the other hand, the variegated colour of 'broken' tulips results from a viral infection.
- Viruses are linked to a significant fraction of human cancers.

Fig. 6.15 Model of tobacco mosaic virus, designed and based on work by R. Franklin, A. Klug, and colleagues. Initially built for and exhibited at the Brussels World's Fair, 1958, as part of a section on 'Large biological structures' organized by W.L. Bragg. After the Fair the model was displayed at the Royal Institution while Bragg was its director. The model now stands in the MRC Laboratory of Molecular Biology, Cambridge, England. A companion model of the shell of an idealized icosahedral virus is in the Science Museum, London.

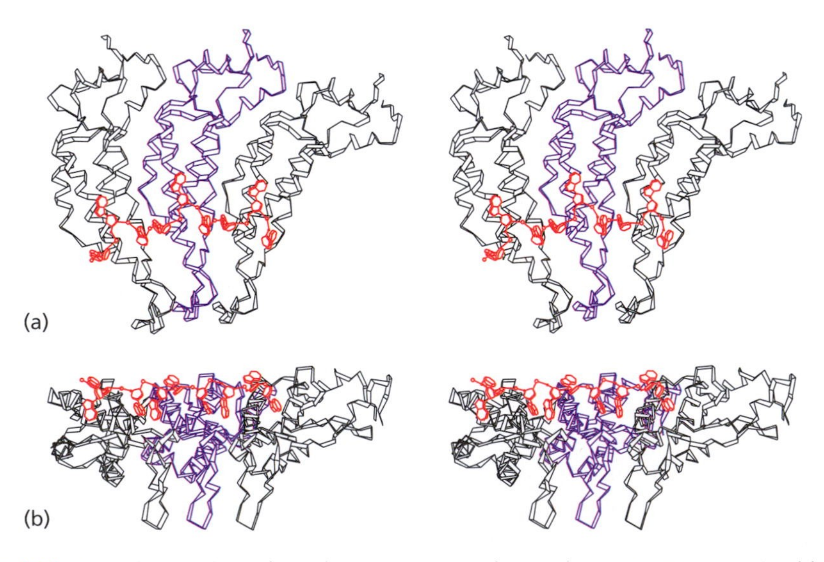

Fig. 6.16 Three adjacent subunits from tobacco mosaic virus, showing the protein–RNA interaction. (a) View parallel to the axis. (b) View approximately perpendicular to the axis, showing how the bases penetrate the protein–protein interfaces.

particle 180 Å in diameter, with icosahedral symmetry, enclosing a single-stranded RNA genome of approximately 1200 nucleotides. Satellite viruses depend on other viruses as well as on their cellular hosts to reproduce. Their genomes are typically shorter than those of complete viruses, and their capsids smaller. Viruses with more than 60 subunits in their capsids encounter a geometrical problem (see Box).

Icosahedral symmetry and quasi-equivalence

Constructing a capsid from multiple copies of a single protein is a sensible strategy for a virus, because it shortens the required length of the genome. The ideal would be to pack identical subunits in identical environments. But available spatial symmetries are limited.

Of the five regular geometric solids—tetrahedron, cube, octahedron, dodecahedron, and icosahedron—the icosahedron allows for the largest number of equivalent subunits, maximizing the interior volume. An icosahedron has 12 vertices, 20 faces, and 30 edges, arranged with twofold, threefold, and fivefold axes of symmetry (see Fig. 6.17). These symmetries will generate, from an object placed at a general position, 59 partners.

It is not possible to place more than 60 identical objects in identical environments in a closed regular array in three-dimensional space. (A helical assembly, as in tobacco mosaic virus, has no such limitations, but the subunits at the ends have different environments from those in the middle.) However, it is possible is to distribute some *multiple* of 60 identical subunits on the surface of an icosahedron in *almost* identical environments.

The geodesic domes of Buckminster Fuller gave the clue to the principle. A regular array of hexagons, as in a common type of fencing material, is planar. If some of the hexagons are replaced by *pentagons* the resulting structure will crimp into a nearly spherical shape (see Fig. 6.18). The environments of different subunits cannot be identical but are *quasi-equivalent*. (see Fig. 6.19). The same symmetry appears in a soccer ball, and in the buckminsterfullerene C_{60} carbon structure.

The number of identical subunits in an icosahedral capsid is limited to a multiple of 60, by a factor of the form $T = h^2 + hk + k^2$. If $h = k = 1$, $T = 3$, to give 180 subunits per capsid.

The theory of quasi-equivalence was seminal in virus structural studies. Some viruses, such as cowpea chlorotic mottle virus, follow it quite closely. Others, such as tomato bushy stunt virus, follow its general ideas, but show greater non-equivalence in the subunit interfaces than anticipated. The polyoma virus capsid, a $T = 7$ structure, was the first to show a qualitative departure from the geometric scheme.

Fig. 6.17 Views of an icosahedron looking down a twofold axis (red), a fivefold axis (blue), and a threefold axis (green).

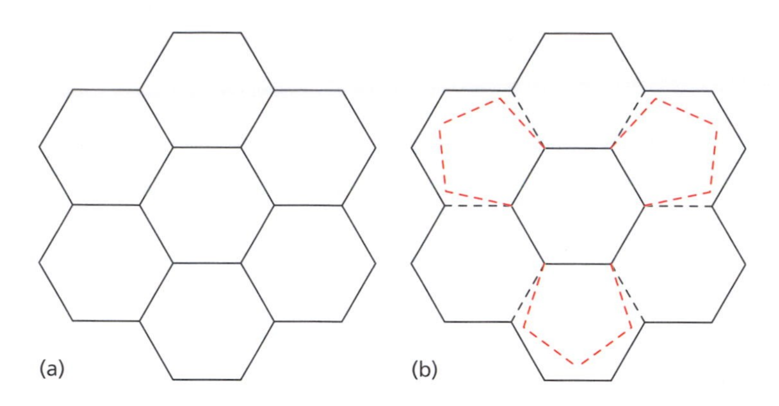

Fig. 6.18 Formation of a structure with icosahedral symmetry starting from a plane hexagonal net. (a) An array of hexagons forms a flat mosaic in the plane. (b) Replacing selected hexagons with pentagons, and attaching the broken red lines to the adjacent broken black lines, requires bending the figure out of the plane, to form part of an object with icosahedral symmetry. Note that the replacement of hexagons with pentagons has reduced the sixfold symmetry of (a) to threefold symmetry in (b). Extension of the pattern will result in a closed figure with fivefold axes at the centres of the pentagons, threefold axes at the centres of the hexagons, and twofold axes halfway between the closest fivefold axes. (To compare the resulting figure with a simple icosahedron, see **www.cs.berkeley.edu/~flab/unfold/unfolding.html**)

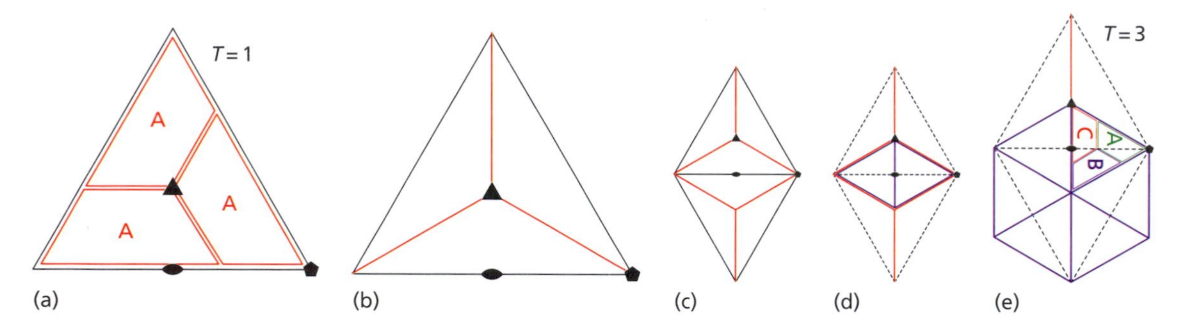

Fig. 6.19 (a) To assemble 60 subunits with icosahedral symmetry, take a simple icosahedron and place three subunits on each of the 20 triangular faces. Each subunit, labelled A, is in an environment identical to that of all others. A filled pentagon represents a fivefold axis perpendicular to the plane of the drawing, a filled triangle represents a threefold axis, and a lozenge represents a twofold axis.

To assemble a *multiple* of 60 subunits in an approximation to icosahedral symmetry, subdivide each face, for example, into three parts, maintaining the symmetry (b, red lines). Doing this with two adjacent faces (c), new triangular faces emerge (d, shown in blue; original triangular faces in broken lines). The new triangular faces correspond to one-third of the original triangular faces, and their centres do not correspond to threefold symmetry axes of the original icosahedron. (e) Placing three subunits in each new triangular face gives nine subunits per original icosahedral face, giving $9 \times 20 = 180$ subunits in all. (To see this, note that the blue hexagon is equivalent to the *two* original triangular faces outlined in broken lines, and that the hexagon contains $3 \times 6 = 18$ subunits). Now, however, the three subunits labelled A, B, and C have slightly different environments (see Problem 1).

Using quasi-equivalence, a virus can assemble a large capsid from many identical proteins, without multiplying the amount of nucleic acid required to encode them. In some viruses, such as cowpea chlorotic mottle virus and tomato bushy stunt virus, all capsid proteins are identical. In others, such as rhinovirus and poliovirus, the capsids have the geometrical organization of a $T = 3$ structure, but the three quasi-equivalent positions contain three different proteins. These require separate viral genes.

Viral capsids have sufficiently rigid structures to form well-ordered crystals, which can be determined by X-ray crystallography. To date, several hundred virus structures are known. They are among the largest macromolecular aggregates solved: Bluetongue virus is over 600 Å in diameter, and contains almost 1000 polypeptide chains.

Because viral capsids have higher symmetry than the nucleic acids they contain, much of the nucleic acid, and in some cases even some of the protein, does not appear in virus crystal structures. The packing of virions in a crystal is determined by the capsids. Because the capsids have full icosahedral symmetry, each virion can take any of 60 equivalent orientations. X-ray crystallography then produces an electron-density map of a virion, averaged over all 60 orientations. The averaging will retain a clear image in the electron-density map of all parts of the virion that have full icosahedral symmetry, because the structure is the same in all the orientations over which the average is taken. Anything of lower symmetry, such as nucleic acid, that is not the same in the different orientations, will, as a result of the averaging, appear fuzzy in the electron-density map.

For an analogy, consider dice packed into a rectangular box. The dice would form a natural cubic lattice. Each die will take one of many possible orientations, at random. A method of structure determination that averaged all the dice in the box would accurately reproduce the cubical size and shape of the dice, but each face would appear as an average of the dot patterns from all possible orientations. (See Exercise 17.)

Tomato bushy stunt virus (TBSV)

Tomato bushy stunt virus is an icosahedral virus with a capsid formed from 180 copies of a 386-residue protein. The capsid is approximately 175 Å in radius, enclosing a single-stranded RNA genome 4776 nucleotides long, which encodes five proteins.

The capsid protein contains three domains (Fig. 6.20). The N-terminal R domain projects into the capsid and interacts with the RNA. The R domains, and the RNA, are

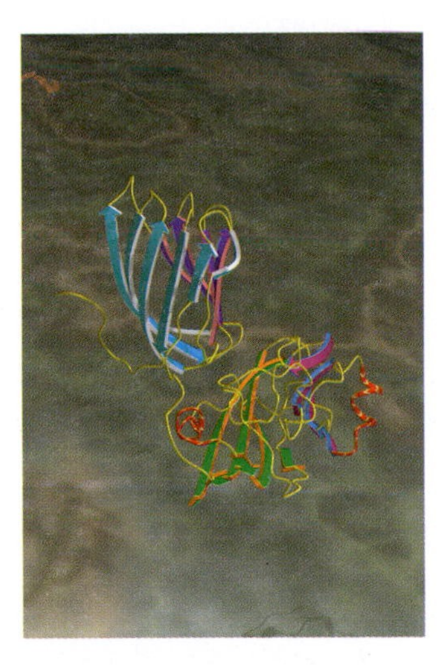

Fig. 6.20 Coat protein of tomato bushy stunt virus [2TBV]. P domain above, S domain below. The R domain is not shown. The general combination of two double-β-sheet domains appears in many virus capsid proteins.

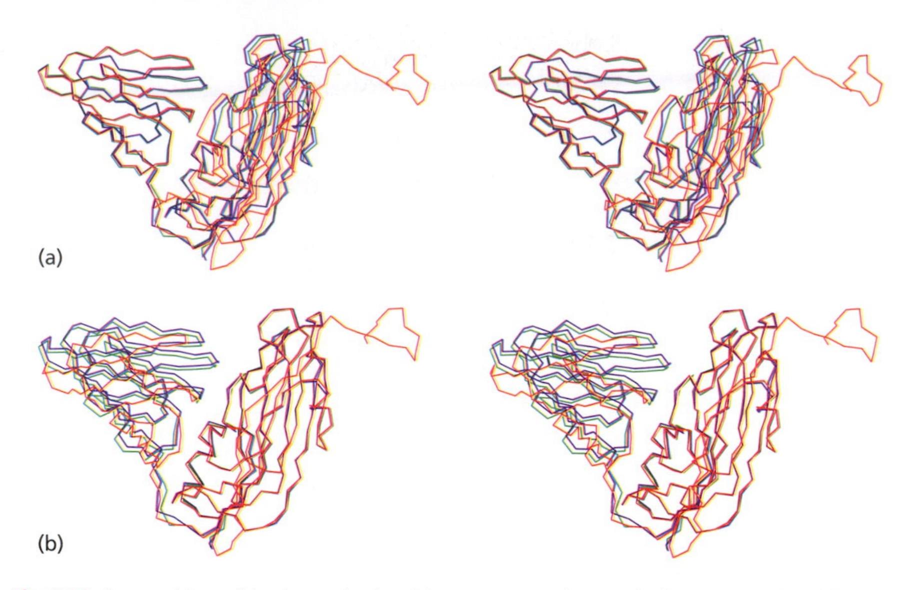

Fig. 6.21 Superpositions of the three subunits of the coat protein of tomato bushy stunt virus [2TBV]. The colours correspond to those in Fig. 6.19: A = green, B = blue, C = red. (a) Structures superposed on the P domain, at the left of the picture. The red and black copies of the S also superpose well, but the S domain of the blue chain does not. (b) Structures superposed on the S domain, at the right of the picture. The red and black copies of the P domain also superpose well, but the P domain of the blue chain does not. The conclusion is that the red and black molecules have similar main chain conformation. The blue molecule differs, by a hinge motion between the domains.

not shown in Fig. 6.20. The central, S, domain, forms the shell. The C-terminal domain, the P or protruding domain, creates prominent surface features. Both S and P domains are double-β-sheet proteins. A short hinge region between the S and P domain mediates the conformational change between the A- and BC-subunits.

Tomato bushy stunt virus is a $T = 3$ structure. The coat proteins have three quasi-equivalent environments, and—despite their identical sequences—differ slightly in structure (Fig. 6.21).

Bacteriophage HK97: protein chain-mail

This section is for those readers who think they've seen everything.

Bacteriophage HK97 is an icosahedral virus with a double-stranded DNA genome. The capsid is 55 nm in diameter, formed of 420 copies of a 281-residue protein. In most viruses, the capsid proteins interact by familiar non-covalent interactions—Van der Waals forces, hydrogen bonding, and salt bridges. In contrast, HK97 forms covalent bonds between the subunits of its capsid.

The side chains of Lys169 and Asn356, in adjacent subunits, form a peptide-like bond:

$$\cdots \mathrm{C\alpha CH_2CH_2CH_2CH_2NH_3^+} + {}^{-}\mathrm{OC(NH_2)CH_2C\alpha}\cdots \rightarrow$$
$$\cdots \mathrm{C\alpha CH_2CH_2CH_2CH_2}\mathbf{NH\text{–}CO}\mathrm{CH_2C\alpha}\cdots + \mathrm{NH_4^+}.$$

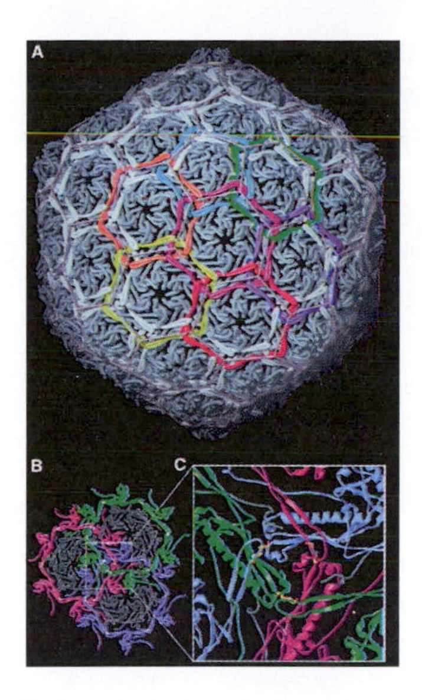

Fig. 6.22 (A) Interlocking chains of subunits in the bacteriophage HK97 capsid. Six hexameric rings are shown in brighter colours. Pentagonal rings are also visible. (B, C) Details of the linkage between adjacent subunits meeting at a threefold axis. (Reproduced with permission from Wikoff, W.R., Liljas, L., Duda, R.L., Tsuruta, H., Hendrix, R.W., and Johnson, J.E. (2000). Topologically linked protein rings in the bacteriophage HK97 capsid. *Science*, **289**, 2129–33. Copyright 2000 AAAS.)

These bonds join sets of monomers into large closed rings. The rings contain five or six subunits, suitable for placement on the threefold or twofold axes, respectively, of the icosahedral symmetry.

Still more bizarre, the rings are threaded through one another, to form a structure akin to chain-mail (Fig. 6.22)!

The photosynthetic reaction centre

The reaction centre from the purple bacterium *Rhodopseudomonas viridis* is the site of the initial step in the capture of light energy in photosynthesis. The reaction centre is a membrane-bound complex of four proteins, binding 14 low molecular weight cofactors. These cofactors include the chromophores that absorb the excitation energy that is converted to electrochemical potential—or redox energy—across the membrane.

Figure 6.23 shows a representation of all atoms of the reaction centre except hydrogens, in front and side views. The assembly as a whole has dimensions 72 Å × 72 Å × 133 Å. The chromophores are embedded in the two central subunits and are only partly visible. The four proteins are the light (L, in blue), medium (M, in purple), and heavy (H, in green) subunits, and a cytochrome (orange).

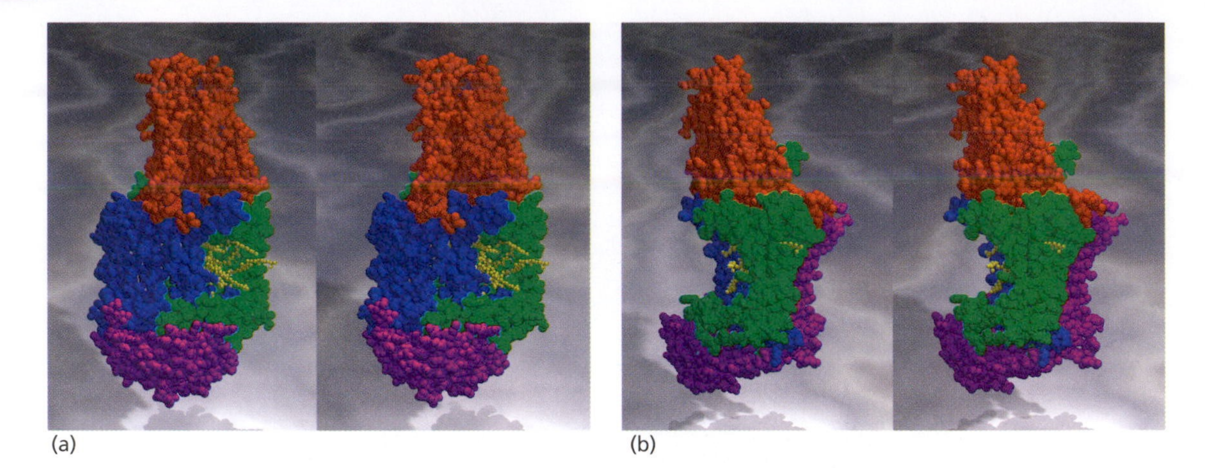

Fig. 6.23 The photosynthetic reaction centre from *Rhodopseudomonas viridis* [1PRC]. This figure shows an 'all-atom' representation. Parts (a) and (b) show two orientations, at right angles to each other. The four proteins are distinguished by colour: light, blue; medium, purple; heavy, green; cytochrome, orange.

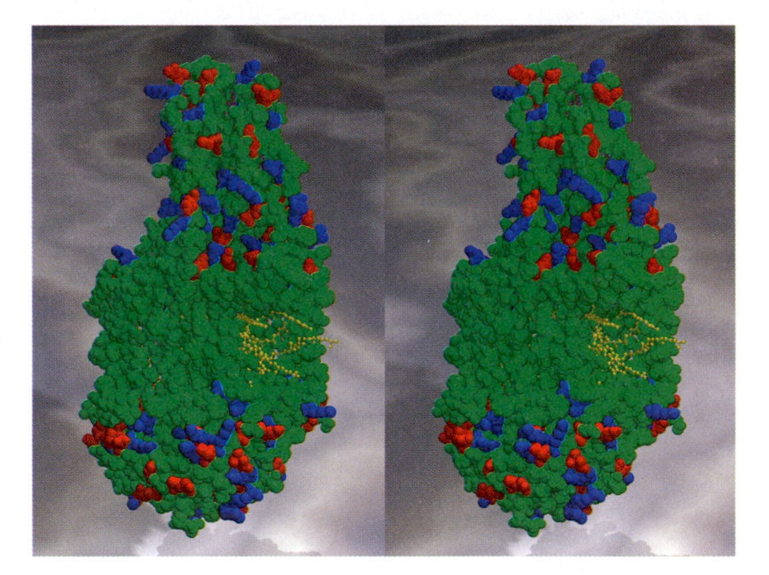

Fig. 6.24 The reaction centre coloured by residue charge (positive, blue; negative, red; neutral, green). The 'green belt' around the waist of the molecule corresponds to the membrane-spanning segment [1PRC].

Figure 6.24 shows the structure with an alternative colour coding. Positively charged residues are coloured blue, negatively charged residues red, the others green. The wide horizontal swathe across the centre of the molecule that is entirely green (except for the cofactors) corresponds to the region that traverses the membrane. The proteins present surfaces of different character to the aqueous environment, at the top and bottom, and to the membrane. The surfaces of the H-subunit and the cytochrome are typical of proteins in aqueous environments: a 'tossed salad' of charged, polar, and uncharged residues. The surfaces presented to the lipid environment of the membrane are devoid of charged residues. These membrane-exposed surfaces resemble the interiors of soluble proteins in physicochemical character.

The schematic diagrams in Fig. 6.25 show that the complex contains transmembrane helices, a structural feature common to many membrane proteins.

There is symmetry in the structure of the L- and M-subunits (Fig. 6.26). They are related by a rotation of approximately 180°, around a vertical axis parallel to the page.

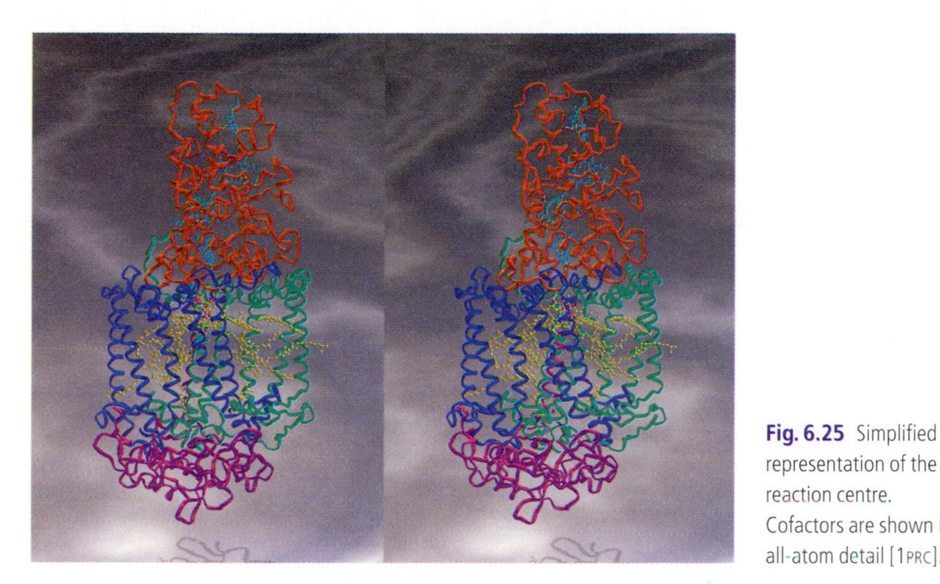

Fig. 6.25 Simplified representation of the reaction centre. Cofactors are shown in all-atom detail [1PRC].

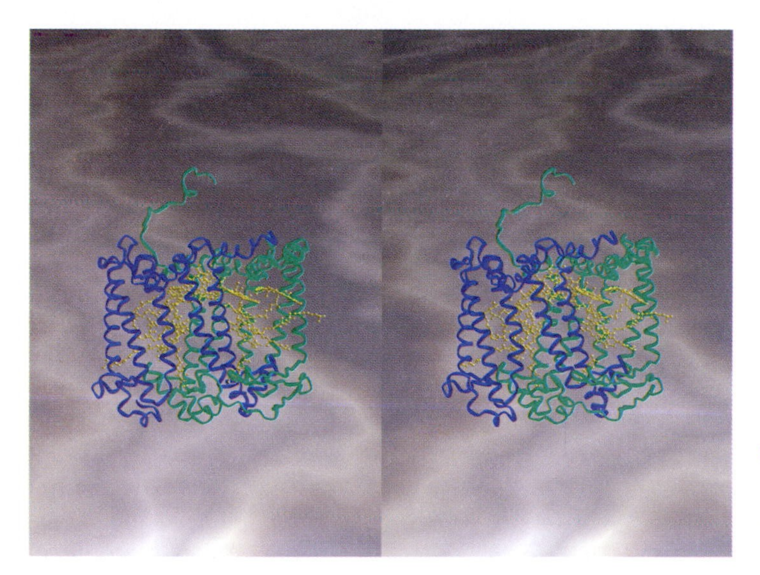

Fig. 6.26 Membrane-spanning subunits L and M [1PRC].

At the heart of the reaction centre is the 'special pair' of bacteriochlorophylls, to which excitation energy is transferred, and which is the site of the ionization—the initial step in electron transfer. Of course, the structure and environment of the special pair is crucial, and we want to understand the nature of the interactions between the proteins and these ligands. Figure 6.27 shows the special pair and its environment. The final picture is a complex one, so we build it up in stages. Figure 6.27(a) shows the special pair on its own. Each bacteriochlorophyll contains a large ring structure—the chlorin ring—and a long phytyl tail. The magnesium atoms at the centres of the chlorin rings are coloured magenta. Figures 6.27(b) and 6.27(c) show the contacts of the chlorin rings with the protein (the contacts to the phytyl tail are not shown). Figure 6.27(b) shows the side chains only; Fig. 6.27(c) shows the backbone as well. Note that most of the contacts are made to side chains; the folding pattern of the backbone places them in the proper position to form these contacts.

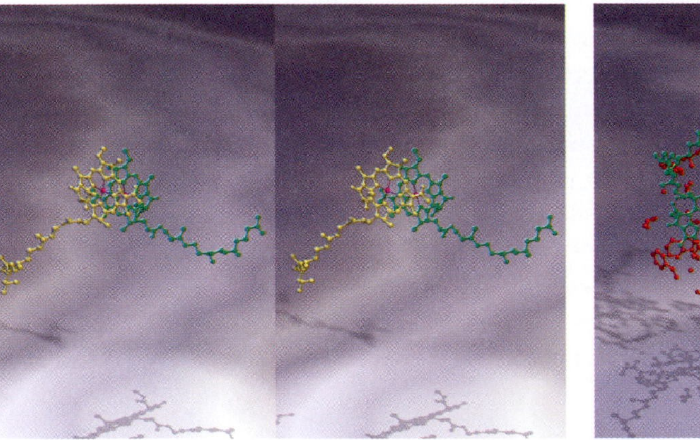

(a) (b)

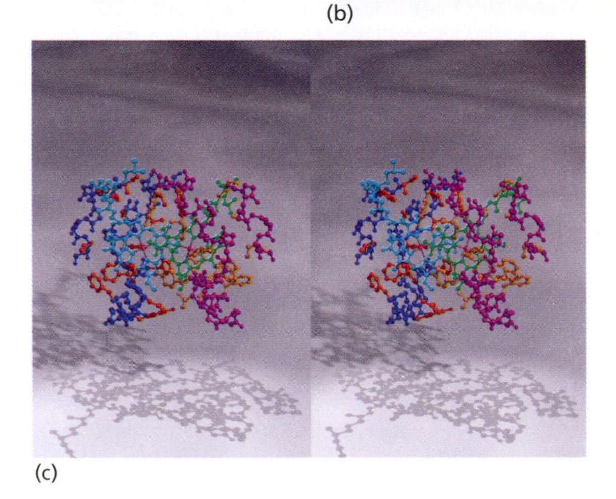

(c)

Fig. 6.27 The environment of the 'special pair': (a) the special pair alone; (b) the special pair plus the side chains that interact with it; (c) the special pair plus the side chains and backbone of the residues that interact with it.

Membrane transport

Membrane proteins mediate the exchange of matter between cell interiors and the surroundings:

- **Channels** or **pores** regulate traffic by: (1) *selectivity*, different channels allowing the passage of specific substances; and (2) *switching* between open and closed states. When open, the channel is only a passive doorway through which allowed solutes diffuse, under the influence of concentration and electrochemical potential.
- **Pumps** transfer substances (in most cases, cations) *against* the prevailing concentration or electrochemical gradient. They must pay the thermodynamic price, generally using energy from ATP or light. For instance, bacteriorhodopsin (Fig. 2.14) is a light-driven proton pump.
- **Carrier proteins** are intermediate between channels and pumps. They provide pathways for specific substances, or classes of substances, to cross membranes down their concentration gradients, like channels. However, rather than acting as simple pores, transport involves conformational changes in the carrier. In one conformation, the carrier exposes a binding site on one side of the membrane, where the passenger ion or molecule embarks. In the other conformation, the carrier exposes a binding site on the other side of the membrane, from which the same passenger is released. Some carriers couple the spontaneous passage across the membrane of one solute, to the driving across the membrane of a different solute *against* its concentration gradient. For instance, a carrier protein in *E. coli* couples the scavenging of glucose-6-phosphate from the medium to export of inorganic phosphate.

Specificity of the potassium channel from *Streptomyces lividans*—room to swing a cation?

The potassium ion channel from *S. lividans* is a pore, specific for K^+ ions, permitting throughput rates of up to 10^8 ions s^{-1}. The ratio of permeability for K^+ and Na^+ is at least 10^4. These facts will appear paradoxical to anyone who has been delayed in a crowd of people passing through an immigration or customs channel—where high selectivity appears to require low throughput.

For the potassium channel, a simple size-exclusion 'sieve' mechanism would not work, because Na^+ is *smaller* than K^+. The radii of hydrated K^+ and Na^+ in aqueous solution are: 1.33 Å and 0.95 Å, respectively. (Ions in aqueous solution bind water tightly, as a result of the strong electrical attraction between a charged ion and the polar water molecules. Have you ever wondered why it is so difficult to break down the crystal lattice of ordinary table salt by heating—the melting point of NaCl is 801 °C—even though it is so easy to take the ions apart simply by adding water? The crystal lattice forces are very strong, *but so are the ion–water forces in the solution.*)

Figure 6.28 shows the structure of part of the *S. lividans* K^+ channel. It is a symmetric tetramer, the subunits packing around a central pore. The pore is 45 Å long. Each subunit contains two transmembrane helices, tilted away from perpendicular to the membrane.

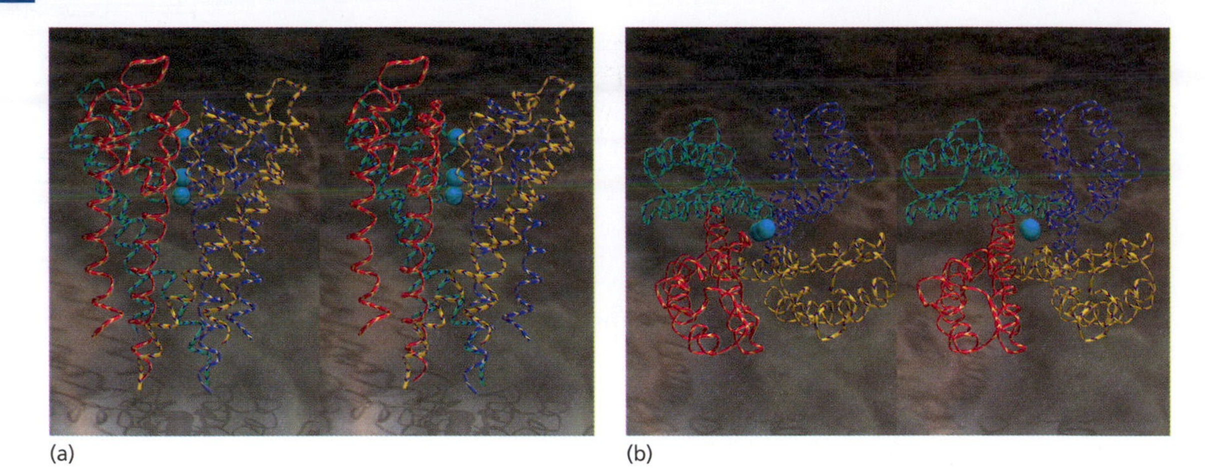

Fig. 6.28 Potassium ion channel from *Streptomyces lividans* [1BL8]. (a) View parallel to membrane. (b) View perpendicular to membrane, looking from outside the cell.

The selectivity filter is formed by a constriction in the pore, 12 Å long and only about 2 Å in diameter. Neither K^+ nor Na^+ can fit through this constriction in their hydrated states. But desolvation of a small ion is energetically costly. The protein lines the constriction in the pore with inward-pointing backbone carbonyl groups, at a radius similar to that of the first hydration sphere of aqueous K^+. Na^+ ions are rejected because these carbonyls are *too far apart* to mimic the first hydration sphere of Na^+. The situtation is like that of a mountain climber, who can ascend a narrow chimney by pushing against its sides, but not a chimney that is wider than the span of his or her arms and legs.

ATPase

ATPase couples the energy *localized* in a proton gradient across a membrane to the high-energy phosphate bond of ATP, which is then distributed and used to drive all manner of life processes.

ATPase comprises a membrane-bound complex, F_0, and a soluble complex, F_1. Separation of the two components revealed that F_0 contains the proton transport machinery, and F_1 the catalytic sites that synthesize ATP. In cow mitochondria, F_1 is a complex of nine polypeptides, containing copies of five different molecules, with the subunit composition: $\alpha_3\beta_3\gamma\delta\epsilon$.

According to the 'binding change' mechanism of ATPase, the enzyme contains three catalytic sites, that cycle among three different states (Fig. 6.29). The *loose* state, L, binds substrates (ADP and P_i) weakly but does not synthesize ATP. The *tight* state, T, binds substrates strongly and converts them to ATP. In the *open* state, O, the affinity for ligands is small, and bound ATP is released.

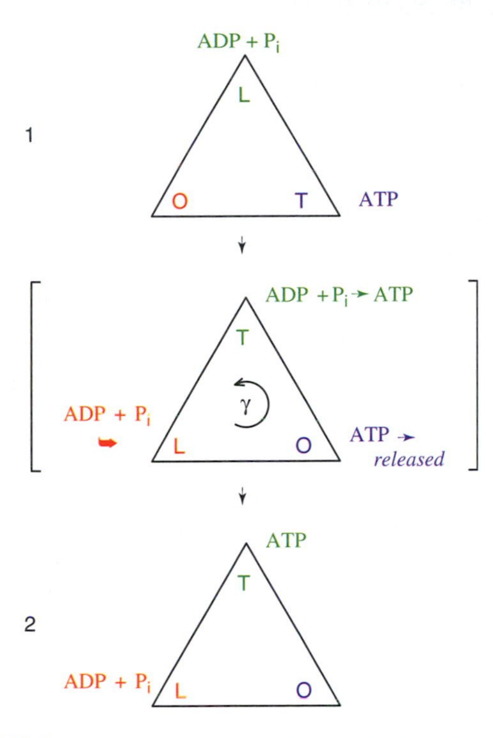

Fig. 6.29 The scheme of P.D. Boyer, confirmed by X-ray crystallography, for the mechanism of ATPase. The molecule contains three binding sites, which interconvert between three conformational states as the molecule rotates.

This diagram shows one stage of the active cycle. The three αβ-dimers have three different states. In 1, the open state O (red) is empty; the loose state L (blue) contains ADP + P_i; and the tight state T (green) contains ATP. In a logical intermediate stage (bracketed), rotation of the γ-, δ-, and ϵ-subunits (not shown) within the $(\alpha\beta)_3$ hexamer converts the L state to a T state, the T state to an O, and the O state to an L. The L state can accept a new charge of substrate. The T state can form ATP. At stage 2, the ATP has fallen out of the O state, new ADP + P_i have bound to the L state, and ATP has been synthesized in the T state. Note that the bracketed state is presented solely for explanation, and *does not* represent a physical trappable intermediate.

Comparison of states 1 and 2 shows that they differ only by a change in colour, signifying that the molecule has been returned to its initial state with *different* subunits in the L, O, and T states.

The crystal structure of the F_1-ATPase subunit from cow heart mitochondria confirmed the binding-change model. Interconversion of the states of the binding sites occurs during the operation of a microscopic motor. The translocation of protons is coupled to synthesis of ATP, through the generation of mechanical energy within the enzyme.

The F_1-subunit has the general shape of a mushroom. The cap of the mushroom contains a ring of three αβ-dimers, forming an approximately spherical assembly ~90 Å in diameter (Fig. 6.30). The binding sites are at the α–β interfaces, with the catalytic sites primarily in the β-subunits. The γ-subunit, containing long coiled helices, is the stem of the mushroom. It penetrates the sphere along a diameter and extends below it to form part of a stalk connecting the F_1- and F_0-subunits (Fig. 6.31). The δ- and ϵ-subunits do not appear in the crystal structure. They are probably disordered in the absence of contacts with the F_0-subunit.

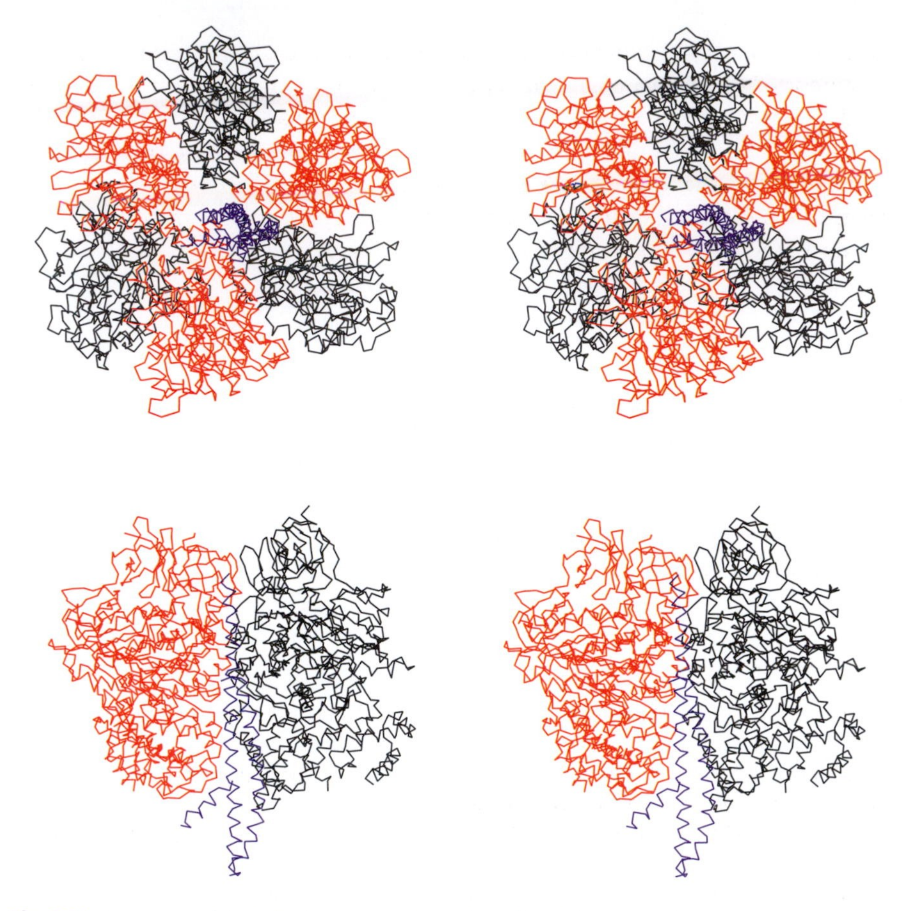

Fig. 6.30 Top view of cow mitochondrial ATPase [1BMF]. α-subunits green, β-subunits red. The α- and β-subunits alternate like segments of an orange.

Fig. 6.31 Penetration of the $(\alpha\beta)_3$ hexamer by the γ-subunit [1BMF]. The γ-subunit breaks the symmetry of the $(\alpha\beta)_3$ hexamer, causing them to adopt three different conformations as in the Boyer model. Rotation of the γ-, δ-, and ϵ-subunits inside the $(\alpha\beta)_3$ hexamer cyclically interconverts the states. The system converts redox energy from the proton gradient across the membrane, to mechanical energy of rotation within the enzyme, to the chemical bond energy in ATP.

The symmetry of the $(\alpha\beta)_3$-subunits is broken by interactions with the asymmetric γ-subunit (Fig. 6.31). The enzyme operates as a microscopic rotatory motor: changing patterns of interaction of the rotating γ-, δ- and ϵ-subunits with the $(\alpha\beta)_3$ hexamer—which remains fixed to the membrane-bound F_0 portion—produce conformational changes, interconverting the states cyclically as in the Boyer model.

The interface between the $(\alpha\beta)_3$- and γ-subunits has features that would be expected from a rotatory mechanism. The internal surface of the $(\alpha\beta)_3$ (the 'bearing') forms a hydrophobic sleeve around the γ-subunit (the 'axle'). The coiled-coil structure of the γ-subunit is typical of proteins that require mechanical rigidity. More subtly, the high overall similarity of the three αβ-dimers suggests that there are no high barriers to the interconversion of the states.

This idea that ATPase functions as a motor, made plausible by the crystal structure, and supported by observations such as inactivation of the enzyme by crosslinking of β- and γ-subunits, was dramatically proved by direct observation. The αβ-subunits are fixed to a solid support, and a fluorescent actin filament attached to the γ-subunit. Adding ATP runs the reaction backwards, coupling ATP hydrolysis to rotation of the enzyme. No membrane, no proton pumping; the energy is lost as heat. Rotation of the actin during activity of the enzyme is seen and recorded.

USEFUL WEB SITES

Information about protein-protein interactions:
http://www.imb-jena.de/jcb/ppi/jcb_ppi_databases.html
Assembly of icosahedron: **www.cs.berkeley.edu/~flab/unfold/unfolding.html**
DNA-protein complexes: **www.biochem.ucl.ac.uk/bsm/prot_dna/prot_dna.html**
Rotation of F1-ATPase: **http//www.k2.ims.ac.jp/F1movies/F1Prop.htm**
Virus particle explorer: **http://mmtsb.scripps.edu/viper/viper.html**

RECOMMENDED READING

Calladine, C.R. and Drew, H.R. (1997). *Understanding DNA: the molecule and how it works*, (2nd edn). Press, London.

Capaldi, R.A. and Aggeler, R. (2002). Mechanism of the F(1)F(0)-type ATP synthase, a biological rotary motor. *Trends Biochem. Sci.*, **27**, 154–60.

Caspar, D.L.D. and Klug, A. (1962). Physical principles in the construction of regular viruses. *Cold Spring Harbor Symp. Quant. Biol.*, **XXVII**, 1–24.

Garvie, C.W. and Wolberger, C. (2001). Recognition of specific DNA sequences. *Mol. Cell*, **8**, 937–46.

Goodsell, D.S. and Olson, A.J. (2000). Structural symmetry and protein function. *Annu. Rev. Biophys. Biomol. Struct.*, **29**, 105–53.

Johnson, J.E. and Speir, J.A. (1999). Principles of virus structure. In *Encyclopedia of virology* (2nd edn) (ed. A. Granoff and R.G. Webster) Vol. 3, pp. 1946–56. Academic Press. London.

Kleanthous, C. (ed.) (1999). *Protein–protein interactions*. Oxford University Press, Oxford.

Reddy, V., Natarajan, P., Okerberg, B., Li, K., Damodaran, K., Morton, R., Brooks, C. III, and Johnson, J. (2001). VIrus Particle ExploreR (VIPER), a website for virus capsid structures and their computational analyses. *J. Virology*, **75**, 11943–7.

Strauss, J.H. and Strauss, E.G. (2002). *Viruses and human disease*. Academic Press, London.

Stuart, D.I., Grimes, J., Mertens, P. and Burroughs, N. (1999). Crystallography and the atomic anatomy of viruses. *Microbiol. Today*, **26**, 59–61.

Tsonis, P.A. (2003). *Anatomy of gene regulation/A three-dimensional structural analysis*. Cambridge University Press, Cambridge.

Yoshida, M., Muneyuki, E., and Hisabori, T. (2001). ATP synthase—a marvellous rotary engine of the cell. *Nat. Rev. Mol. Cell Biol.*, **2**, 669–77.

EXERCISES, PROBLEMS, AND WEBLEMS

Exercises

1. Find illustrations in this book of isologous and heterologous dimers.
2. **(a)** Does the dimeric globin from *Scapharca inaequivalvis* (Fig. 2.22(b)) appear to have an axis of twofold symmetry? One way to check is to photocopy Fig. 2.22(b) onto a transparency, rotate by 180°, and try to superpose the copy on the original.
   (b) Does the interface in this dimer appear to be isologous?
3. **(a)** Why is an isologous open structure impossible?
   (b) Must a dimer with an isologous interface have a twofold axis of symmetry?
4. On a photocopy of Fig. 6.4(a), indicate: (1) a phosphate group, (2) a sugar. On a photocopy of Fig. 6.4(b), indicate: (3) a phosphate group (shown as a single sphere), (4) a sugar, (5) a purine, and (6) a pyrimidine.
5. From inspection of Fig. 6.4(b):
   (a) How many base pairs are visible in the major groove?
   (b) Compare the accessibility of bases through the major and minor grooves.
6. Draw a picture analogous to Fig. 6.5 showing what amino acid side chain would readily form two hydrogen bonds to an A–T base pair.
7. From Fig. 6.6(a), estimate the angle through which binding of λ cro bends DNA.
8. On a photocopy of Fig. 6.6(a), indicate: (1) the helix-turn-helix motif, (2) the four-stranded β-sheet, (3) the N-terminal tail in the minor groove.
9. On a photocopy of Fig. 6.9(b), indicate: (1) the three side chains primarily responsible for recognition, and (2) the four side chains that ligate the zinc ion. (One side chain does both).
10. At which positions does the sequence to which λ cro binds differ from an exact palindrome? Which, if any, of these positions make contacts with the protein?
11. On a photocopy of Fig. 6.16(a), show the approximate position of the intersection of the helix axis with the plane of the picture.
12. On three separate photocopies of Fig. 6.32, indicate by a '5' each of two sites related to the red dot by fivefold symmetry. Indicate by a '2' each of two sites related to the red dot by twofold symmetry. Indicate by a '3' each of two sites related to the red dot by threefold symmetry.

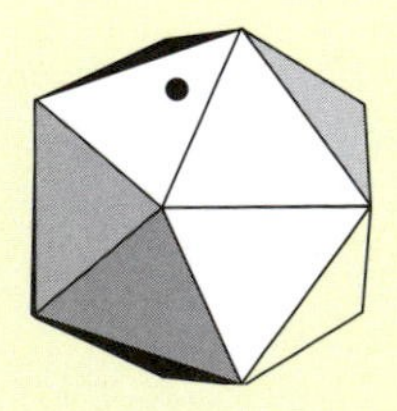

Fig. 6.32 A point on the surface of an icosahedron. Find symmetry-related points. (Ex. 12).

13. On a photocopy of Fig. 6.22, circle the region of part A that appears in part C.
14. On a photocopy of Fig. 6.22, circle two pentagonal rings.
15. On a photocopy of Fig. 6.19(e), indicate the positions of: (1) four additional twofold axes, (2) one additional threefold axis, and (3) four additional fivefold axes.
16. What fraction of a closed figure with icosahedral symmetry will be formed the faces outlined in thick lines after attaching broken red and black lines in Fig. 6.18(b)?
17. What would be the appearance of the face of a die, averaged over all possible orientations, (a) if all six faces of the die were copies of the face containing a 'four' on a normal die? (b) if all six faces of the die were copies of the face containing a 'two' on a normal die? (c) if the die had the normal distribution of dots on all faces?

Problems

1. (First do Exercise 15.) (a) On a photocopy of Fig. 6.19(e), draw in the positions of subunits of type A, B, and C in the rest of the area of the figure. (b) Which subunits meet at twofold axes? at threefold axes? at fivefold axes? (c) How many different types of intersubunit contacts are formed?
2. (a) Compare the mass ratios of nucleic acid:protein for tobacco mosaic virus and tomato bushy stunt virus. (b) Compare the capsid encoding efficiency—the ratio of the length of the capsid protein to the total genome length—for these two viruses.
3. Satellite tobacco necrosis virus is an icosahedral virus with a $T = 1$ structure of inner radius 60 Å, containing one molecule of single-stranded RNA. If one ribonucleotide occupies ~600–700 $Å^3$ in a virion, estimate the maximum length of the RNA molecule that could fit inside the capsid, and compare with the observed genome size of 620 nucleotides.
4. *Bacillus subtilis* phosphocarrier protein interacts with a kinase from *Lactobacillus casei*. The crystal structures of these proteins have been solved both separately ([1JB1] and [1SPH]) and in the complex ([1kkl]). (a) Calculate the accessible surface area of the two components separately, and of the complex. From these results calculate the surface area buried in the complex. (b) Extract the coordinates of each protein from the complex, and calculate their accessible surface areas. Recalculate the buried surface area from these results, and compare with (a). The difference arises because the kinase changes conformation upon forming the complex. If the structures of the components individually had never been determined, how accurate a value of the buried surface area can be calculated from the coordinates of the complex alone?
5. You are asked to write a computer program to identify crevices in the surface of a protein, as candidates for active sites. You have available a standard program for the calculation of accessible surface area from the coordinates of a protein structure, which permits you to specify the radius of the probe sphere. Suppose that you decide to proceed by calculating the accessible surface area of a protein with different probe radii—the standard value, 1.4 Å, and a succession of larger values. How would you analyse and interpret the results?
6. Compare the interactions of the recognition helix in the major groove in λ cro and antennapedia: in each structure, how many hydrogen bonds are formed to (1) the sugar–phosphate backbone of DNA, and (2) the bases of DNA?

Weblems

1. Find oligomeric proteins that in their natural assembly have the following symmetries: (a) an axis of threefold symmetry; (b) an axis of fourfold symmetry; (c) an axis of sixfold symmetry; (d) an axis of fourfold symmetry *and* a perpendicular axis of twofold symmetry (group D4); (e) cubic symmetry; (f) octahedral symmetry.
2. **(a)** Write the sequence of reactions for the synthesis of methionine and *S*-adenosyl-methionine from aspartate in *E. coli*.

   (b) Which genes coding for enzymes in this pathway have expression regulated by the Met repressor?

Chapter 7

PROTEINS IN DISEASE

LEARNING GOALS

1. **To understand some basic ideas about the interaction between protein science and medicine**, and to recognize the genetic and functional components of the ways in which proteins contribute to disease states.
2. **To appreciate the role of the aggregation of misfolded proteins in several diseases**, including sickle-cell anaemia, amyloidoses, Huntington, Parkinson, and Alzheimer diseases, and prion diseases such as 'mad-cow disease'.
3. **To differentiate the conformational states of the serpins**; and to know how they contribute to the mechanism of inhibition, and how they participate in the formation of aggregates.
4. **To appreciate the very great variety of immunoglobulin structures required to recognize the entire organic world**, and to understand the genetic mechanisms for generating antibody genes by combining segments.
5. **To understand the distinction between variable and constant domains, the significance of hypervariable regions, and the distinction between framework and complementarity-determining regions (CDRs).**
6. **To be able to relate the structures of antibodies to their function of binding antigens**: the antigen-binding site is formed from loops from variable domains; to know the canonical structure model of the structures of antigen-combining sites, its successes and limitations.
7. **To understand structure–function relationships in proteins of the MHC complex, and T-cell receptors**, and how they interact during the immune response.
8. **To understand some of the basic molecular biology of cancer**, appreciating the contribution that protein science can make to diagnosis and treatment.

Introduction

With life so dependent on proteins, there is ample opportunity for things to go wrong. Some general types of problems include (1) an absent or dysfunctional protein, (2) protein aggregation, and (3) infection by pathogenic organisms.

(1) **An absent or dysfunctional protein.** This may arise from a mutation. If a normal allele on the partner chromosome provides adequate healthy protein, abnormal symptoms may be a recessive trait. Mutations on the X chromosome may lead to sex-linked abnormalities such as some forms of colour blindness.

Some missing proteins can be supplied, such as insulin for diabetes, Factor VIII for the most common form of haemophilia, and human growth hormone for children who produce inadequate amounts. Attempts to repair or replace damaged genes using viruses as vectors are in the experimental stage and show promise.

- *Cutting a metabolic pathway* by knocking out an enzyme can lead to (a) loss of product, (b) accumulation of precursor.
  - Phenylketonuria is a hereditary disease resulting from the loss of phenylalanine hydroxylase, the enzyme that converts phenylalanine to tyrosine. If untreated, phenylalanine accumulates in the blood, to toxic levels. Phenylketonuria is relatively easy to detect, and patients can live on diets restricted in phenylalanine.
  - Certain enzyme deficiencies can more easily be accommodated by changes in lifestyle, and they can show complex interactions with other diseases. Glucose-6-phosphate dehydrogenase (G6PDH) is the most common enzyme deficiency, affecting over 400 million people worldwide. It is a recessive X-linked genetic defect, affecting up to 10% of populations in which mutations are common.

See Box: Glucose-6-phosphate deficiency, food taboos, folk medicine, pharmacogenomics, and mosquito breeding seasons.

  - In some cases, species-wide loss of biosynthetic enzymes contributes to the list of essential nutrients. For instance, most animals can synthesize vitamin C, but we must provide it in our diet.
  - Loss of some proteins is surprisingly innocuous. Mice lacking myoglobin thrive, and even show athletic performance comparable to that of normal mice.
- *Dysfunction of a regulatory protein or receptor* can disorganize the operation of a pathway even if all components are normal. Some abnormal regulatory proteins cannot be activated at all, others are constitutively activated and cannot be shut off. The results include:
  - Physiological defects: a number of diseases are associated with mutations in **G-protein-coupled receptors**. Some mutations in **opsins** are associated with colour blindness. Certain mutations in the common G-protein target of olfactory receptors lead to loss of the sense of smell.

Glucose-6-phosphate deficiency, food taboos, folk medicine, pharmacogenomics, and mosquito breeding seasons

Glucose-6-phosphate dehydrogenase (G6PDH) catalyses the reaction:

$$\text{glucose-6-phosphate} + \text{NADP} \rightarrow \text{6-phosphogluconate} + \text{NADPH},$$

the first step in the pentose phosphate shunt.

This reaction is required to produce the reduced glutathione needed to dispose of hydrogen peroxide (H_2O_2). It is particularly important in mature red blood cells, which, lacking nuclei and mitochondria, are metabolically impoverished, and have no alternative mechanism for detoxifying H_2O_2. Without active G6PDH, a build-up of H_2O_2 will oxidize and denature haemoglobin, leading to the destruction of red blood cells and producing a condition called haemolytic anaemia.

Eating fava beans, especially uncooked, can induce anaemic episodes in people deficient in G6PDH. The danger of eating fava beans has been recognized since antiquity, and has been associated with food taboos, and with preparation techniques designed to reduce toxicity. Pythagoras, for example, banned the eating of fava beans in his school. We now know that fava beans contain the compounds vicine and convicine. These are metabolized in the intestine to isouramil and divicine, which react with oxygen to produce hydrogen peroxide, subjecting cells to oxidative stress.

Other chemicals, including certain drugs, present the same danger to G6PDH-deficient people. During World War I, some patients were observed to suffer dangerous side effects of the antimalarial drug primaquine. Many drugs, including sulphonamides, are now contraindicated for use in G6PDH-deficient patients, as is the taking of large doses of vitamin C. The observation of variations in the effectiveness and toxicity of different drugs in different people has developed into the new field of **pharmacogenomics**, the tailoring of drug treatments to the genotype of the individual patient.

Why does primaquine produce haemolytic anaemia in G6PDH-deficient patients, and does this have a relationship to its antimalarial activity? Why have dysfunctional G6PDH genes remained at such a high level in the population? And why have fava beans continued to be grown if non-toxic alternatives are available?

The malarial parasite invades the red blood cell of its host, and competes metabolically with normal activity. Primaquine and related drugs, such as **chloroquine**, subject the red blood cells to oxidative stress. Cells stressed by *both* parasite and drug are the most vulnerable, and if they die the parasite dies with them. Because consumption of fava beans subjects cells to oxidative stress, they also provide an antimalarial effect, recognized in folk medicine. Indeed, fava beans have some effect against malaria even for people with normal G6PDH activity; those with abnormal G6PDH have an advantage, until the maturing *Plasmodium* produces its own G6PDH.

The link with malaria is the likely explanation of the persistence of the gene in the population and the fava bean in agriculture. A final clue appears in the the calendar: there is good overlap between the fava bean harvest period and the peak *Anopheles* breeding season.

- Developmental defects: several types arise from mutations in hormone receptors.
- Cancer: mutations in *ras* appear in 30% of human cancers. A cell containing p21 Ras trapped in a constitutively active state is continuously triggered to proliferate.

(2) **Protein aggregation**. Many diseases are associated with the formation of insoluble aggregates, usually of misfolded proteins. These include classical **amyloidoses**, Alzheimer and Huntington disease, aggregates of misfolded **serpins**, and **prion** diseases. Aggregation also creates problems in the production, storage, and delivery of insulin preparations used in the therapy of diabetes.

(3) **Infection by pathogenic organisms**. This is part of the ongoing wars between species. To avoid the necessity of developing *ad hoc* defences against every new threat, vertebrates developed a generalized approach to detect and repel any foreign invader—the immune system. If works amazingly well—except in the case of AIDS, which attacks the immune system itself.

Diseases of protein aggregation

The first disease attributable to protein aggregation was sickle-cell anaemia, the first recognized 'molecular disease'. Sickle-cell anaemia arises from a single site mutation, converting normal haemoglobin, HbA, to HbS: β19Gln → Val. HbS/HbA heterozygotes suffer a milder form of the disease than HbS/HbS homozygotes.

The mutation creates a sticky hydrophobic patch on the surface of the red cell, normally interrupted by the polar Gln side chain. As a result, the deoxy form of haemoglobin forms polymers that precipitate within the red blood cell. Red blood cells must deform to squeeze through fine capillaries. In sickle-cell anaemia, the precipitates of HbS make red cells too rigid to deform adequately. In the traffic jam building up behind a plugged capillary, arriving red cells release their oxygen to surrounding hypoxic tissues, become deoxygenated, and thereby aggravate the problem.

Many diseases are now recognized to arise from protein aggregation (see Box). Misfolded proteins are more prone to aggregate, and mutated proteins are more prone to misfold. Overproduction of proteins as a result of the breakdown of control mechanisms, as in myelomas that overproduce immunoglobulin light chains, also aggravates the threat of aggregation.

Amyloidoses

Classical **amyloidoses** involve extracellular accumulations of insoluble fibrils, 80–100 Å in diameter. Although the name amyloid implies polysaccharide, the aggregates contain protein. (Virchow, in the mid-nineteenth centure, named them amyloid because they could be stained with iodine, like starch.) Common features include:

- Characteristic microscopic appearance seen after staining with haematoxylin/eosin.
- Straight, unbranched regular fibrils seen in electron micrographs.
- β-Sheet structure indicated by X-ray fibre diffraction, with the strands perpendicular to the fibre axis—the 'cross-β' structure

Diseases associated with protein aggregates

Disease	Aggregating protein	Comment
Sickle-cell anaemia	Deoxyhaemoglobin–S	Mutation creates a hydrophobic patch on the protein surface
Classical amyloidoses	Immunoglobulin light chains, transthyretin, and many others	Extracellular fibrillar deposits
Emphysema associated with Z-antitrypsin	Mutant α_1-antitrypsin	Destabilization of structure facilitates aggregation
Huntington	Altered huntingtin	One of several polyglutamine repeat diseases
Parkinson	α-Synuclein	Fibrillar deposits found in Lewy bodies
Alzheimer	Aβ	Aβ = 42 residue fragment in amyloid plaques
	Tau	In neurofibrillary tangles
Spongiform encephalopathies	Prion proteins	Amyloid deposits, protein infectious

- Bright green fluorescence seen under polarized light after staining with the dye Congo Red, which is specific for β-sheet structures.
- Solubility at low ionic strength.

Many proteins are known to form amyloid fibrils, including immunoglobulin light chains and their fragments, transthyretin, and lysozyme. Although these proteins do not have a common folding pattern in their native state, they can apparently adopt a common structure in the amyloid state. It has been suggested that *all* proteins can form this common cross-β fibrillar structure, under suitable conditions.

Alzheimer disease

This is a neurodegenerative disease common in the elderly. It is associated with two types of deposits:

1. dense insoluble extracellular protein deposits, called senile plaques. These contain the Aβ fragment (the N-terminal 42 residues) of a cell-surface receptor in neurons, the β-protein precursor (βPP) or APP (amyloid precursor protein) (Fig. 7.1).

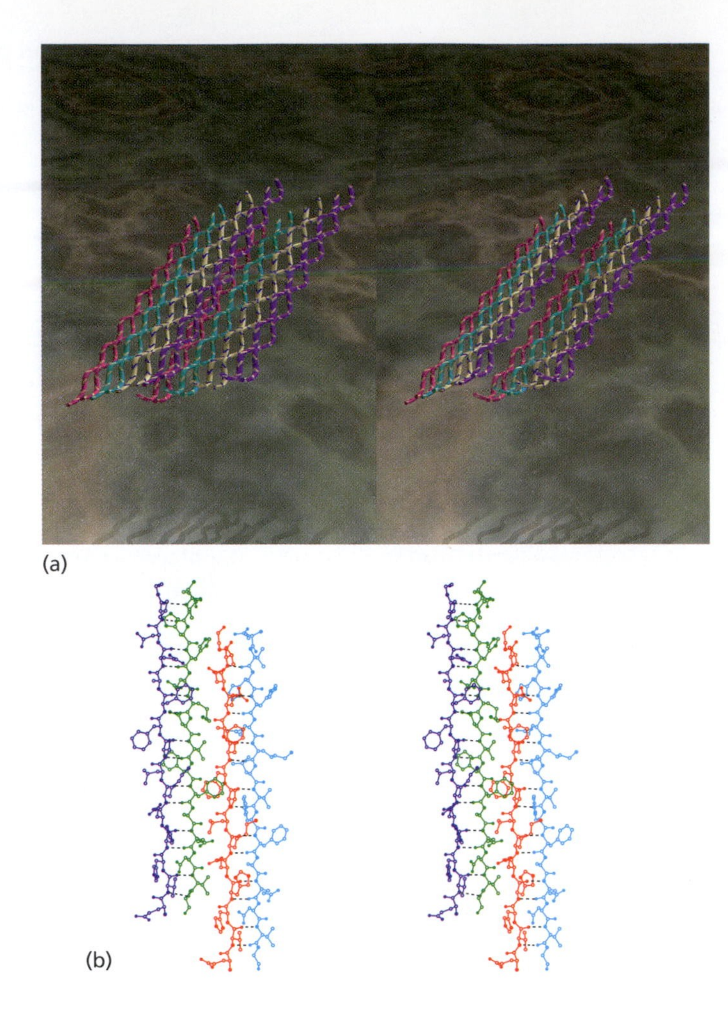

Fig. 7.1 Aβ fragments of the amyloid precursor protein form deposits in the brain associated with Alzheimer disease. This structure of a smaller fragment of Aβ (residues 11–25) forms a cross-β-structure similar to that of classical amyloid protein aggregates. (a) The fragment forms extended stacks of antiparallel β-sheets (Each set of four strands of the same colour corresponds to one unit cell of the crystal.) The axis of the corresponding fibre is in the plane of one of the sheets, perpendicular to the stand direction—not very far from parallel to the view direction. These pictures show repeating units. In the crystal or fibre the sheets contain more strands, and the stacks contain more sheets.

Aβ fragments containing residues 1–40 (very close to the molecule in deposits isolated from the brains of Alzheimer patients) form a fibre of diameter ~70 Å. The 15-residue fragment in this crystal structure also forms fibres, with diameter ~50 Å.

The interresidue distance along the strands is ~3.3 Å. Therefore the 15-residue extended strand is close to 50 Å long; that is, the length of the strands is close to the diameter of the fibre. The distance between the stacked sheets is ~10.6 Å. As a rough estimate the fibre contains 4 or 5 sheets stacked as shown in this picture. These sheets are extended to macroscopic dimensions, in a direction in the plane of any of the sheets and perpendicular to the strand direction.

(b) The atomic structure of one set of four strands, showing both the interstrand hydrogen bonding and the stacking of the sheets. This figure shows two pairs of hydrogen-bonded strands. In their packing together there has been a displacement of two residues in the strand direction—perpendicular to the corresponding fibre axis. Formation of such a structure from longer protein chains could be possible by connecting the hydrogen-bonded strands into a succession of hairpins.

(I thank Drs L. Serpell, E.D.T. Atkins, and P. Sikorski for the coordinates of their structure.)

2. neurofibrillary tangles, or paired helical filaments, inside neurons, containing microtubule-associated protein tau. There is some evidence that amyloid deposits promote tangle formation.

Aggregates of tau appear in other neurodegenerative diseases, the '**tauopathies**'.

Parkinson disease

Parkinson disease, also common in the elderly, is a neurodegenerative disorder commonly characterized by its effects on movement. Symptoms include tremor and rigidity. Parkinson disease results from reduction in levels of the signal transmitter molecule dopamine, caused by the death of neurons that produce it. Cytoplasmic inclusions in brain cells, called *Lewy bodies*, are associated with Parkinson disease. Lewy bodies contain aggregates of the protein α-synuclein.

Huntington disease

This disease, one of several polyglutamine repeat diseases, is associated with aggregates of a protein containing expanded blocks of glutamine residues. The longer the block, the greater the severity of the disease and the lower the age of onset. A threshold in the number of repeated Gln residues required for the appearance of symptoms—for Huntington disease about 40 Gln—suggests that a stable nucleus is required for aggregate formation.

Prion diseases—spongiform encephalopathies

Prion diseases are a set of neurodegenerative conditions of animals and humans, associated with the deposition of protein aggregates in the brain (see Box). The characteristic sponge-like appearance of the brains of affected individuals in postmortem investigation reflects the holes left behind after the death of neurons.

They are unusual among protein deposition diseases in that they are transmissible, and unsusual among transmissible diseases in that the infectious agent is a protein. Moreover—another unusual feature—some prion diseases are hereditary, for example, *familial* **Creutzfeldt-Jakob disease (CJD)**. (Distinguish between a disease transmitted perinatally from mother to baby by passage of an infectious agent—as in many AIDS cases—with a truly hereditary disease depending on parental genotype.) All hereditary human prion diseases involve mutations in the same gene, that encodes the protein found in the aggregates deposited in the brains of sufferers of *both* hereditary and infectious prion diseases.

We mentioned, in connection with circular dichroism, that the prion protein can exist in two forms: the normal PrP^{C} and the dangerous PrP^{Sc}. PrP^{Sc}, but not PrP^{C}, can (1) form aggregates, (2) catalyse the conversion of additional PrP^{C} to PrP^{Sc} within the brain of an individual person or animal, and (3) infect other individuals, by various routes including ingestion of nervous tissue from an affected animal (or person, in the case of **kuru**).

PrP^{C} = prion protein—*cellular*; PrP^{Sc} = prion protein—*scrapie*.

Prion disease presents widespread health problems for humans and animals. In 2001 a serious epidemic of bovine spongiform encephalopathy (BSE; colloquially, '**mad-cow**

Some diseases associated with prion proteins

Disease	Species affected	Symptoms
Scrapie	Sheep	Abnormal gait and behaviour, tremor, weight loss
Bovine spongiform encephalopathy, 'mad-cow disease'	Cow	Similar to scrapie
Kuru	Human	Loss of coordination, weakness, dementia often late or absent
Creutzfeldt–Jakob disease (CJD)	Human, age > 55 years	Impaired vision and motor control, rapidly progressive dementia: median survival time 19 weeks
variant CJD	Human, age usually 15–30 years	Psychiatric and sensory abnormalities preceding a less-rapidly progressive dementia: median survival time 14 months
Gerstmann–Straüssler–Scheinker syndrome	Human	Impaired coordination, slurred speech, mild dementia
Fatal familial insomnia	Human	Insomnia, dementia

disease') devastated the United Kingdom countryside. There was an apparent association with the appearance of human cases of *variant* CJD (vCJD). In the hereditary disease, familial CJD, symptoms begin to appear in people aged 55–75, whereas variant CJD affected people in their twenties. It is hypothesized that these outbreaks were associated with transmission of prion protein infections across species barriers: sheep to cows for BSE, and cows to humans for vCJD.

Prion proteins form a family of homologous proteins in different species of higher animals, and also in yeast, but apparently not in *Caenorhabditis elegans* or *Drosophila*. The normal human prion protein is synthesized as a 253-residue polypeptide. This comprises: an N-terminal signal peptide, followed by a domain containing ~5 tandem repeats of the octapeptide PHGGGWGQ (in mammals), a conserved 140-residue domain, and a C-terminal hydrophobic domain. The signal domain and the C-terminal domain are cleaved off, and the protein is anchored to the extracellular side of the cell membrane of neurons by a GPI (glycosylphosphatidylinositol) group bound to the C-terminal residue Ser231 of the mature protein.

The normal role of PrP^C is not clear. Mice in which PrP^C has been knocked out develop normally for a time, and eventually die of apparently unrelated developmental defects. In fact, PrP^{C}-knockout mice are not susceptible to infection with PrP^{Sc}, an observation important in proving the mechanism of the disease.

The nature of the conformational change is still not entirely clear. The change from an α to β structure shown by CD is one clue. In principle, prion proteins show multiple structures from one polypeptide sequence. However, differences in glycosylation patterns between PrP^{C} and PrP^{Sc} have been reported; these may play a role in determining the conformation.

See Chapter 3, p. 93

The mechanism by which PrP^{Sc} catalyses the transformation of additional PrP^{C} to PrP^{Sc} is also not clear. Inherited prion diseases are associated with mutants, presumably increasing the tendency for conformational mobility. A related question concerns the kinetics of the process—what governs the rate of accumulation of aggregates that causes many prion diseases to appear only among the elderly?

Serpins: SERine Protease INhibitorS—conformational disease

The **serpins** are a family of proteins with a variety of biological roles, not limited to the inhibition of proteases. They are of interest: (1) for their medical importance—clinical consequences of serpin dysfunction include blood clotting disorders, emphysema, cirrhosis, and mental illness; (2) for their very unusual conformational changes between states of different folding pattern; and (3) for their unusual mechanism of action. Unlike inhibitors that bind tightly—but in principle reversibly—to active sites, serpins perform a suicide mission: the serpin is cleaved to form a stable protease–inhibitor complex in which the protease is partially denatured, not only rendering it inactive but increasing its susceptibility to proteolysis.

That serpins must undergo dramatic conformational changes when cleaved by proteases was recognized from the first serpin crystal structure, cleaved α_1-antitrypsin. The residues adjacent to the cleaved peptide bond are 65 Å apart! The region N-terminal to the scissile bond, called the 'reactive centre', formed a strand at the centre of a large six-stranded β-sheet (it is the red strand within the green sheet in Fig. 7.2). In the uncleaved structure, these residues, separated by a major element of the secondary structure of the protein in the cleaved form, must somehow come together.

How do you think this can happen?

Conformational states

Several conformational states of serpins are known including:

- In the *active* or '*native*' form of serpins, the polypeptide chain is intact. The reactive centre—the segment that interacts with the protease—forms an exposed loop at one end of the main β-sheet. In the native form this β-sheet is *five-stranded* (Fig. 7.3). Comparison with the cleaved form shows that the central strand has been pulled out, and the surrounding strands have closed up around it. Figure 7.4 shows a superposition of the native and cleaved forms of α_1-antitrypsin.

- In the *cleaved* form the conformational change integrates the reactive centre loop into the body of the molecule, as a sixth strand in the central β-sheet (Fig. 7.3). A free cleaved serpin cannot inhibit protease.
- The *latent* state is another uncleaved form, in which the reactive centre loop is inserted into the main β-sheet as in the cleaved form, but the chain is intact

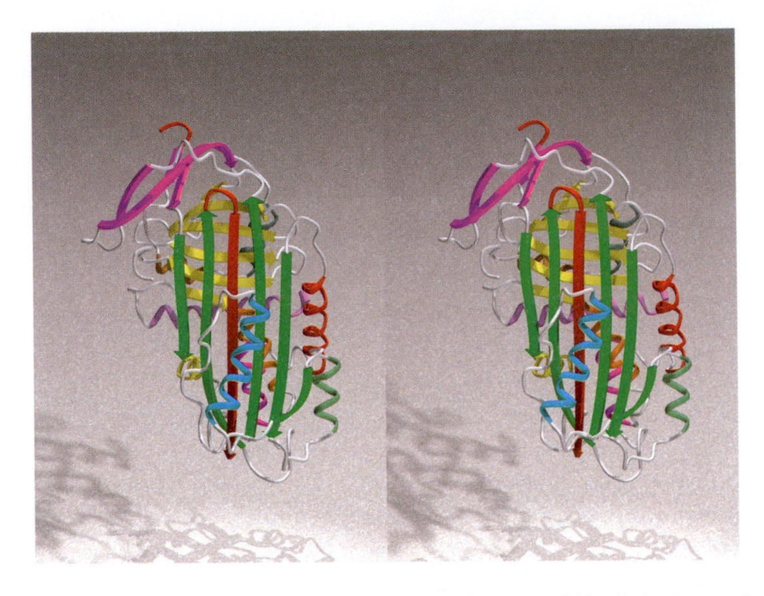

Fig. 7.2 Cleaved α_1-antitrypsin [7API]. The secondary structure of the serpin fold includes three β-sheets and nine α-helices. The natural role of α_1-antitrypsin is to inhibit elastase in the lung. In this conformation the reactive centre forms the red strand within the green β-sheet.

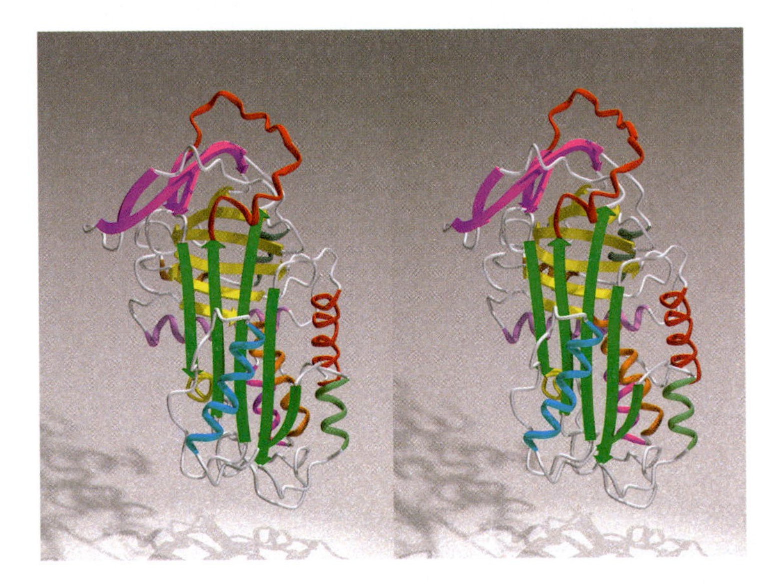

Fig. 7.3 The native form of α_1-antitrypsin [2PSI].

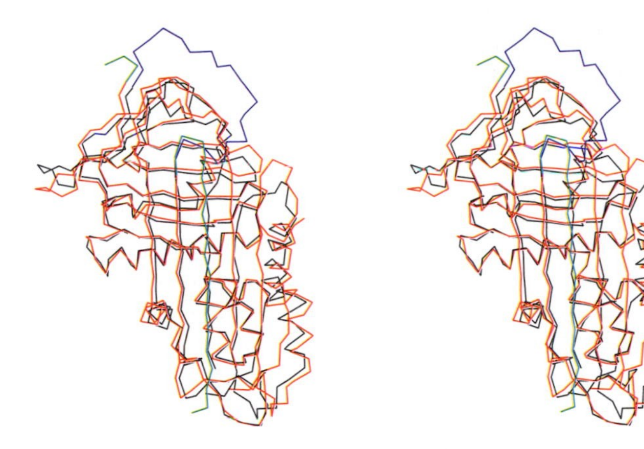

Fig. 7.4 The native and cleaved forms of α_1-antitrypsin, superposed on the largest common substructure [2PSI, 7API].

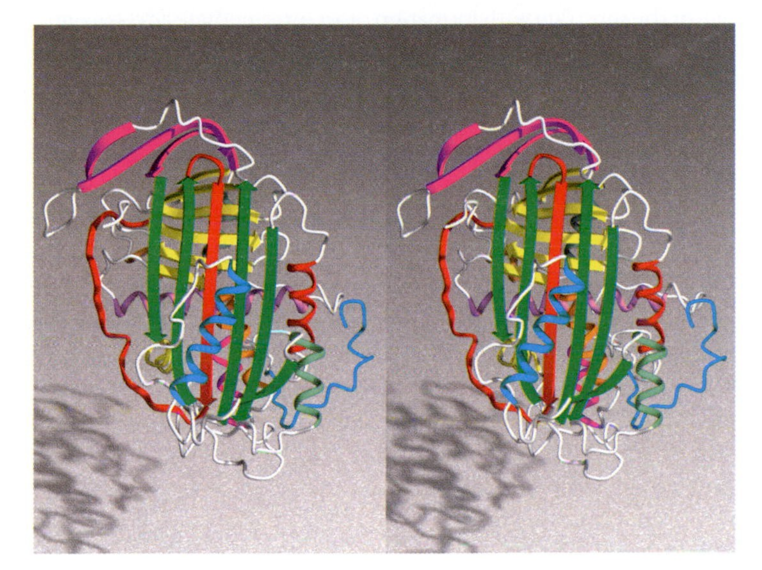

Fig. 7.5 The latent state of human antithrombin [1ATH]. Antithrombin plays a role in controlling the proteases involved in blood coagulation.

(Fig. 7.5). Under physiological conditions the native states of inhibitory serpins are metastable, converting spontaneously to the latent state at different rates (within about 2 hours for plasminogen activator inhibitor-1). The multiple conformational states observed among the serpins are an exception to the rule that homologous proteins have structures containing a core with the same folding pattern.

- Serpins form *polymers* by a mechanism akin to domain swapping. If the reactive centre loop of a serpin inserts, not into the sheet of the same molecule as in the cleaved and latent structures, but into the sheet of a second serpin, and the process continues, a polymer results. Polymerization is particularly common in certain mutant serpins, such as Z α_1-antitrypsin (E342K), in which the structure is destabilized. The polymers form deposits in the liver, eventually leading to cirrhosis. Lack of functional α_1-antitrypsin also causes lung damage as a result of inadequate inhibition of elastase, often leading to emphysema. The combination of a genome homozygous for Z α_1-antitrypsin and smoking cigarettes is a guarantee of death from emphysema at a young age. This is a brutally simple illustration of the maxim, 'Genetics loads the gun and environment pulls the trigger' (Prof. J. Stern).

In relying on structural mobility for their mechanism of action—necessarily involving at least partial unfolding—the serpins have rendered themselves vulnerable to aggregation. From the idea that the lesion involves the dynamics of interconversion of structural states emerged the concept of *conformational disease*, embracing conditions produced by serpin mutants, prions, and other aggregating proteins.

Mechanism of protease inhibition by serpins

Free serpin in the native state presents an exposed reactive-centre loop to protease attack. The normal mechanism of action of trypsin-like serine proteases involves cleavage of the substrate with formation of an intermediate acyl–enzyme complex, with the amino-terminal part of the substrate linked to the enzyme. Hydrolysis of the acyl–enzyme complex restores the original state of the enzyme. This mechanism is described in detail in the first book of this trilogy, *Introduction to protein architecture*.

A protease that studied that or another textbook is in for a nasty surprise! Cleavage of α_1-antitrypsin by trypsin triggers the serpin conformational change: the reactive-centre loop inserts into the main β-sheet, and the (hapless) enzyme, bound in the acyl–enzyme complex, is dragged to the opposite pole of the serpin. Steric clashes break up the protease structure. The catalytic triad is forced apart, preventing release of the enzyme by normal deacylation. The salt bridge between the free amino terminus and the side chain of Asp194, involved in the activation of trypsin, is broken, an alternative salt bridge forming between Asp194 of the proteinase and Lys328 of the inhibitor. This reverses the conformational change that activated the protease. In fact, almost 40% of the protease structure becomes disordered, rendering it susceptible to proteolytic attack. A brutal scenario.

The immune system

The vertebrate immune system has the job of identifying foreign pathogens, and defending the body against them. To do this, it:

- must recognize foreign molecules, and accurately distinguish between 'self' and 'non-self';
- destroy the invaders: molecules, viruses, or bacterial cells;
- remember foreign substances previously encountered, and mobilize more-rapid and specific responses to subsequent encounters with them.

The immune system has evolved specific weapons at the genetic, protein, and cellular level.

In *cellular immunity*, a macrophage ingests an **antigen**, digests it, and displays fragments on its surface, in complex with proteins of the major histocompatibility complex (MHC). T cells recognizing these complexes are induced to mature and proliferate. Cytotoxic T cells will destroy cells that bear their target antigen.

In *humoral immunity*, a B cell will engulf and digest molecules that bind to the immunoglobulin on its surface. It will display the fragments produced, on its surface, in complex with MHC proteins. These complexes are then recognized by T-cell receptors on T-helper cells, leading to activation of B cells to proliferate, and to differentiate them into a factory to produce and excrete the specific **antibody** elicited. Some B cells become 'memory cells' that remain poised to respond to subsequent challenges from the same antigen by copious synthesis of secondary antibodies. This is the mechanism that keeps us from getting certain childhood diseases more than once, and why a second pregnancy involving rhesus (Rh) factor incompatibility is more dangerous than the first.

Antibody structure

The immune system has evolved to be able to recognize the entire organic world. It is designed to generate diversity: during our lifetimes we synthesize about 10^{10} antibodies.

Antibodies contain multiple polypeptide chains, distinguished by size into light (L) chains (r.m.m. about 23 000) and heavy (H) chains (r.m.m. about 50 000–70 000). Each chain is itself modular, containing a series of homologous domains, each with a characteristic double-β-sheet structure. Light chains contain two domains, and heavy chains contain four or five domains. Light chains are distinguished into κ and λ classes or *isotypes*. In the human, κ and λ light chains are present in comparable proportions; in the mouse κ light chains predominate.

Different classes of immunoglobulins—IgG, IgA, IgM, IgD, and IgE—differ in the assembly of their chains and domains (Fig. 7.6). IgGs usually have two heavy chains each containing four domains, and two light chains each containing two domains (Fig. 7.7). Figure 2.23 showed the structure of a complete IgG.

How is antibody diversity created? The genes for antibodies are assembled from a combination of DNA segments. The immunoglobulin loci in vertebrate genomes contain tandem segments that join in different combinations to generate immunoglobulin genes (see Box). Each light-chain gene contains a selection: a *V* segment, a *J* segment, and a *C* segment; each heavy-chain gene contains a *V* segment, a *D* segment, a *J* segment, and several *C* segments. Additional variability is introduced by imprecise joining of the segments.

V (variable) segments code for the major, N-terminal, part of the V_L and V_H domains. *J* (joining) segments code for the C-terminal part of the V_L and V_H domains. *D* (diversity) segments code for a region in the V_H domain between those encoded by the V_H and J_H segments.

Upon challenge by a pathogen, the products of these genes mount the initial *primary immune response*, mobilizing antibodies with dissociation constants in the

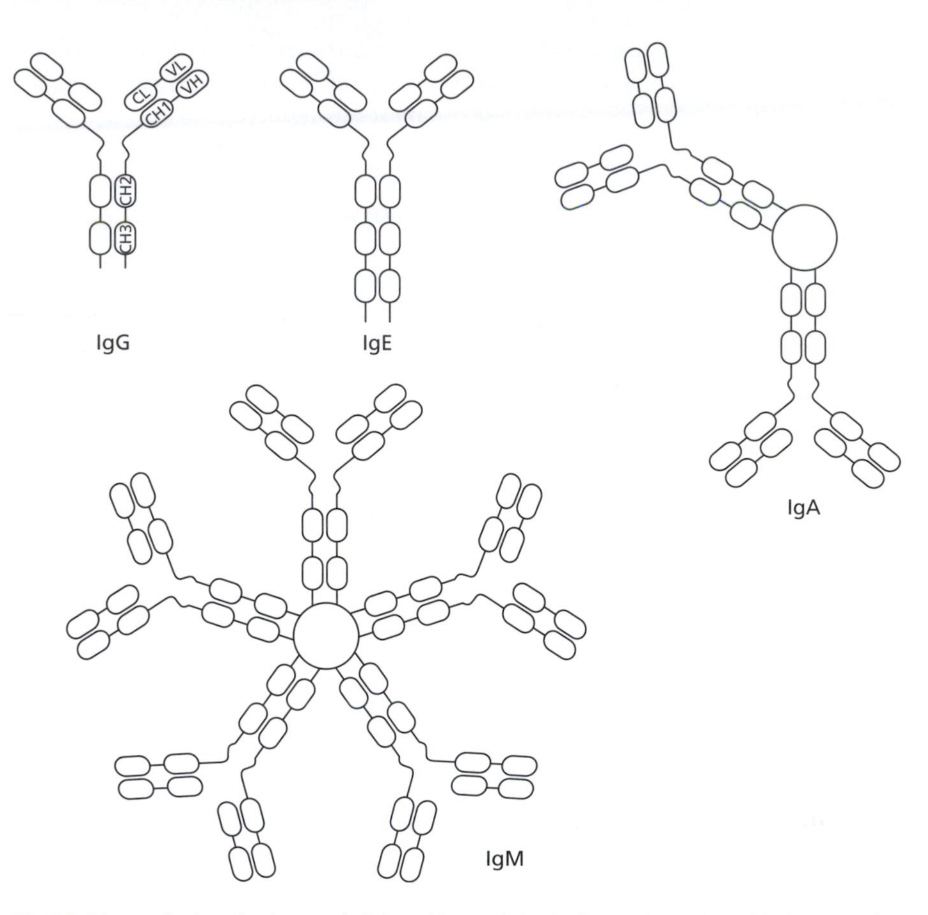

Fig. 7.6 Most antibody molecules contain light and heavy chains. Each comprises one variable domain and different numbers of constant domains. The combination of 2 light + 2 heavy chains is a higher order building block. Immunoglobulins of different classes (IgG, IgA, IgM, IgD, and IgE) show different states of oligomerization, with additional chains where necessary serving as linkers.

The IgG molecule contains four polypeptide chains: two identical light chains each containing one variable and one constant domain, denoted V_L and C_L; and two identical heavy chains each containing one variable and three constant domains, denoted V_H, C_H1, C_H2, and C_H3. The C_H1 and C_H2 domains are linked by a peptide called the 'hinge region'. The angle between the V_L–V_H domain pair and the C_L–C_H1 domain pair is called the 'elbow angle'. In solution, these joints are probably flexible. Immunoglobulins also contain carbohydrate moieties not shown in this figure.

micromolar range. Subsequently, *somatic mutation* tunes the affinity and specificity of the antibodies for the *secondary response*, typically achieving nanomolar dissociation constants. Artificial selection of antibodies, using phage display techniques, can also routinely produce molecules of comparable affinity to those arising naturally from somatic mutation.

Immunoglobulins have specific binding sites that interact with ligands, with lock-and-key complementarity. Molecules related to antibodies appear on cell surfaces, to mediate cell–cell recognition and signalling processes, triggering the proliferation of

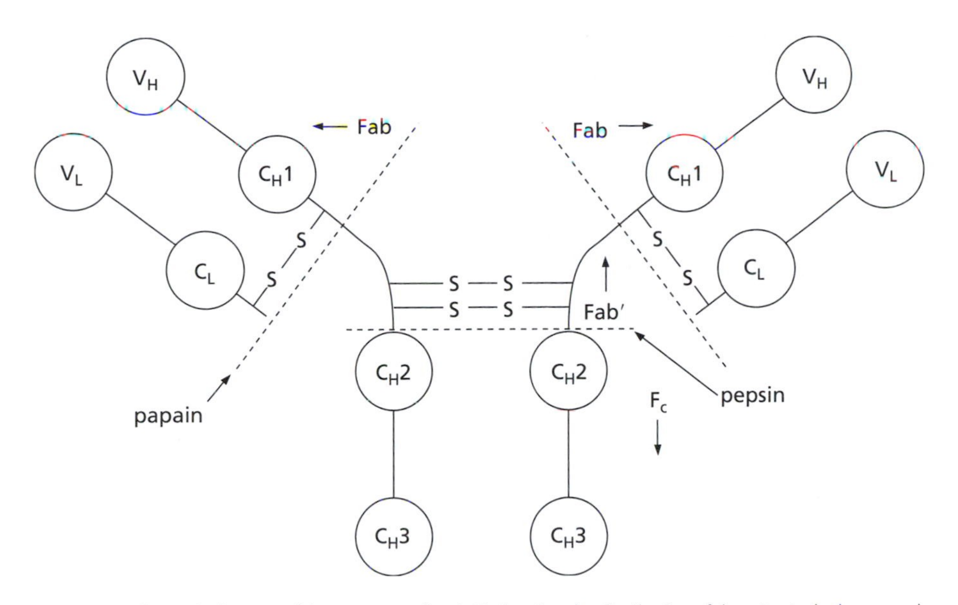

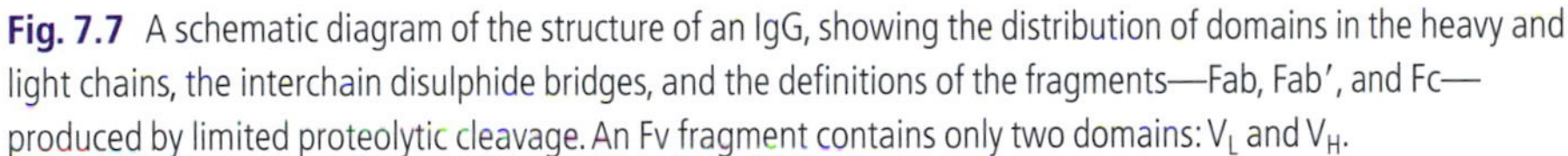
Fig. 7.7 A schematic diagram of the structure of an IgG, showing the distribution of domains in the heavy and light chains, the interchain disulphide bridges, and the definitions of the fragments—Fab, Fab′, and Fc—produced by limited proteolytic cleavage. An Fv fragment contains only two domains: V_L and V_H.

Functional immunoglobulin gene segments in the human genome

κ Light chains		λ Light chains		Heavy chain	
Segment	Multiplicity	Segment	Multiplicity	Segment	Multiplicity
V_κ	30-35	V_λ	29-33	V_H	38-46
				D	23
J_κ	5	J_λ	4-5	J_H	6

particular cells in response to antigenic challenge. Many related proteins are known; antibodies are members of a large superfamily of proteins, most, but not all, of which are involved in molecular recognition.

The constant, the variable, and the hypervariable

Patterns in multiple sequence alignments of antibodies distinguished two types of antibody domains: variable (V) and constant (C), on the basis of the extent of sequence conservation. (A third type of domain, the I type, occurs in other molecules of the immunoglobulin superfamily.)

Within the variable domains are regions of still higher variability that determine antibody specificity. These **hypervariable** segments are called **complementarity determining regions (CDRs)**. The structurally conserved regions of the variable domains outside the CDRs are called the **framework**.

Constant and variable domains have a similar, but not identical, framework structure. The hypervariable regions correspond to surface loops in the variable domains, and do indeed interact with antigens. Each IgG has two copies of the antigen-binding site, one from each light–heavy chain pair. This duplication permits multiple interactions with antigens to form aggregates.

The interactions between the different chains in an IgG include disulphide bridges (see Fig. 7.7); and interfaces between corresponding domains—V_L–V_H, C_L–C_H1, and C_H3–C_H3 but not C_H2–C_H2—that pack together to form extensive Van der Waals contacts. Like other oligomeric proteins, the interior interfaces are formed by the packing of complementary surfaces. The tendency to conserve the residues involved in these interfaces explains why different light and heavy chains can pair fairly freely to form complete immunoglobulins. In the case of the V_L–V_H interaction, the conservation of the relative geometry implies that, to a good approximation, *the double-β-sheet frameworks of V_L and V_H domains form a scaffolding of nearly constant structure on which the antigen-binding site is erected.*

The antigen-binding site

CDR = complementarity-determining region.

The antigen-binding sites of most antibodies are formed primarily from six loops—three from the V_L domain and three from the V_H domain. The three loops from the V_L domain are called L1, L2, and L3, (or CDR1, CDR2, and CDR3) in order of their appearance in the amino acid sequence (see Fig. 7.8); V_H domains contain three corresponding CDRs, H1, H2, and H3. (Some of the antibodies of the camel and related animals contain only heavy chains, and three V_H CDRs suffice to create the antigen-binding site.) Four of the loops, L2, H2, L3, and H3, are β-hairpins. (Recall that a β-hairpin is a loop that links successive antiparallel strands of a single β-sheet.) In contrast, L1 and H1 form bridges from a strand in one of the two β-sheets to a strand in the other.

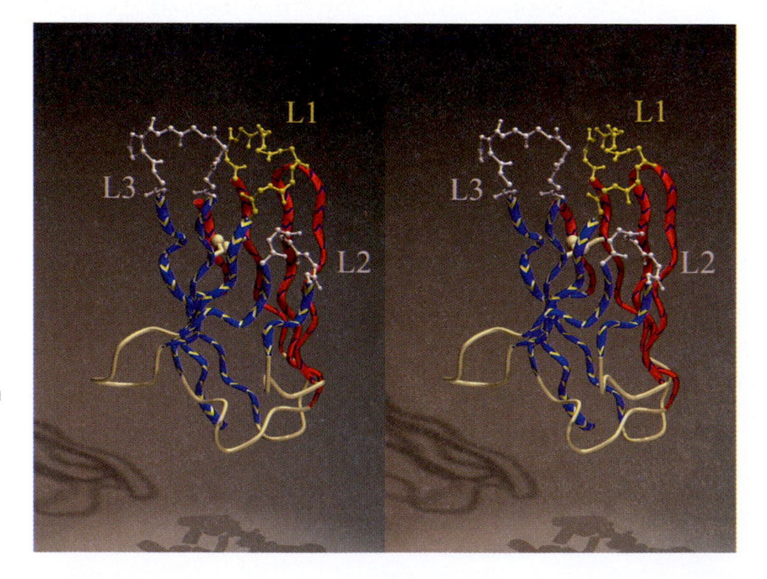

Fig. 7.8 The V_κ domain REI, a variable domain from a light chain, indicating the antigen-binding loops L1, L2, and L3 [1REI].

Figures 7.9 and 7.10 illustrate the construction of the antigen-binding site. Figure 7.9 shows the binding of a small organic ligand, or hapten—phosphorylcholine—by the immunoglobulin McPC603 and TE33 finding a peptide. An orientation was chosen looking down onto the antigen-binding site—an 'antigen's-eye view'. Figure 7.10 shows the complex of Fab HyHEL-5 and hen egg-white lysozyme.

In the complex between Fab D1.3 and hen egg-white lysozyme, the interface is a relatively flat patch of dimensions 20 × 30 Å, although one glutamine side chain from the lysozyme inserts into a pocket in the antibody. A total of 17 residues from the antibody make contact with a total of 17 residues from the lysozyme. There are 32 pairs of residues in contact. Of the 17 residues from the antibody, 14 are from the CDRs. The other three are framework residues, but they are adjacent to CDRs.

Conformations of antigen-binding loops of antibodies

Of the six antigen-binding loops of immunoglobulin structures, five have only a small discrete repertoire of main chain conformations, called **canonical structures**. These conformations are determined by a few particular residues within the loop, or outside the loop but interacting with it. Among corresponding loops of the same length, only these residues need to be conserved to maintain the conformation of the loop. Other residues in the sequences of the loops are thus left free to vary, to modulate the surface

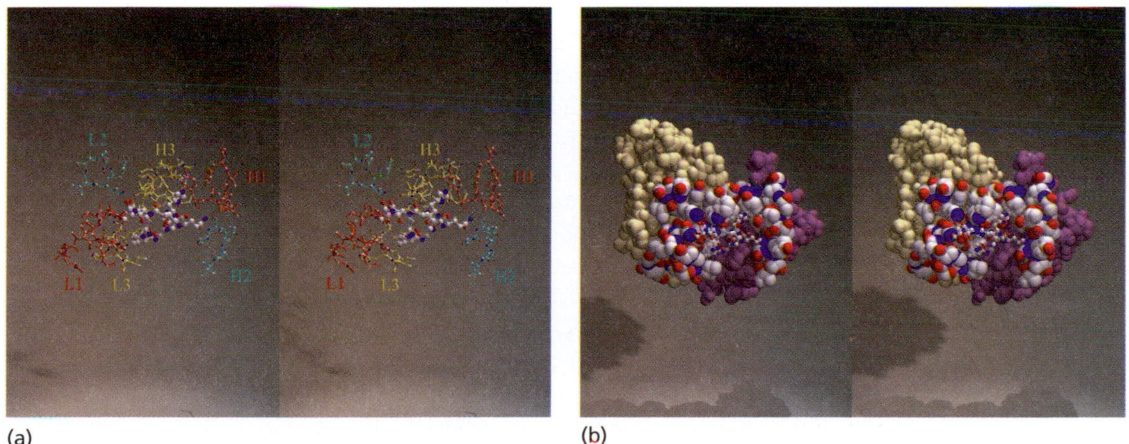

Fig. 7.9 (a) The spatial distribution of antigen-binding loops, or CDRs, in the antigen-binding site of McPC603 [1MCP]. The ligand is phosphorylcholine. The light-chain loops are at the left, the heavy-chain loops are at the right. Reading clockwise starting at '7 o'clock', the loops appear in the order L1, L2, H3, H1, H2, L3. There is a rough symmetry in the arrangment of the loops: L1 is opposite H1, L2 is opposite H2, and L3 is opposite H3. The central position of H3 is noteworthy. (b) Space-filling picture of antibody TE33, binding a 12-residue peptide from cholera toxin [1TET]. Smaller atoms represent the ligand. Larger atoms represent the antigen-binding loops of the antibody. For the ligand and the antigen-binding loops, atoms are coloured as follows: carbon, white; nitrogen, blue; oxygen, red. The framework residues of the light chain are shown in yellow. The framework residues of the heavy chain are shown in mauve.

In both (a) and (b), the orientation is chosen to give a view down onto the antigen-binding site, an 'antigen's-eye view'.

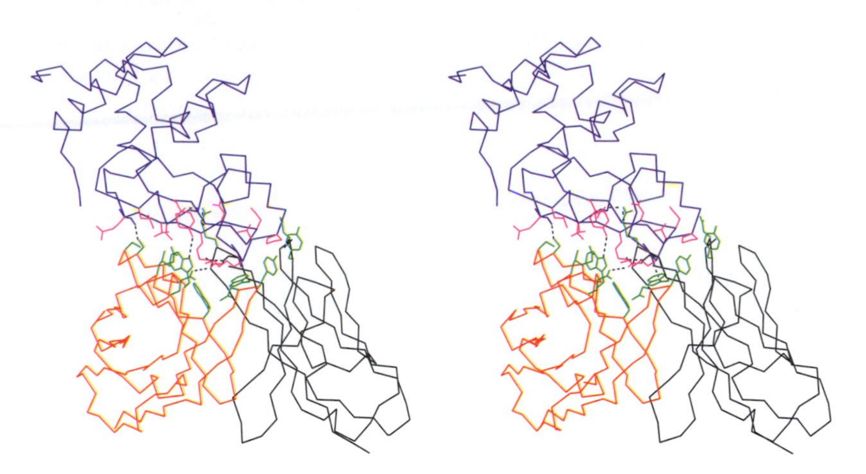

Fig. 7.10 Interactions between antibody HyHEL-5 and hen egg-white lysozyme [1BQL]. Chain trace of light chain in black, chain trace of heavy chain in red, chain trace of lysozyme in blue. Side chains of residues in the lysozyme that make contact with the antibody are shown in magenta. Residues in the antibody that make contact with the lysozyme are shown in green. Broken lines indicate hydrogen bonds between antigen and antibody.

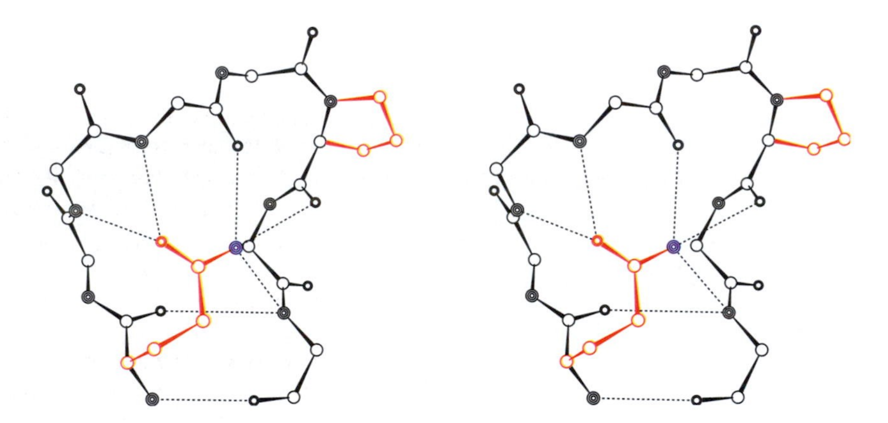

Fig. 7.11 The canonical structure of L3 from Bence-Jones protein REI [1REI]. This is the most common canonical structure of the κ L3 loop. Residues responsible for this conformation are shown in red: the peptide preceding a proline can adopt a relatively low-energy *cis* conformation; A glutamine residue at the position just N-terminal to the loop can form hydrogen bonds to inward-pointing main chain atoms in the loop. Asn and His can also make these hydrogen bonds. Therefore any Vκ L3 loop seven residues long with a proline at the fifth position, preceded by a Gln, Asn, or His, would be expected to adopt this conformation. Other residues in the loop are free to vary.

topography and charge distribution of the antigen-binding site. The signature patterns of the different loop conformations in the sequence provide useful methods for predicting the structure of antibody combining sites.

An example of a canonical structure is the L3 loop from V_κ REI is a β-hairpin, containing a *cis*-proline at position 95, and stabilized by hydrogen bonds between the side chain of the residue at position 90, just N-terminal to the loop, and main chain atoms of residues in the loop (Fig. 7.11). The side chain at position 90 is a Gln in REI (the residue at this position can also be an Asn or a His in other V_κ chains). The requirement for Gln, Asn, or His at position 90 arises from the necessity to form hydrogen bonds to inward-pointing main chain atoms in the loop. The combination of loop length, one of these polar side chains at position 90, and the proline at position 95 constitute the 'signature' of the conformation of the loop, from which it can be predicted in a sequence of an immunoglobulin the structure of which has not been experimentally determined.

What is the biological significance of this robustness of the main chain conformation of the antigen-binding loops? Perhaps it plays a role in the tuning of antibodies by somatic mutation. That is, in the system as it has evolved, a change in a residue in the antigen-combining site will, in most cases, produce only a conservative structural change in the main chain—this is what one would want in a system that already shows affinity and requires only 'fine-tuning'. Consider the alternative: if most mutations completely altered the main chain conformation of the loops, the effect would be to produce a succession of independent primary responses, rather than a secondary response with structures perturbed in only minor (but crucially important) ways from a set of already selected primary antibodies. The reader should not underestimate the power of very tiny structural changes to produce very large effects on binding affinity.

Greater variability in the H3 loop

H3, the third hypervariable region of the heavy chain, is far more variable in length, sequence, and structure than the other antigen-binding loops. It cannot therefore be included in the canonical-structure description of the conformational repertoire of the three hypervariable regions of V_L chains and the first two of V_H chains. Because the H3 loop falls in the region of the *V–D–J* join in the assembly of the immunoglobulin heavy-chain gene, several mechanisms contribute to generation of its diversity, including combinatorial choice of V_H, D, and J_H gene segments, and alternative splicing patterns at the junctions.

In expressed antibodies, H3 appears prominently at the centre of the antigen-binding site (Fig. 7.9). Given this central position, H3 makes significant interactions—with other loops, with the framework, with the light-chain partner, and with ligands—that influence its conformation. Thus H3—in contrast to the other five antigen-binding loops—has a conformation that depends strongly on its molecular environment. Indeed, structures containing the V_H domain of antibody B1–8 combined with two different V_L domains show two very different conformations of H3. This important observation implies that (unlike the other five antigen-binding loops) general rules governing the conformation of H3 *must* involve interactions outside its local region in the sequence.

Somatic mutation and the maturation of the antibody response

Somatic mutation is the process whereby antibodies active in the primary response are tuned in affinity and specificity by tinkering with their sequences to produce the 'mature' antibodies of the secondary response. Whereas the diversity in the primary

CASE STUDY Antibody maturation

P.G. Schultz and co-workers have studied the maturation of an anti-nitrophenyl phosphonate catalytic antibody by X-ray crystallography. They solved the structures of the primary (germ line) and somatically mutated Fab fragment, each with and without ligand. Their results reveal both the sites and the structural effects of mutations.

There are nine amino acid sequence changes between germ-line and mature antibody; three in the light chain and six in the heavy chain. One of the light-chain mutants appears within the L1 loop; the other two are in positions in the sequence near L1 and L2. One of the heavy-chain mutants appears in H2, three are in regions adjacent to or in contact with the antigen-binding loops, and two are surface residues at the opposite ends of the domains and would appear to have little effect on the antigen-binding site. No mutant is at a position directly in contact with the antigen. No mutant appears in H3, even though there are extensive contacts between H3 and the antigen. The conformations of two of the antigen-binding loops (H1 and H2) differ between germ-line and mature antibody.

The antigen-binding site of the mature antibody has a similar conformation in ligated and unligated states; a fixed conformation complementary to the antigen. This structure is shared by the ligated state of the germ-line antibody, but the unligated state of the germ-line antibody shows differences in conformation. That is, the germ-line antibody *can* adopt the antigen-binding conformation, and is *induced* to do so by the ligand; the mature antibody adopts this conformation even in the absence of ligand.

antibodies is highest in H3 and L3, around the centre of the antigen-binding site, somatic mutations spread the diversity to its periphery. Usually, somatic mutations are isolated point substitutions, and not insertions or deletions.

The immune system is one of many biological examples of molecular recognition. Individual antibody–antigen complexes fit the classical ideas of lock-and-key complementarity and induced fit; but the system as a whole is more complex than, for example, enzymes, in which the lock and key are in most cases fixed and unchanging within the lifetime of the organism. In the primary response, the immune system achieves affinity for many keys by providing many locks; but achievement of the spectacular affinity of the secondary response requires a mechanism that perturbs the locks for better fit. It is the integration of the immune response over the system as a whole, involving both genetic and structural diversity, that adds the dimensions of complexity.

Proteins of the major histocompatibility complex (MHC)

Surgical patients, if not immunosuppressed by drugs, will reject transplanted organs—unless the donor is an identical twin—because the transplant is recognized as foreign. The immunological distinction between 'self' and 'non-self' resides in the proteins of the major histocompatibility complex (MHC) and their interaction with T-cell receptors. MHC proteins bind intracellularly produced peptides and present them on cell surfaces. The triggering event in alerting the immune system to the presence of a foreign protein is the recognition, by a T-cell receptor, of a complex between an MHC protein and a peptide derived from the foreign protein.

MHC proteins fall into two classes, with related but different structures and functions. The two classes function as parallel systems, to produce different immune responses appropriate to intracellular and extracellular pathogens, respectively (Fig. 7.12). Class I

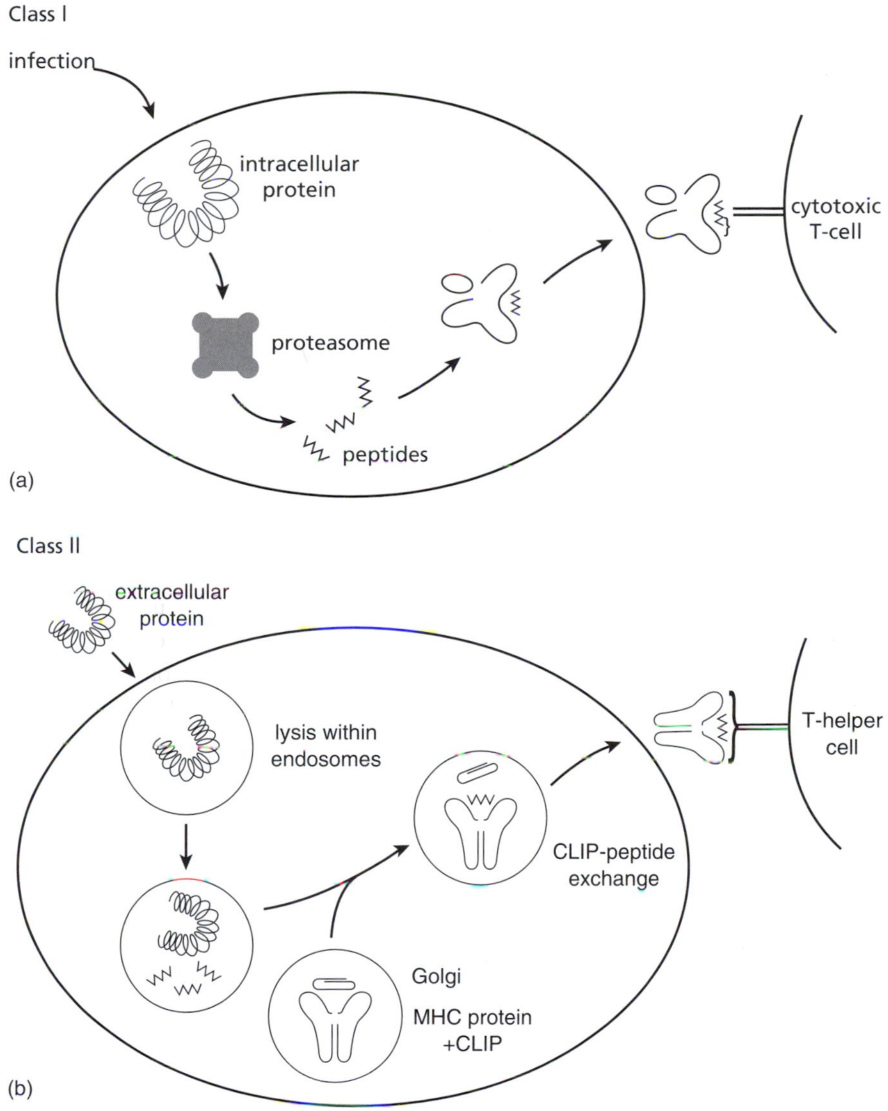

Fig. 7.12 (a) Class I and (b) class II MHC molecules participate in two parallel systems to trigger the immune response to foreign proteins originating inside and outside cells. Peptides derived from foreign proteins are loaded intracellularly and transported to the surface where they are presented to T cells. The representation of the class II invariant peptide (CLIP) is an icon not a drawing of its structure.

MHC molecules appear on the surfaces of most cells of the body, and present peptides derived from proteins degraded in the cytosol. These peptide–MHC complexes alert the body to intracellular pathogens. They interact with cytotoxic T cells, and direct the immune response to the presenting cell and those in its vicinity. Class II MHC molecules appear on the surfaces of specialized cells of the immune system: B lymphocytes and antigen-presenting macrophages. They present oligopeptides derived from exogenous antigens (which have been endocytosed and chopped into peptides). These peptide–MHC complexes interact with helper T cells, mediating the proliferation of cells synthesizing antibodies that circulate in the blood, and the activation of macrophages.

In addition to triggering immune responses in mature individuals, MHC–peptide complexes are also involved in the removal of self-complementary T cells in the thymus during development, at the stage when the distinction between self and non-self is 'learnt'.

Each individual in vertebrate species expresses a set of MHC proteins selected from a diverse genetic repertoire in the species. In humans the MHC complex is a set of linked genes on chromosome 6. The system is highly polymorphic, with 50–150 alleles per locus, showing greater sequence variation than most polymorphic proteins. Each of us produces six class I molecules and a somewhat higher complement of class II molecules. Each MHC protein must therefore be able to bind many peptides, if ~30 MHC proteins are to present the large number of possible antigens.

The set of MHC proteins expressed defines the **haplotype** of an individual. The number of possible haplotypes has been estimated to be of the order of 10^{12}, although, because of linkage, the combinations are non-random, and because of selective pressure fewer combinations appear than expected. Like fingerprints and restriction fragment length polymorphisms, our haplotypes are a personal identification code. Haplotypes have been used extensively in anthropology in measuring quantitatively the relationships among human populations, and in tracing paths of migration.

Structures of MHC proteins

MHC proteins are modular proteins containing several characteristic domains. These include peptide-binding domains with folds special to the MHC system, and immunoglobulin-like domains (Fig. 7.13).

Class I MHC proteins contain two polypeptide chains (Fig. 7.14). The longer chain (r.m.m. 44 000 in humans; 47 000 in mice) has a modular structure of the form α_1–α_2–α_3– followed by a short hydrophobic membrane-spanning segment and a 30-residue cytoplasmic tail. The α-domains are approximately 90 residues in length. The second chain is β_2-microglobulin, a non-polymorphic structure (that is, constant within the species), the gene for which is unlinked from the MHC complex.

The α_1- and α_2-domains of class I MHC proteins have a common fold, and interact to form a symmetrical combined structure. They bind peptides in a groove between them, created by two long curved α-helices (Fig. 7.15). The variability in the amino acid sequence of MHC proteins is high in the regions that surround the groove, to create variety in specificity. The α_3-domain and β_2-microglobulin do not interact with bound peptides. They are double-β-sheet proteins with topologies of the immunoglobulin superfamily.

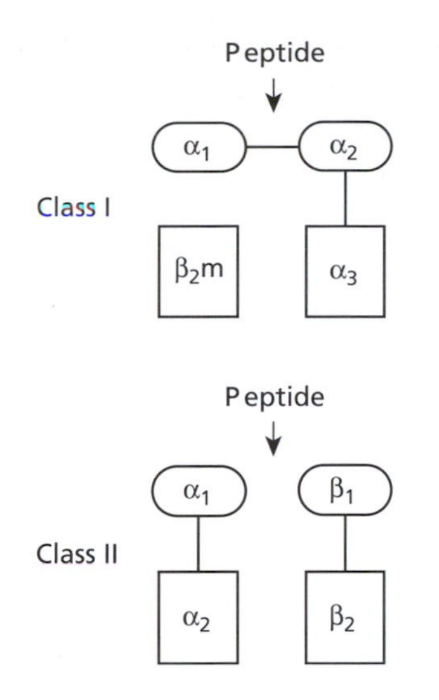

Fig. 7.13 MHC proteins are modular proteins containing two peptide-binding domains (rounded cartouches) and two immunoglobulin-like domains (rectangles). Peptides bind in a groove created by the α_1/α_2-domains in class I MHC proteins and by the α_1/β_1-domains in class II MHC proteins.

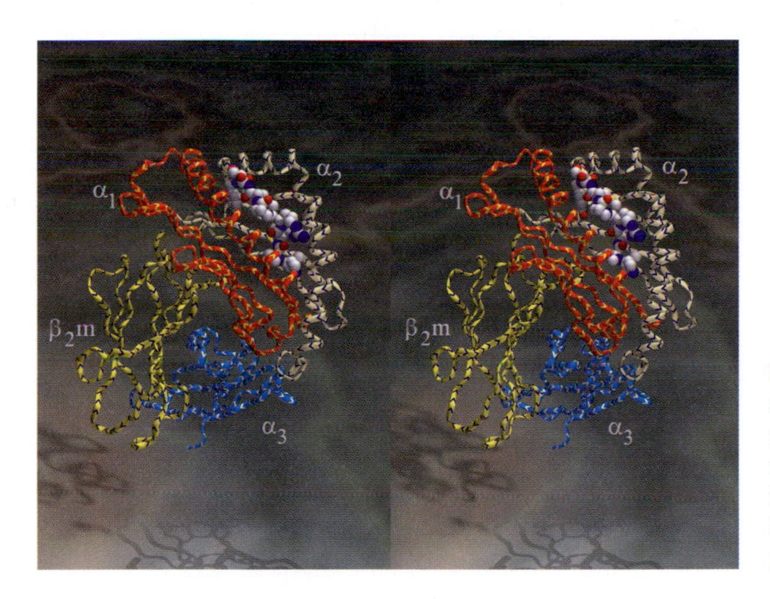

Fig. 7.14 The structure of a class I MHC protein, B35, binding the peptide VPLRPMTY from the nef protein of HIV-1 [1A1N]. (β_2m = β_2-microglobulin.)

The α_1- and α_2-domains of class I MHC proteins, and the α_1- and β_1-domains of class II, have a common folding pattern. Each has a four-stranded β-sheet at the N-terminus, followed by a short bridging helix, and then a long C-terminal helix that lies across the β-sheet. The strands from each domain interact laterally to form an 8-stranded β-sheet, positioning the C-terminal helices from the two domains to form the sides of the peptide-binding cleft. Each helix has a pronounced curvature. Because

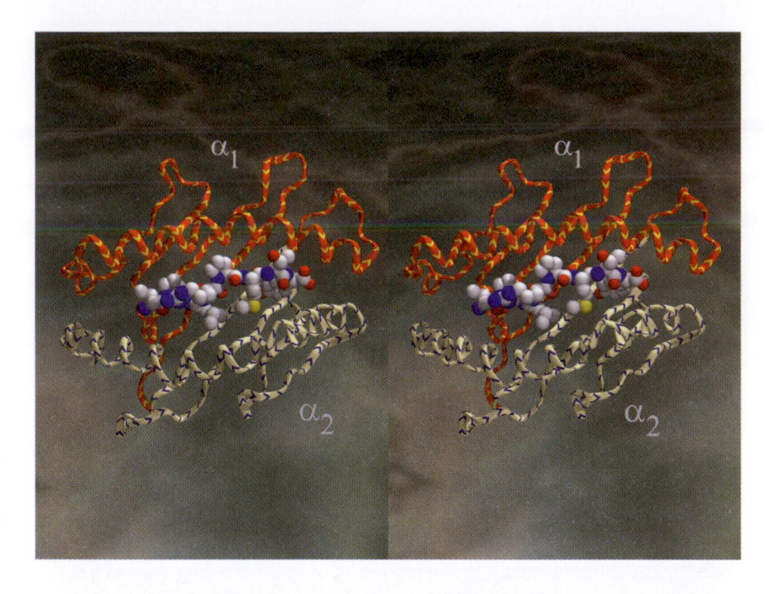

Fig. 7.15 The peptide-binding domains and ligand from class I MHC protein B35 [1A1N].

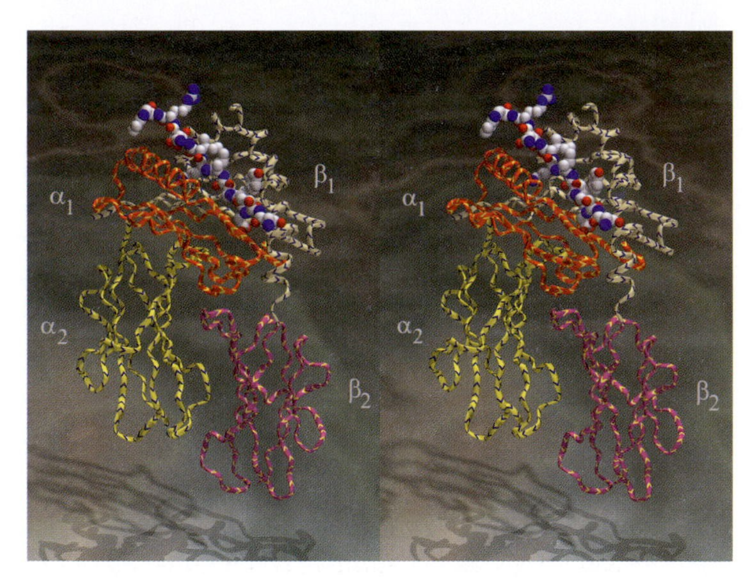

Fig. 7.16 The structure of a class II MHC protein, I-Ak, binding the peptide STDYGILQINSRW from hen egg-white lysozyme [1IAK].

of the approximate twofold symmetry of the α_1–α_2 unit, the long C-terminal helices run antiparallel to each other. Peptides bind in an orientation parallel to that of the α_1-helix.

Class II MHC proteins contain an α-chain (r.m.m. 34 000) containing two domains, $\alpha_1 + \alpha_2$, a homologous β-chain (r.m.m. 29 000) containing two domains, $\beta_1 + \beta_2$, and an invariant chain I1 (r.m.m. 31 000) (Fig. 7.16; the invariant chain does not appear). The α_1- and β_1-domains pack together to make a structure similar to that formed by the α_1- and α_2-domains of class I MHC proteins, and they bind peptides in a similar mode (Fig. 7.17).

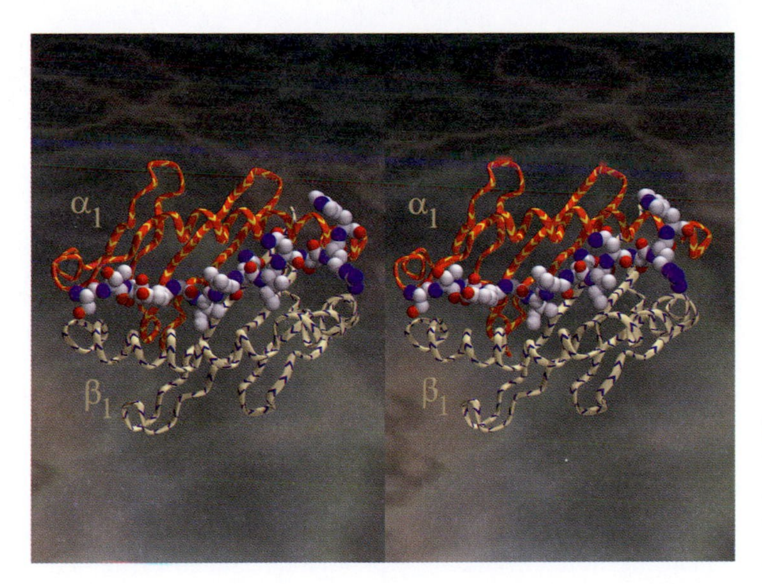

Fig. 7.17 The peptide-binding domains and ligand from class II MHC protein, I-AK [1IAK].

Class I molecules can bind peptides of limited length—from about 8 to 11 residues—but most commonly 8–9. Two factors impose the limits: the closure of the cleft at either end, and a salt bridge to the C-terminal carboxyl group of the peptide. The cleft will accommodate 9-residue peptides in a nearly extended conformation; longer peptides can bulge out or zig-zag, or their C-termini can extend out beyond the end of the pocket. In class II molecules the cleft is also closed at the right, but open at the left.

Specificities of the MHC system

Two aspects of MHC specificity are: (1) the self-foreign distinction, which depends on T-cell scrutiny of MHC–peptide complexes; and (2) the selection of different types of immune reponse to different categories of threats—extracellular vs. intracellular. This selection is accomplished by directing peptides derived from extracellular and intracellular sources through the class I and class II systems, to form complexes recognized by different types of T cells.

The MHC proteins themselves have broad specificity, each binding many peptides, including those of self and non-self origin. Cell surfaces contain large numbers of MHC–peptide complexes, among which those binding foreign peptides are a small minority. T-cell receptors, in contrast, have narrow specificity, and pick out the complexes containing foreign peptides, like a professional antiques dealer spotting a valuable item in a rummage sale. T-cell diversity is created by the same types of genetic recombination mechanisms that produce the diversity of antibodies in the primary immune response. (There is no analogue of somatic mutation in the T-cell system, however.)

T cells 'see' both the MHC and peptide components of the complex they recognize. The variability in MHC proteins extends to the residues that interact with T cells but do not interact with the bound peptides. Two MHC proteins binding the same peptide could be recognized by different T cells.

Class I and class II MHC proteins function in parallel, selecting different immune responses to extracellular and intracellular pathogens

Challenge by internally synthesized foreign proteins, as in virus-infected cells, leads to peptide cleavage by cytosolic proteasomes. They encounter class I MHC proteins in the endoplasmic reticulum, and the peptide complexes are moved to the cell surface.

Challenge by external foreign proteins leads to cleavage in endosomes. Class II MHC proteins start out in the endoplasmic reticulum associated with the Invariant (Ii) chain. In this complex the peptide binding site is blocked, to suppress flooding the system with self-derived peptides, and to prevent picking up peptide fragments from internally synthesized proteins, which are to be bound to the class I MHC molecules. The complex moves to specialized organelles where it is exposed to peptide fragments of imported proteins. Cleavage of the Ii chain leaves the CLIP (CLass II Invariant chain Peptide) in the binding site. A molecule related to the MHC proteins—HLA-DM in humans and H-2M in mice—catalyses peptide exchange, removing the CLIP and enhancing the rate of binding. Peptide binding stabilizes the α-chain–β-chain dimers, which are brought to the cell surface.

Peptide binding

The crystal structures of both class I and class II MHC proteins, with and without bound peptides, reveal: (1) the nature of the peptide–MHC protein interactions; (2) the mechanism of the broad specificity; and (3) the nature of the surface presented to T-cell receptors.

- **Class I:** The complexes of the class I MHC protein HLA-B53 show the conformation and interactions of the ligand. Figure 7.18 shows the shape of the cleft from the complex of HLA-B53 and the peptide LS6 (KPIVQYDNF) from the malarial parasite *P. falciparum*. The cleft is a deep groove, with a hump in the middle, and pockets at either end which receive inward-pointing side chains from near the chain termini of the peptide. The middle part of the floor of the groove is lined by polar side chains: 9Tyr, 70Asn, 74Tyr, and 97Arg. The peptide is in an extended conformation, the conformational angles of all residues in the β-region of the Ramachandran plot. The main chain atoms (N, Cα, C, O) of all but two central peptide residues make contact with the MHC protein, and every side chain of the peptide makes contact with the MHC protein.
- **Class II:** In the structure of the class II MHC protein I-Ak binding the peptide STDYGILQINSRW, the peptide lies in the cleft with its ends protruding from both sides of the MHC protein (Fig. 7.19). All residues of the ligand are in the β-conformation. As the binding groove is pinched off at the right, in the orientation of this picture, the ligand turns up, out of the cleft.

There is a difference between class I and class II in the interactions between peptide side chains and binding groove. In class I the terminal side chains of the ligand sit in pockets at the bottom of the groove, and the central side chains do not point down into

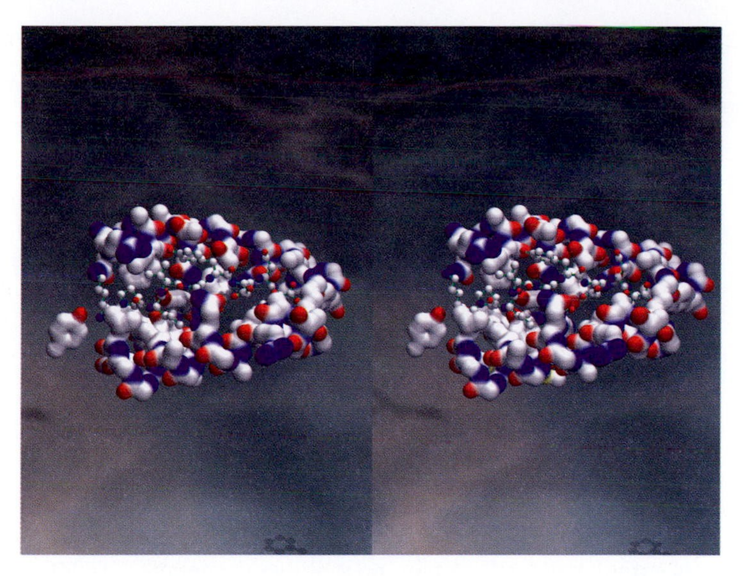

Fig. 7.18 Space-filling picture of the cleft in class I protein B53 [1A1O].

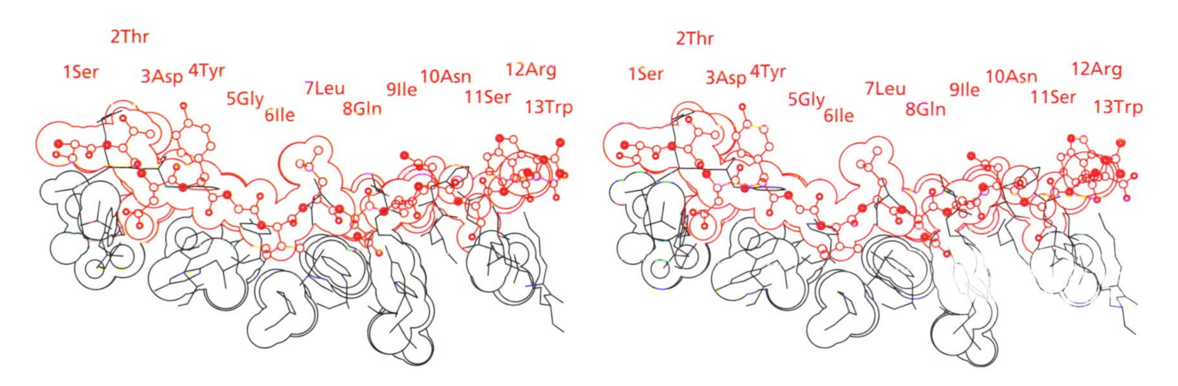

Fig. 7.19 Interaction of class II MHC protein I-Ak (black) and ligand (red) [1IAK].

the groove but at right angles to it. In I-Ak, class II, the central side chains are buried in pockets in the base of the groove.

Class I and class II molecules bind peptides in ways that are generally similar but different in detail. (The peptide-binding domains of MHC proteins have been compared to a mouth, the two long helices corresponding to the lips. In binding peptides, class I and class II molecules employ different embouchure.) The most obvious difference is the interaction of the terminal residues of peptide ligands with class I molecules in a way that limits the length of peptides that can be bound. MHC proteins achieve the goal of broad specificity but fairly tight binding by numerous contacts with the main chain of the peptide, which are peptide-sequence independent, and by conformational changes that allow tuning of the crevice to bind several peptides.

T-cell receptors

T-cell receptors (TCRs) are proteins related to immunoglobulins, that interact with MHC-presented peptides to trigger immune responses. Like immunoglobulins, they are composed of V and C domains homologous to immunoglobulin domains. They also show a diversity generated by combinatorial assembly of genes from segments, but there is no subsequent development by somatic mutation. The binding sites of TCRs are created by six loops, three from each V segment, homologous to the CDRs of antibodies. However, unlike antibodies which must bind molecules of very different sizes and shapes, the targets of TCRs are exclusively MHC–peptide complexes, the *overall* sizes and shapes of which are fairly well circumscribed. (MHC-peptide complex abbreviates MHC-protein-peptide complex.)

Binding of an MHC–peptide complex by a TCR on the surface of both cytotoxic T cells and T-helper cells initiates a signal transduction cascade. Because of the high concentration of MHC proteins presenting self peptides, T cells that would bind them are destroyed in the thymus early in life, to prevent autoimmune responses.

TCRs are dimers of proteins built of domains in a manner similar to immunoglobulins. Most TCRs contain an α- and a β-chain, each consisting of a variable (V) domain (bearing the binding loops), a constant (C) domain, a transmembrane segment, and a cytoplasmic tail. (The α, β notation is independent of the nomenclature of the chains in MHC proteins.) The α-chains have r.m.m. ~40 000–50 000, and the β-chains have r.m.m. ~40 000–45 000. Neither chain has additional constant domains, as do immunoglobulin heavy chains.

The TCR–MHC–peptide complex

The **ternary complex** of a T-cell receptor with a class I MHC protein presenting a 9-residue viral peptide, shows how the TCR recognizes the MHC–peptide complex (Fig. 7.20).

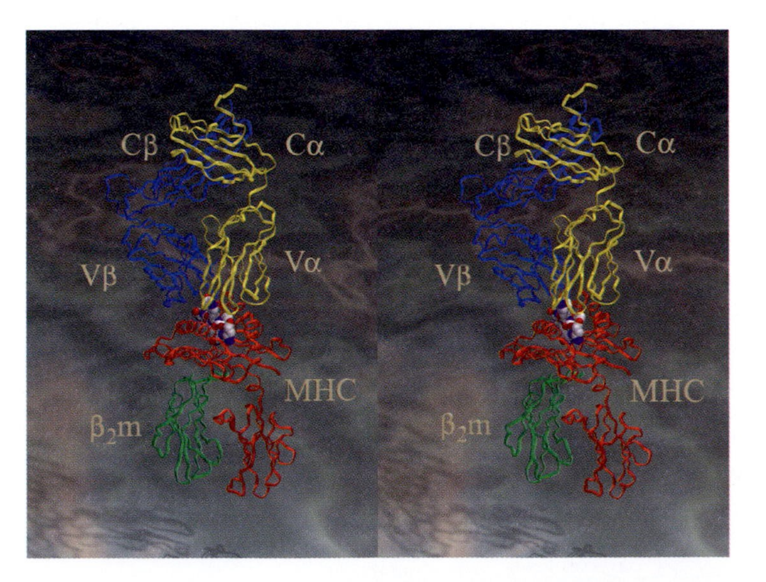

Fig. 7.20 A human T-cell receptor in complex with a class I MHC protein and a viral peptide [2CKB]. ($β_2m$ = $β_2$-microglobulin.)

A total of 11 residues from the TCR are in contact with the peptide. A total of 17 residues from the TCR are in contact with the MHC protein, only 6 of which also make TCR–peptide contacts. The remaining 11 recognize the MHC protein, or, at least, the conformation of the MHC protein that appears in this peptide complex. Conversely, 26 residues from the MHC protein are in contact with the peptide. A total of 16 residues from the MHC protein are in contact with the TCR, of which 5 are also in contact with the peptide.

Cancer

Biologically, cancer is a problem of genetic and cellular regulation. However, much clinical diagnosis and treatment of cancer, other than by surgery or radiation, involves protein science. Proteins and their interactions underly the mechanisms of cellular transformation and proliferation in cancer, which must ultimately be understood at the protein level. Protein expression patterns allow distinctions to be made between different varieties of cancer affecting the same tissue. This permits both more accurate prognosis and selection of optimal treatment. Emerging structures of proteins involved in cancer open new approaches to drug development.

Normal growth and development involve the orderly division and differentiation of cells. As our bodies develop, tissues and organs reach defined sizes. After we reach adulthood, cell division is matched to the needs for cell replacement, to maintain constant size. Different tissues have different rates of turnover. Red blood cells have a lifespan of about 120 days and must continually be replaced. Liver cells divide, on average, once every 2 years. Neurons, on the other hand, are not replaced in the adult. Wound healing is a special case of temporary local active cell division. This shows that cells are still receptive to stimuli to proliferate. When the wound is healed, they stop. Other localized and temporary growth spurts occur in the uterus and mammary glands during pregnancy.

Release of checks on cell division can cause a cell to replicate itself in an unconstrained way, in cancer. An adult contains roughly 2^{47} cells. Only a few successive rounds of cell division are required for a single rampant cell to attain a size that can interfere with normal activity of the organ in which it is growing. Conversely, the fact that for most people, over tens of years *none* of the 2^{47} cells escapes the controls, is a tribute to the power and robustness of the system.

Differences between normal and cancer cells may include:

1. Cancer cells fail to be controlled by normal mechanisms that regulate proliferation. Whereas normal cells can undergo only a limited number of cell divisions, controlled in part by telomere shortening, cancer cells are 'immortalized' and can divide indefinitely. The result is a tumour, a clonal mass of cells.

   Benign tumours are characterized by excessive growth only. Although it may invade surrounding tissues, the tumour nevertheless shares with normal cells the property of adhesion to like cells. The cancer remains compact and localized.

2. Far more dangerously, cancer cells may also lose their normal response to regulatory signals from neighbouring cells (*contact inhibition* of cell movement and

growth), and the normal requirement for a surface on which to grow (*anchorage-dependent* growth).

Malignant tumours arise when cells lose their normal properties of adhesion to like cells, becoming unresponsive to contact inhibition, and/or achieving anchorange-independent growth. Cells may break loose and seed secondary tumours elsewhere in the body, a process called metastasis.

Malfunctions in two classes of genes contribute to the development of cancer

Oncogenes

Recall: Ras–GDP, inactive; Ras–GTP, activated.

Oncogenes encode regulators of cell growth. Activation of an oncogene can transform a normal cell into a cancerous one. Mutations in Ras proteins are very common in cancers, as Ras is involved in signalling pathways triggering cell proliferation. Mutations that inactivate GTP hydrolysis leave Ras proteins in a constantly activated state.

Many oncogenes are altered forms of **protooncogenes**, that encode proteins with normal roles in healthy cells. Conversion of a protooncogene to a cancer-producing oncogene can occur via a point mutation, a chromosomal rearrangement, an increase in copy number, or viral insertion of an abnormally active promotor.

Tumour suppressor genes

These encode proteins that oppose cell proliferation. Some are inhibitors of the products of oncogenes. Inactivation of tumour suppressor genes weakens the regulation of cell division.

A person may have two copies of a tumour suppressor gene, on paired chromosomes. One normal allele may suffice. People with a single defective allele are predisposed to develop cancer should a mutation occur in the normal allele. A person with two normal alleles would require two separate mutations *in the same cell* to silence the gene completely and lose its protection. This is the 'two hit' theory.

For example, **retinoblastoma**, a cancer of the retina, presents clinically in two forms. In *familial retinoblastoma*, tumours develop very early in infancy, in both eyes, and at many sites. *Sporadic retinoblastoma* develops somewhat later, in early childhood, usually with a single tumour in a single eye. What is the difference?

The retinoblastoma susceptibility protein is a tumour suppressor. Familial retinoblastoma is associated with the inheritance of one defective gene, in *all* somatic cells. *Any* retinal cell suffering a mutation in the healthy allele that inactivates the protein can develop a tumour. In sporadic retinoblastoma, both inherited alleles are normal in the parents. It requires mutation in *both genes in the same cell* of the child, to deprive that cell completely of functional protein and to develop a tumour. In summary:

familial retinoblastoma = 1 germ-line mutation (affecting all cells)
+ 1 somatic mutation (in any cell or cells)

sporadic retinoblastoma = 2 somatic mutations (in the same cell)

Viruses and cancer

A link between viruses and cancer first appeared almost a century ago, with the discovery that extracts of chicken tumours were infectious. The infectious agent, Rous sarcoma virus, is a **retrovirus,** capable of reverse-transcribing its RNA into the host genome, at random loci. (The virus then uses the host protein-synthesizing machinery to transcribe and translate the DNA copy of its genome.)

Reverse transcriptase is an RNA-directed DNA polymerase.

Viruses implicated in human cancers include: Epstein–Barr virus (Burkitt's lymphoma and infectious mononucleosis = glandular fever), hepatitis virus B and C (liver cancer), and **papillomavirus** (cervical cancer). The association of AIDS (HIV-1 infection) with **Kaposi's sarcoma** is not the result of the virus introducing an oncogene, but with the loss of the immune system surveillance against cancer.

Mechanisms by which viral infections may lead to cancer include:

- The viral genome may itself contain an oncogene, expressed by the host.
- The virus may insert a promotor or enhancer next to a normal stimulator of cell division, causing abnormal expression levels.

In the case of Rous sarcoma virus, the virus contains a gene v-*src* (viral src) homologous to a eukaryotic gene c-*src* (cellular src). c-Src is a protein tyrosine kinase involved in pathways that control cell proliferation. The viral form v-Src is locked in the activated state, resulting in uncontrolled cell proliferation. Rous sarcoma virus particles without a functional v-*src* gene do not cause cancer.

The virus originally must have picked up the c-*src* gene by a 'read-through' error in replication. The mutations that cause the protein encoded by the viral gene to induce copious proliferation of infected host cells conferred an obvious selective advantage on the virus.

Cancer and protein structures

The identification of genes associated with cancer—oncogenes and tumour suppressor genes—has provided targets for crystallographers to determine the structures of the proteins that such genes encode. Several recently solved crystal structures illuminate the biological roles of these proteins, help us to understand the effects of mutations in them, and provide targets for rational drug design.

The breast cancer 1 (**BRCA1**) protein is a tumour suppressor. It is mutated in approximately 50% of patients who have a familial predisposition to breast or ovarian cancer. A single defective *BRCA1* allele is sufficient to increase the risk, even though heterozygotes may have one normal gene, which becomes lost or inactivated in a tumour.

BRCA1 is a 1863-residue protein. It has an N-terminal ring-finger domain, followed by a predicted helical coiled-coil region, followed by two tandem BRCT domains. Structures of the N- and C-terminal domains are available. Ring fingers contain cysteine-rich sequences binding Zinc. The BRCT domain (standing for BRCA C-Terminal domain) is a small α/β protein with a central four-stranded β-sheet flanked by helices. The tandem BRCT domains form an asymmetric dimer. Each domain has homologues in other proteins in different species.

See Weblem 7.

BRCA1 interacts with many other proteins and is implicated in several different functions. These include:

- *Sensing and signalling of lesions in DNA*: BRCA1 responds to several types of DNA damage, and activates different repair mechanisms appropriate to each.
- *Preserving chromosome structure*: As chromosome integrity may suffer *as a consequence of* inaccurate repair of DNA damage, these functions are related.
- *Mediating checkpoint tests at points in the cell cycle,* in part at least by regulation of the transcription of genes encoding proteins involved in checkpoint enforcement. Microarray experiments have identified sets of genes, the expression of which is regulated by BRCA1.

The unifying idea is that BRCA1 acts to respond to DNA damage by eliciting repair mechanisms and, in case repair is unsuccessful, checkpoint mechanisms that stop unrepaired damage from propagating. Loss of its function leads to the accumulation of damaged DNA in cells, resulting in the transition to the cancerous state.

The BRCA1 protein is at a nexus of regulatory pathways. Understanding its function on the basis of its amino acid sequence and crystal structure is a goal unattainable without a detailed knowledge of its patterns of interaction in space and time within a living cell. We have reached the shores of new dimensions of complexity, and this is a probably a good place to stop.

USEFUL WEB SITES

Site compiling information on inborn errors in metabolism:
http://web.indstate.edu/thcme/mwking/inborn.html
Searching the Kabat databank of antibody sequences and specificities:
http://immuno.bme.nwu.edu/
Compilation of links to on-line databases and resources of immunological interest, as well as much other material related to computational molecular biology:
www.infobiogen.fr/services/deambulum/english/db5.html
Immunogenetics database, with links to sequence analysis tools:
www.genetik.uni-koeln.de/dnaplot/

RECOMMENDED READING

General

Lindenbaum, S. (2001). Kuru, prions, and human affairs: thinking about epidemics. *Annu. Rev. Anthropol.*, **30**, 363–85.

Protein aggregation

Dobson, C.M. (2002). Protein-misfolding diseases: getting out of shape. *Nature*, **418**, 729–30.
Serpell, L.C., Sunde, M. and Blake, C.C. (1997). The molecular basis of amyloidosis. *Cell Mol. Life Sci.*, **53**, 871–87.
Sipe, J.D. and Cohen, A.S. (2000). Review: history of amyloid fibril. *J. Struct. Biol.*, **130**, 88–98.

Serpins

Carrell, R.W. and Lomas, D.A. (2002). α_1-Antitrypsin deficiency—a model for conformational diseases. *N. Engl. J. Med.*, **346**, 45–53.

Irving, J.A., Pike, R.N., Lesk, A.M. and Whisstock, J.C. (2000). Phylogeny of the serpin superfamily: implications of patterns of amino acid conservation for structure and function. *Genome Res.*, **10**, 1845–64.

Whisstock, J.C., Skinner, R. and Lesk, A.M. (1998). An atlas of serpin conformations. *Trends Biochem. Sci.*, **23**, 63–7.

Immunoglobulin structure and superfamily

Al-Lazikani, B., Lesk, A.M. and Chothia, C. (1997). Standard conformations for the canonical structures of immunoglobulins. *J. Mol. Biol.*, **273**, 927–48.

Neuberger, M. (2002). Antibodies: a paradigm for the evolution of molecular recognition. *Biochem. Soc. Trans.*, **30**, 341–50.

Padlan, E. (1996). X-ray crystallography of antibodies. *Adv. Prot. Chem.*, **49**, 57–133.

Wedemayer, G.J., Patten, P.A., Wang, L.H., Schultz, P.G. and Stevens, R.C. (1997). Structural insights into the evolution of an antibody combining site. *Science*, **276**, 1665–9.

Proteins of the Major Histocompatibility Complex (MHC) and T-cell receptors (TCR)

Bongrand, P. and Mallissen, B. (1998). Quantitative aspects of T-cell recognition: from within the antigen-presenting cell to within the T cell. *BioEssays*, **20**, 412–20.

Klein, J., Takahata, N. and Ayala, F.J. (1993). MHC polymorphism and human origins. *Sci. Am.*, **269**(6), 46–51.

Molecular biology of cancer

Huyton, T., Bates, P.A., Zhang, X., Sternberg, M.J. and Freemont, P.S. (2000). The BRCA1 C-terminal domain: structure and function. *Mutat. Res.*, **460**, 319–32.

Venkitaraman, A.R. (2002). Cancer susceptibility and the functions of BRCA1 and BRCA2. *Cell*, **108**, 171–82.

Venkitaraman, A.R. (2003). A growing network of cancer-susceptibility genes. *N. Engl. J. Med.*, **348**, 1917–19.

Vogelstein, B. and Kinzler, K.W. (ed.) (2002). *The genetic basis of human cancer* (2nd edn.). McGraw-Hill Professional, New York.

EXERCISES, PROBLEMS, AND WEBLEMS

Exercises

1. Draw a strip diagram (similar in format to those on p. 21) of the human prion protein as it is synthesized, showing the cleavages involved in its maturation. Indicate any significant landmarks.
2. Knock-out mice lacking the gene for prion protein cannot develop spongiform encephalopathy even if inoculated with PrP^{Sc}. How does this suggest a way to breed strains of sheep and cows that would be safe to eat?
3. Insertions that lengthen the region of the reactive centre loop of serpins convert them from inhibitors to substrates. Explain.
4. A complete IgG—as shown in Fig. 7.7—has an r.m.m. of about 170 000. Estimate the r.m.m. of (a) a Fab fragment, (b) the Fc fragment, (c) an Fv fragment.
5. Estimate the minimum size of an antigen or hapten that could make contacts with all six CDRs of the antibody shown in Fig. 7.9. (The total width of the figure is 40 Å.)
6. The end-to-end distance of an N-residue peptide in a nearly extended conformation is about 3.8 Å × N. The rise per turn of an α-helix is 3.6 Å. Estimate the number of residues in an α-helix needed to flank a groove designed to bind a 9-residue peptide in a nearly extended conformation. Compare with the lengths of the long helices in MHC molecule-binding domains.

Problems

1. The homologous antigen-binding loops of V_κ and V_λ domains of immunoglobulins have different repertoires of canonical structures. Suppose the antigen-binding loops from a V_κ antibody are 'transplanted' into the homologous positions in a V_λ domain. Would you expect affinity to be retained? Explain your reasons.
2. **(a)** Assume the following estimates: (1) stimulation of a T-cell requires approximately 100 MHC–peptide complexes per cell (as a consequence of the kinetics of dissociation of MHC–peptide complexes and of cell–cell encounters); and (2) an antigen-presenting cell expresses in the order of 10^5 MHC molecules on its surface. How many peptide species can an antigen-presenting cell effectively display at any time?

   (b) A typical eukaryotic cell synthesizes significant amounts of about 2000 proteins, average length 300 residues. How many 8-residue peptides can theoretically be generated from them?

   (c) What fraction of these peptides can a single antigen-presenting cell effectively display at any time?
3. By DNA sequencing, it is relatively easy to detect mutations in the gene for BRCA1. Those that cause dysfunction of the protein raise the chances of early development of breast or ovarian cancer. In some cases, such as a mutation that causes a phase shift near the N-terminus, it is easy to say that the protein will not be functional. But what about substitution mutations? How could one hope to use the known structures of domains

from BRCA1, and their complexes with other proteins, to try to predict whether a mutation will seriously interfere with function or not? What experiments not involving human patients could be carried out to test such predictions?

Weblems

1. What chicken sequence corresponds to the tandem octapeptide repeat PHGGGWGQ in the N-terminal domain of mammalian prion proteins?

2. **(a)** Find the sequence of an antibody or antibody fragment against the HIV gp120 protein. Use: **http://immuno.bme.nwu.edu/antibody_spec.html** Report the name of the antibody.
(b) What species is it from?

3. In what tissues are proteins of the immunoglobulin superfamily found in (a) vertebrates, (b) invertebrates?

4. Find PDB entries of five immunoglobulins, or fragments of immunoglobulins, determined by X-ray crystallography at better than 2.3 Å resolution.

5. Retrieve the DNA sequence of the heavy chain of the human antirabies antibody MAb105 (EMBL Data Library accession number L08089). Submit the sequence to a search of the databank V BASE of human antibody genes. Use: **www.genetik.uni-koeln.de/dnaplot/**
(a) From what germ-line V sequence is this domain derived?
(b) What are the differences between the amino acid sequence coded by the germ-line segments and that of the corresponding region of the expressed antibody?

6. **(a)** Calculate the surface area buried in forming the hen egg-white lysozyme–D1.3 complex. (Use PDB data set 1vfb.)
(b) What fraction of the area of the antibody buried in the complex is non-polar?
(c) What fraction of the area of the antibody buried in the complex is polar?

7. The Protein Data Bank contains structures of influenza haemagglutinin, and of a complex between Fab 17/9 and an octapeptide cut out from influenza haemagglutinin. What is the root-mean-square difference in main chain atom positions between the octapeptide in the Fab complex and within the complete haemagglutinin structure? Does the peptide adopt its native conformation—that is, its conformation within the complete haemagglutinin structure—in the antibody complex?

8. Draw a picture of the tandem BRCT domains from human BRCA1. Identify 7 point substitution mutations that are known to confer a predisposition to breast or ovarian cancer, and map them onto the structure.

9. Identify five proteins that bind to BRCA1, and the specific function of BRCA1 in which they are involved.

EPILOGUE

Many years ago, James Watson asserted that molecular biologists were spoiled by the structure of DNA: the logic of the activity of DNA was so obvious, that when protein structures came along their complexity was frustrating in comparison. That was when haemoglobin was people's idea of a protein with a complicated mechanism of action.

The perspective has shifted. For one thing, although unquestionably the basic idea of how DNA stores and replicates information is implicit in the double helix, the mechanisms of transcription and translation are anything but simple. One can't even blame it all on proteins: remember that the ribosome contains a ribozyme!

It is undeniable that there has been progress in understanding the activity of more-and-more complex combinations—larger macromolecular aggregates, and more densely reticulated regulatory networks. Again the ribosome is a prime example. But the real difficulty is that the problems lying at the heart of life processes involve complicated patterns of interaction among the components, in space and in time. Individual sequences and structures give us pieces of the puzzle. Knowing full genomes and thousands of protein structures denies us the excuse that we're missing too many pieces. But these are only the static and isolated data. To integrate them will require new kinds of information, developments from what is already being supplied by microarrays, and by methods for detecting patterns of interaction, and networks of control mechanisms.

In the last century molecular biologists took living things apart. Our task now is to put them back together.

ABBREVIATIONS

A site	aminoacyl site on ribosome
Å	angstrom unit (=10^{-10} metres, i.e. 0.1 nm)
AIDS	acquired immunodeficiency syndrome
Ala	alanine
ANS	8-anilino-1-naphthalenesulphonic acid (see Glossary)
APP	amyloid precursor protein
Arg	arginine
ASA	accessible surface area
Asp	aspartic acid
atm	atmosphere (=101 325 pascal)
ATP	adenosine triphosphate
ATPase	adenosine triphosphatase
β_2m	β_2-microglobulin
bp	base pair
βPP	β-protein precursor
BRCA1	(see Glossary)
BRCT	(see Glossary)
BSE	bovine spongiform encephalopathy
bZIP	basic zipper
C	constant (region of an immunoglobulin)
CAFASP	Critical Assessment of Fully Automated Structure Prediction (see Glossary)
CAPRI	Critical Assessment of Predicted Interactions (see Glossary)
CASP	Critical Assessment of Structure Prediction (see Glossary)
CD	circular dichroism
CDR	complementarity-determining region
CFTR	cystic fibrosis transmembrane conductance receptor
CID	collision-induced dissociation
CJD	Creutzfeldt–Jakob disease
CLIP	Class II invariant chain peptide
Cro	C-repressor and other things
c-*src*	normal cellular gene (or protooncogene; see Glossary)
Cys	cysteine
D	diversity (region of an immunoglobulin)
D_2O	deuterium oxide (=heavy water)
DPG	diphosphoglycerate
E site	exit site on ribosome
E	energy
EC	Enzyme Commission
ELC	essential light chain
EM	electron microscopy

EMBL	European Molecular Biology Laboratory
ENCODE	Encyclopedia of DNA Elements
ESI	electrospray ionization
EST	expressed sequence tag
F6P	fructose-6-phosphate
Fab	antigen-binding fragment of an immunoglobulin
FAD	flavin adenine dinucleotide
Fc	crystallizable fragment of an immunoglobulin
FISH	fluorescent *in situ* hybridization
FMN	flavin mononucleotide
Fos	Finkel osteosarcoma
FRAP	fluorescence recovery after photobleaching
FRET	fluorescence resonance energy transfer
FSSP	Fold classification based on Sequence–Structure alignment of Proteins
Fts	filamentous temperature sensitive
Fv	antibody fragment containing V_L and V_H domains
G	Gibbs free energy
G6PDH	glucose-6-phosphate dehydrogenase
GdmCl	guanidinium chloride (= guanidine hydrochloride)
GDP	guanosine diphosphate
GFP	green fluorescent protein
Gln	glutamine
Glu	glutamic acid
Gly	glycine
GO	Gene Ontology™
GPCR	G-protein-coupled receptor
GPI	glycosylphosphatidylinositol
GTP	guanosine triphosphate
GTPase	guanosine triphosphatase
H	enthalpy
H	hydrophobic
h	Planck's constant
Hb	haemoglobin
HbA	normal haemoglobin
HbS	sickle-cell haemoglobin
His	histidine
HIV	human immunodeficiency virus
HLA	human leucocyte antigen
HPLC	high-performance liquid chromatography
Hz	Hertz
Ii	invariant chain
Ile	isoleucine
IUB	International Union of Biochemistry
IUPAC	International Union of Pure and Applied Chemistry
J	joining (region of an immunoglobulin)
J	joule
Jun	*junana* (the Japanese for 17)

κ	immunoglobulin light-chain type
K	kelvin temperature
K_d	dissociation constant for the reversible breakdown of a complex or chemical compound
k_i, k_{-i}	forward and reverse rate constants for the *i*th step in an enzyme reaction
K_M	Michaelis constant
λ	wavelength
λ	immunoglobulin light-chain type
LDH	lactate dehydrogenase
Leu	leucine
LINUS	Local Independently Nucleated Units of Structures
LWS	long-wavelength sensitive
Lys	lysine
MAD	multiwavelength anomalous dispersion
MALDI	matrix-assisted laser desorption ionization
MAP	microtubule-associated protein
MAP	mitogen-activated protein
Mb	megabase
MDH	malate dehydrogenase
Met	methionine
MHC	major histocompatibility complex
μm	micrometre ($=10^{-6}$ metre)
mmHg	unit of pressure = 133.322 Pa
mRNA	messenger RNA
MS	mass spectrometer/spectrometry
MS/MS	tandem mass spectrometry
MWS	medium-wavelength sensitive
NAD	nicotinamide–adenine dinucleotide
NADH	reduced form of NAD
NADP	nicotinamide–adenine dinucleotide phosphate
NADPH	reduced form of NADPH
NMR	nuclear magnetic resonance
NO	nitric oxide
ompa	outer membrane protein-A
ORF	open reading frame
P site	peptidyl site on ribosome
P	polar
PCR	polymerase chain reaction
PDB	Protein Data Bank
Phe	phenylalanine
P_i	inorganic phosphate
PIR	Protein Identification Resource
pK_a	measure of the strength of an acid
pm	picomole ($= 10^{-9}$ mole)
pO_2	partial pressure of oxygen
Pro	proline
PrP^c	prion protein, cellular

PrP^{sc}	prion protein, scrapie
r.m.m.	relative molecular mass
Ras	rat sarcoma (virus)
RGS	regulator of G-protein signalling
RLC	regulatory light chain of myosin
rps	ribosomal protein from small subunit
s	second
S	entropy
S	Svedberg unit ($=10^{-13}$ second)
SCOP	Structural Classification of Proteins
sdh	succinate dehydrogenase
Ser	serine
Serpin	serine proteinase inhibitor
SIV	simian immunodeficiency virus
SNP	single-nucleotide polymorphism
SWISS-PROT	a protein sequence database
SWS	short-wavelength sensitive
T	temperature
TBSV	tomato bushy stunt virus
TCR	T-cell receptor
Thr	threonine
TMV	tobacco mosaic virus
TOF	time of flight
TrEMBL	translated EMBL
tRNA	transfer RNA
Trp	tryptophan
Tyr	tyrosine
UniProt	United Protein Database
UVS/VS	ultraviolet sensitive/violet sensitive
V	variable (region of an immunoglobulin)
Val	valine
vCJD	variant form of CJD
V_{max}	maximum velocity (of a reaction)
v-*src*	a viral oncogene first isolated from the Rous sarcoma virus
Zif	zinc finger
ZIP	zipper

GLOSSARY

2,3-diphosphoglycerate	An allosteric effector for haemoglobin
3-hydroxyretinal	Visual pigment
4-hydroxyproline	An unusual amino acid, found in collagen
abzyme	A catalytic protein based on an antibody structure
accessible surface area	The computed area of the boundary between protein and water
accessory pigment	A molecule that absorbs light and transfers the excitation energy to an active site, for instance in photosynthesis or vision
achiral	Having no centre of symmetry; among the natural amino acids only glycine is achiral
actinidin	A cysteinyl proteinase from kiwi fruit
activation barrier	The energy difference between the reactants and the highest-energy point on a trajectory; the higher the activation barrier, the slower the reaction; some enzymes speed up reactions by lowering activation barriers
active site	A specific location on a protein that is the focus of biological activity; for instance a crevice in the surface of an enzyme that binds substrate specifically and juxtaposes catalytic residues
acylenzyme	An intermediate in the cleavage of proteins by chymotrypsin-like serine proteinases
acylphosphatase	A small protein containing α-helices and β-sheets
adaptor hypothesis	The conjecture that there must be molecules that mediate information transfer between DNA and the ribosome, discovery of tRNA to fill this role proved the truth of the conjecture
adhesion	Property of sticking together; normal cells adhere to like cells; loss of this property in cancer is involved in metastasis
aerobic	Metabolism in the presence of oxygen (compare anaerobic)
affinity	Measure of the tenacity with which two molecules stick together
aggregate	A grouping of objects; protein aggregates are often insoluble, arise under unnatural conditions, and are associated with diseases
agonist	A substance that stimulates an activity in a different molecule
algorithm	A computational procedue for a calculation or other analysis
alignment	The assignment of residue–residue correspondences in a protein or nucleic acid sequence
aliphatic	Having the properties of hydrocarbons
allele	One possible sequence at any genetic locus
allosteric	A change in one part of a protein affecting activity at another

α-helix	A hydrogen-bonded structure of a localized region of the **backbone** of a protein
α_1-antitrypsin	A serine protease inhibitor (serpin; see below) protecting the lungs against damage from elastase
alternative splicing	Variation in the assembly of exons in forming messenger RNA from genome sequence
Alzheimer disease	A neurodegenerative disease, primarily of the elderly, associated with deposition of protein aggregates in the brain
amino acid	One of the constituents of proteins; proteins are polymers of amino acids
amyloidosis	One of a class of diseases characterized by extracellular deposits of protein aggregates
anaemia	Deficiency of red blood cells
anaerobic	Metabolism in the absence of oxygen
anchorage-dependent growth	Requirement for cell multiplication of attachment to a suitable substrate
anisotropic	With properties that differ depending on orientation
ANS or 8-anilino-1-naphthalene-sulphonic acid	A dye specific for hydrophobic residues
antennapedia	A mutation in a gene in *Drosophila* causing developmental defects
antibody	One of a large set of proteins capable of defending the vertebrate body by specific binding to foreign substances
anticodon	A triplet of bases within tRNA that binds to a complementary triplet of bases in messenger RNA
antigen	A foreign substance recognized by an antibody
apoptosis	Programmed cell death
Archaea	One of the basic divisions of life, contains numerous simple organisms adapted to unusual environments such as high temperatures
aromatic	A type of hydrocarbon
aspartate carbamoyl-transferase	An allosteric protein, catalysing the first step in pyrimidine biosynthesis
atomic resolution	A very detailed structure determination, in which the experimental method shows the positions of individual atoms
azurin	An electron transport protein in the bacterial respiratory chain
backbone	The repetitive part of protein structures, a linear chain linked by peptide bonds
bacteriorhodopsin	A molecule related to G-protein-coupled receptors that functions as a light-driven proton pump
barnase	Bacterial ribonuclease, a protein studied for the properties of its folding pathway
Bence-Jones protein	An overproduced immunoglobulin light-chain dimer

β-barrel	A type of assembly of regions in a protein in which strands of β-sheet wrap around into a closed cylindrical structure
β-hairpin	Structure of a local region of a protein in which a short loop connects two antiparallel strands of a β-sheet
β-sheet	A common structure in proteins in which several segments interact laterally by hydrogen bonding
BPTI	Bovine pancreatic trypsin inhibitor, a small protein
BRCA1	A protein, mutations in which are associated with a familial predisposition to breast and ovarian cancer
BRCT	The C-terminal domains of BRCA1
Burkitt's lymphoma	A tumour of B lymphocytes generally associated with infection by Epstein–Barr virus
CAFASP	A blind test of protein structure prediction by computer programs
canonical structure	One of a roster of discrete conformations of antigen-binding loops of antibodies
CAPRI	A blind test of attempts to predict the mode of association of two proteins from the structures of the two partners separately
capsid	The assembly of proteins that forms the outer shell of a virus
CASP	A blind test of attempts to predict the structures of proteins
catalysis	Speeding up of the rate of a reaction
central dogma	Originally, 'DNA makes RNA makes Protein' (Francis Crick, 1958)
chaos	A dynamical state in which small changes in initial conditions may make large changes in the subsequent behaviour
chaperone	A protein that catalyses refolding of misfolded proteins
checkpoint	Mechanism whereby cells containing damaged DNA are prevented from continuing with the cell cycle
chemical shift	Modification of resonant frequency of a nucleus in NMR by the molecular environment
chevron plot	A graph of rates of folding and unfolding as a function of the concentration of denaturant; reveals the nature of interactions in the transition state
chlorin ring	Heterocyclic organic ring system related to porphyrin (see below), binds magnesium to form a chromophore in chlorophylls
chloroquine	An antimalarial drug
chromophore	A molecule or portion of a molecule that absorbs light
CID	Collision-induced dissociation, a technique for forming ions from a vaporized protein
circadian rhythm	Natural periodicities in biological activity, deriving from the 24-hour day
circular dichroism (CD)	Effect on the propagation of polarized light through a solution, indicating the secondary structure content of a protein
codon	Triplet of nucleotides in genome that is translated into one amino acid

coiled-coil Structure in which two α-helices wind around each other to form a supercoil

collagen A structural material in our bodies

complementarity-determining region (CDR) A region in an antibody contributing to the specificity of antigen binding

conformation Spatial disposition of atoms in a molecule

contact inhibition Regulatory mechanism by which cell–cell adhesion inhibits proliferation

cooperative A dose–response curve in which binding of some ligand alters the binding constant for additional ligand

Creutzfeldt–Jakob disease A neurodegenerative disease associated with the prion protein

cryoEM Structure determination by low-temperature electron microscopy

crystal structure Structure determination from the diffraction of X-rays by a crystal

cytokine Substance produced by the immune system, affecting the immune response

cytoskeleton Structural elements within the cell giving shape and rigidity to the cell and providing tracks for intracellular transport proteins

DALI Distance ALIgnment—a program to align the amino acid sequences of proteins based on their structures

dehydrogenase An enzyme that reduces a substrate

denaturant A substance that destabilizes a native protein structure

denaturation Loss of native protein structure

depolarization of fluorescence Partial or total loss of polarization between absorption and emission, as a result of molecular motion or reorientation during the interval

detoxification Reaction to change a dangerous substance to a harmless one

dielectric constant Measure of how far electrostatic effects can be felt

dihedral angle A quantitative description of the conformation of a molecule in terms of of rotation around bonds

diphosphoglycerate An allosteric effector for haemoglobin

dissociation constant A measure of the affinity of two molecules

disulphide bridge A post-translational modification in which two cysteine residues, which may be far apart in the sequence, come together in the structure and form an S–S bond

divergence Progressive loss of similarity during evolution

docking Prediction of the mode of association of two proteins from knowledge of the separate structures

domain A compact subunit of a protein structure

domain swapping A mechanism of dimer formation in which two domains exchange regions of the chain

dose–response curve	The variation of the effect with the concentration of a molecule
ectopic	In the wrong place
effector	A molecule that enhances the activity of another molecule
Ehlers–Danlos syndrome	An inborn error of collagen formation
ellipticity	A measure of the degree of polarization of light
enantiomorph	One of two mirror-image structures
equilibrium	A state which is the initial state of no spontaneous process
erythrocyte	Red blood cell
Eukaryotes	Major grouping of life forms, those that have a nucleus
excited state	State of an atom or molecule with higher than minimum energy
exocytosis	Transport process in which substances are exported from a cell
exon (expressed region)	A transcribed region of a gene
Φ-value	Relation between interactions of a residue in the native state of a protein and the interactions in the transition state for folding
Fab	A fragment of an antibody, including the antigen-binding site
fluorescence	Re-emission of absorbed radiation without change in spin state
fluorophore	A molecule or portion of a molecule from which radiation is emitted during fluorescence
folding funnel	A pictorial concept describing the folding of a protein from a denatured state with many conformations, to a native state with only one
free R-factor	Measure of agreement between protein structure determined by X-ray crystallography and experimental data on which it is based
genomics	analysis and comparison of the complete genetic material of organisms of different species
Gibbs free energy	A measure of spontaneity and equilibrium under conditions of constant temperature and pressure
GPCR	G-protein-coupled receptor, a protein that transmits external signals across a cell membrane, received inside the cell by a G-protein
ground state	State of an atom or molecule with minimum energy
haemophilia	Disease involving defects in the mechanism of blood clotting
hairpin (see β-hairpin)	Two successive strands of an antiparallel β-sheet linked by a short peptide.
haplotype	Individual distribution of MHC alleles
heterodimer	An association of two different proteins
homeodomain	A protein involved in organizing the development of the body plan
homeotic	A gene involved in the development of the body plan
homologues	Genes or proteins related by evolution
Huntington disease	A hereditary, adult-onset, polyglutamine-repeat disease characterized by movement disorder and dementia
hydrophilic	Preferring an aqueous environment to an organic one
hydrophobic	Preferring an organic environment to an aqueous one

hydrophobicity scale A measure of the relative preferences of amino acids for aqueous and organic solvents

hyperthermophile An organism that can survive very high temperatures

hypervariable A region in an antibody responsible for antigen-binding specificity

inhibitor A molecule that reduces or destroys the activity of another

interface The region of abutting of two or more proteins in a multimer or aggregate

intein A protein that autocatalyses the removal of an internal peptide

intron Intervening region, non-coding, removed from RNA before translation

isomorphous replacement Method of solving protein crystal structures, by measuring the differences in the diffraction pattern caused by the addition of a heavy metal without changing the structure of the protein or its packing in the crystal

Kaposi's sarcoma Cancer of connective tissues common in AIDS patients

kinase A protein that phosphorylates another protein

kuru Adult-onset neurodegenerative disease, formerly observed in Papua New Guinea, caused by conformational change and aggregation of prion protein

leucine zipper Mode of dimerization of two protein helices based on interaction between periodically spaced leucine residues

LINUS A Monte Carlo simulation program for *ab initio* protein folding calculations

lymphoma Cancer of lymphanic issue, including Burkett's lymphoma caused by Epstein-Barr virus

lysis Destruction of a cell, for example by bursting out of the products of a viral infection

lysogeny Incorporation of bacteriophage genome into that of the host cell, the virus thereby entering a quiescent state

mad-cow disease Bovine spongiform encephalopathy, a prion (see below) disease

main chain The common backbone of the polypeptide chain in proteins

mass spectrometry Identification or partial sequencing of proteins by measurements of masses of ions

MHC protein Protein of the immune system governing individual sensitivity to different antigens

microarray Device to measure protein expression patterns, or genotypes

MOD-BASE database of homology models of proteins

mmHg Unit of pressure: 760 mmHg = 1 atm = 101 325 Pa

MODELLER A program to predict the structure of a protein knowing the structures of one or more close relatives

modular protein A protein containing a concatenation of regions, each of which folds into a compact unit

module A unit of protein structure that appears in different contexts in different proteins

molecular replacement Method of solving protein crystal structures, based on the relationship of the unknown structure to a protein of known structure

molten globule A partially-structured stage in the folding of a protein

multiwavelength anomalous dispersion (MAD) Method of solving protein crystal structures, based on measurements of X-ray absorption by specific atoms at specific wavelengths

native state The compact, active conformation of a protein

oncogene A gene involved in the development of cancer

open reading frame A region of DNA without Stop codons, a candidate for being a protein-coding gene

opsin A protein containing a chromophore active in the visual process

orthologous Relationship between two proteins in different species

palindrome A character string that is identical with its reverse; for instance 'MAXISTAYAWAYATSIXAM'. For nucleic acids, a sequence complementary to its reversal.

papillomavirus virus associated with cervical cancer

paralogous Relationship between two proteins that have diverged as a result of a gene duplication

Parkinson disease Neurodegenerative disease common in the elderly, characterized by effects on movement, caused by reduction in dopamine levels in the brain as a result of neuronal death

peptide bond Link between two amino acids to form protein backbone

peptidomimetics Molecules with structural similarities to parts of proteins that can compete with proteins for binding sites

pharmacogenomics Tailoring of drug treatment to the genotype of the patient

phase problem The loss of information essential for structure determination in measuring X-ray diffraction patterns

phenylketonuria A enzyme deficiency disease leading, if untreated, to the toxic accumulation of phenylalanine

phylogenetic tree Set of ancestor–descendent relationships among a set of species or proteins

plasmid A small extra segment of DNA, often circular, external to the normal chromosome

plastocyanin An electron-transport protein in photosynthesis

porphyrin ring Heterocyclic organic ring system containing four pyrrole groups; the haem group found in globins, cytochromes c, and certain other proteins contains an iron-bound porphyrin

post-translational modification Chemical change in a protein after synthesis by the ribosome

primary structure The set of chemical bonds in a protein

prion A protein that, if converted to a dangerous form, can cause disease

prokaryote A relatively simple form of life in which cells lack a nucleus

proteasome	A large protein aggregate responsible for selective proteolysis
proteomics	Measurement of the spatial and temporal distribution of protein expression patterns and interactions
protooncogene	A gene with a normal function which if mutated can cause cancer
quantum yield	The fraction of photons absorbed that is released in an ensuing processes, such as fluorescence
quaternary structure	The assembly of a multichain protein
quenching	Reduction or elimination of fluorescence resulting from non-radiative modes of de-excitation
racemic	Mixture of left- and right-handed molecules
reaction coordinate	An idealized depiction of the course of a reaction
reading frame	One of the ways to 'parse' a region of DNA into successive triplets, that may potentially code for protein
reticulation	Network formation
retinal	The chromophore in visual pigments
retinoblastoma	A form of cancer of the eye, propensity for which is associated with a known oncogene
retrovirus	A virus that can insert its genetic material into the host genome
reversible	Thermodynamically, an idealized process that proceeds through a succession of equilibrium states
R-factor	A measure of the agreement between a molecular model and the X-ray diffraction data on which it is based
ribosome	The large ribonucleoprotein molecular machine that synthesizes proteins at the direction of messenger RNA sequences
ribozyme	An RNA molecule with catalytic activity
ROSETTA	A state-of-the-art program for the prediction of protein structure from an amino acid sequence
rotamer library	A set of commonly observed conformations of side chains
sarcoma	Tumour of connective tissue, Rous sarcoma was the first cancer observed to be caused by a virus
Sasisekharan–Ramakrishnan–Ramachandran plot	A graph of the relative energies of different backbone conformations of an amino acid
sarcomere	The structural unit of a muscle fibre
scrapie	A prion disease of sheep
secondary structure	hydrogen-bonded, main-chain conformations, including α-helices and β-sheets, common to many proteins
serpin	A serine proteinase inhibitor or other protein related to α_1-antitrypsin
sickle-cell anaemia	A disease caused by a mutation in haemoglobin and resulting in polymerization of deoxyhaemoglobin in the red blood cell
side chain	The atoms attached to the polypeptide chain at each residue; the succession of side chains determines the amino acid sequence of a protein

sorting signal Short peptide specifying the destination of an expressed protein

spliceosome A large macromolecular nucleoprotein complex responsible for removing introns from RNA

spongiform encephalopathies Progressive degenerative neurological diseases characterized by cell death resulting from conformational change and aggregation of prion protein

steady-state A non-equilibrium condition in which the amounts of certain substances remain constant

structural genomics The programme to determine enough crystal structures of proteins to permit at least rough modelling of almost all proteins from their sequences

strange attractors A dynamical state of a chaotic system, that gives some appearance of stability

supersecondary structure A local structure in proteins involving standard assemblies of α-helices and strands of β-sheet

synchrotron A particle accelerator capable of producing a very bright, tuneable, X-ray source

tauopathies A group of diseases involving aggregates of tau protein

TCR T-cell receptor

ternary complex A combination of three molecules

tertiary structure The assembly of helices and sheets in a protein

TIM barrel A very common type of β-barrel structure in proteins containing an eight-stranded β-sheet and eight α-helices

time-of-flight Measurement of the charge-to-mass ratio of an ion from the time it takes to travel a known distance with a known kinetic energy

topology The spatial aspects of the assembly of helices and sheets in a protein structure

torsional Internal rotation

transition A mutation in which a purine (A or G) is replaced by another purine or a pyrimidine (T or C) by another pyrimidine

transition state The highest-energy point on a reaction pathway (see 'activation barrier')

translocation Moving of a gene, or larger section of a chromosome, from one place in the genome to another

transversion A single-site mutation in which a purine (A or G) is replaced by a pyrimidine (T or C) or a pyrimidine is replaced by a purine

'wobble' hypothesis The suggestion, later confirmed, that the rules of base-pairing might be less stringent in the interaction between the third position of a codon and tRNA, than at the other two positions

X-ray crystallography A method of structure determination based on measurements of X-ray diffraction by a crystal

zinc finger A type of transcriptional regulator that binds zinc

zymogen An inactive precursor of an active enzyme

INDEX

E

F

G

H

I

K

L

M

N

O

P

Q

R